CAMBRIDGE LIB

Books of endu~~ring scholarly~~

MW00835448

Technology

The focus of this series is engineering, broadly construed. It covers technological innovation from a range of periods and cultures, but centres on the technological achievements of the industrial era in the West, particularly in the nineteenth century, as understood by their contemporaries. Infrastructure is one major focus, covering the building of railways and canals, bridges and tunnels, land drainage, the laying of submarine cables, and the construction of docks and lighthouses. Other key topics include developments in industrial and manufacturing fields such as mining technology, the production of iron and steel, the use of steam power, and chemical processes such as photography and textile dyes.

The Electric Arc

An electric arc is formed when a current passes between two conductors through a non-conducting medium like air. Although the phenomenon was discovered during early electrical experiments and utilised widely in lighting by the end of the nineteenth century, its problems were not fully understood. First published in 1902, this book represents one of the first systematic investigations of the electric arc, and the best-known work of suffragist and electrical engineer Hertha Ayrton (1854–1923). It includes a chapter on the history of the discovery, over a hundred illustrations and tables, and Ayrton's explanation of the enduring problem of arc instability. As a result of her research, she went on to patent anti-aircraft lights and new arc-lamp technology. She later became the first female recipient of the Royal Society's Hughes Medal. Remaining relevant to students of electrical engineering and the history of science, this book shares her insights and expertise.

The Electric Arc

HERTHA AYRTON

CAMBRIDGE UNIVERSITY PRESS

Cambridge, New York, Melbourne, Madrid, Cape Town,
Singapore, São Paolo, Delhi, Mexico City

Published in the United States of America by Cambridge University Press, New York

www.cambridge.org
Information on this title: www.cambridge.org/9781108052689

© in this compilation Cambridge University Press 2012

This edition first published 1902
This digitally printed version 2012

ISBN 978-1-108-05268-9 Paperback

TO

MADAME BODICHON,

WHOSE CLEAR-SIGHTED ENTHUSIASM FOR THE FREEDOM

AND ENLIGHTENMENT OF WOMEN ENABLED HER TO STRIKE

AWAY SO MANY BARRIERS FROM THEIR PATH ;

WHOSE GREAT INTELLECT, LARGE TOLERANCE AND

NOBLE PRESENCE WERE AN INSPIRATION TO ALL

WHO KNEW HER ;

TO HER

WHOSE FRIENDSHIP CHANGED AND BEAUTIFIED MY

WHOLE LIFE, I DEDICATE THIS BOOK.

THE ELECTRIC ARC.

By HERTHA AYRTON,

MEMBER OF THE INSTITUTION OF ELECTRICAL ENGINEERS.

LONDON:
"THE ELECTRICIAN" PRINTING AND PUBLISHING COMPANY,
LIMITED,
SALISBURY COURT, FLEET STREET, E.C.

Printed and Published by
"THE ELECTRICIAN" PRINTING AND PUBLISHING CO., LIMITED,
1, 2, and 3, Salisbury Court, Fleet Street,
London, E.C.

PREFACE.

THIS book, which owes its origin to a series of articles published in *The Electrician* in 1895-6, has attained to its present proportions almost with the growth of an organic body. In experimenting on the arc, my aim was not so much to add to the large number of isolated facts that had already been discovered, as to form some idea of the bearing of these upon one another, and thus to arrive at a clear conception of what takes place in each part of the arc and carbons at every moment. The attempt to correlate all the known phenomena, and to bind them together into one consistent whole, led to the deduction of new facts, which, when duly tested by experiment, became parts of the growing body, and, themselves, opened up fresh questions, to be answered in their turn by experiment. Thus the subject grew and developed in what might almost be called a natural way.

From the first it seemed to me that the fact that the resistance of the material in the gap between the carbons must not only depend upon the current, but that it must depend upon it in many apparently contradictory ways, could not but lead to curious complications

A

in the relation between the P.D. and the current—quite apart from any back E.M.F. that the arc might possess. In the attempt to disentangle the various effects on this resistance that a change of current must produce, and to see how far all that was apparently mysterious in the arc might be the natural result of such complexity in the resistance of a portion of the circuit, the theory presented in Chapter XII. gradually evolved itself. This theory, whatever may be its shortcomings, has at least not been hastily built up to fit a few of the more salient characteristics of the arc ; it has literally *evolved* itself, during the course of a detailed study, from many points of view, of each separate phenomenon.   For although the central idea, that the carbon vapour changed into mist at a short distance from the crater, occurred to me at a very early period of the work, its complete application to the whole series of phenomena, and the full recognition of all that it entailed, followed but slowly, as each part of the subject was considered in turn.

The experiments of other observers have been employed in two ways: (1) In *confirmation* of theory developed from my own experiments, and (2) as the *basis* of theory, for which further tests were devised. The law connecting P.D., current, and length of arc, for instance, was first constructed from my own results, and then was shown to be applicable to those obtained much earlier by Messrs. Edlund, Peukert, Cross and Shepard, and Ayrton.   The theory concerning the light, on the other hand, was entirely deduced from the experiments of others.   M. Blondel's interesting and systematic researches, the admirable work of Mr. Trotter, and Prof. Ayrton's Chicago Paper were all laid under contribution, and the deductions drawn from them were then tested by new experiments.

In seeking to compare my results with those of other observers, and in searching for accounts of experiments that might furnish material for theory, I have often been struck with the excellent work that has been done by men whose names are quite unfamiliar to us in England. There are admirable Papers on the arc, for instance, by Nebel, Feussner, Luggin, Granquist, and Herzfeld, to which reference is seldom seen in any English publication ; while other work, which is in some cases far inferior, is constantly quoted. I have, therefore, given in Chapter II. short abstracts of most of the important Papers on the direct-current arc that appeared up to the end of the nineteenth century, while those referring principally or entirely to the light are discussed in Chapter XI. At the end of Chapter II. is a chronological list of *all* the original communications that I could find when that chapter was written ; but the names of a few to which my attention has since been directed, and of some that appeared after the list was made, together with the dates of my own contributions to *The Electrician* and to various societies, are added at the end of the Appendix. The latest paper of all— an extremely interesting one "On the Resistance and Electromotive Forces of the Electric Arc," read by Mr. Duddell before the Royal Society in June last— I should much have liked to discuss in connection with this book, but, as it is not yet published in full, that is unfortunately impossible.

As it seemed better not to wait till the whole book was ready, before publishing the most important of the new results obtained, some part of almost every chapter has been made the subject of a Paper that has been read before one or other of the societies interested in such work. These Papers generally covered but a portion of the ground, however, giving the main experiments and conclusions only, without following them up, or showing how they bore upon one another. In the book these are all connected together, and many new results are set forth which have been developed during the process. At the end of each chapter is a summary of the most important conclusions reached in it, which, it is hoped, may be found useful in making each step perfectly clear before the next is taken.

Besides the light experiments already mentioned, all those on the time-change of P.D. immediately after starting the arc, and after sudden changes of current, originally formed part of Prof. Ayrton's ill-fated Chicago Paper, which, after being read at the Electrical Congress in 1893, was accidentally burnt in the Secretary's office, whilst awaiting publication. These highly important experiments were not only the first of their kind, but, as far as I know, they still remain unique. Most of the figures in the first chapter, all the experiments and curves that relate to cored carbons in the fourth and fifth, and some of those on hissing in the tenth, also belonged to this Chicago Paper, which was as full of suggestion as it was rich in accomplished work.

Although the book is concerned entirely with the arc itself, and does not touch at all upon lamps and their devices, it is hoped that it may appeal to the practical man as well as to the physicist. For not only the cause but the practical bearing of each peculiarity of the arc has been considered ; the directions in which improvements may be hoped for have been pointed out, and the conditions requisite to secure the maximum production of light from a given expenditure of power in the generator have been fully discussed.

In conclusion, I have to thank Prof. Blondel, Prof. Fleming and Mr. Trotter for kind permission to use figures from their Papers; Mr. Fithian for taking the beautiful photographs of the Hissing Arc reproduced in Fig. 81; Mr. Mather for much valuable advice and assistance with experiments, and Mr. Maurice Solomon for his suggestive criticism of the MS. and careful revision of the proofs.

HERTHA AYRTON.

TABLE OF CONTENTS.

CHAPTER I.

CHAPTER II.

TABLE OF CONTENTS.

CHAPTER III.

Impossibility at First of getting Definite P.D. with Fixed Current and Length of Arc.—Cause of Difficulty.—Low P.D. and Subsequent Rapid Rise on Striking Arc with Cored Positive Carbon.—Influence of Current, Length of Arc, and Shapes and Temperatures of Carbons on Time Required for P.D. to become Constant after Striking Arc.—First *Rise* of P.D. with *Increase* of Current with Cored Carbons.—Peculiar Changes of P.D. with Sudden Changes of Current, and their Causes.—Summary.

CHAPTER IV.

General Character of Curves for P.D. and Current with Constant Length of Arc for Solid Carbons.—Same with Positive Carbon Cored.—Discussion of Variations Caused by Core.—Different Positions of Hissing Points with Solid Carbons and with Positive Carbon Cored.—Influence of Strength of Current on Diminution of P.D. due to Core.—Hypothesis as to Action of Core in Modifying P.D.—Curves showing Straight Line Law connecting P.D. with Length of Arc, for Constant Current, with Solid Carbons.—Curves for Same Connection with Cored Positive Carbon, showing P.D. Practically Independent of Current for One Length of Arc.—Discussion of Differences between the Two Sets of Curves, and Explanation, on Abovementioned Hypothesis.—Soft and Hard Crater Ratios.—Deductions from them as to Influence of Current and Length of Arc on Area of Crater.—Summary.

CHAPTER V.

Measurements of Diameter of Crater with Arc Burning.—Curves of Area of Crater and P.D. between Carbons for Various Currents and Lengths of Arc.—Curves of Area of Crater and Length of Arc, of Soft Crater Ratio and Length of Arc, and of

CHAPTER VI.

The Equation for P.D., Current, and Length of Arc, with Solid Carbons, and its Application to the Results of Earlier Experimenters

CHAPTER VII.

The P.D. between each Carbon and the Arc, and the Fall of Potential through the Arc

CHAPTER VIII.

CHAPTER IX.

CHAPTER X.

CHAPTER XI.

CHAPTER XII.

APPENDIX.   445

LIST OF ILLUSTRATIONS.

———◆———

A *

LIST OF TABLES.

ERRATA.

Page 13, line 1, *for* Fig. 9 *read* Fig. 7.

Page 14, line 3, *for* 5 *read* 6.

Page 50, line 7, *for* Fig. 13 *read* Fig. 16.

Page 51, line 8 from end, *for* back E.M.F. of the arc *read* P.D. between the carbons.

Page 62, line 6 from end, *for* 1890 *read* 1891.

Page 64, line 5 from end, *for* Capt. Abney *read* Sir W. de W. Abney.

Page 64, line 3 from end, *for* Mr. Crookes *read* Sir. W. Crookes.

Page 88, line 11, *for* air *read* arc.

Page 95, line 7 from end, *for* p. 2 *read* p. 227.

Page 207, line 3, *for* Twelve *read* Fourteen.

CHAPTER I.

THE APPEARANCE OF THE ARC.

SINCE the discovery of the electric arc early in the present century, Nature has been subjected to a series of questions with the object of extracting from her a statement of the mysterious laws that govern it. These questions—which we call experiments—she has, so far, answered but sparingly. They have been repeated again and again, but, even when replies have been vouchsafed, they have been couched in such ambiguous terms that one experimenter has interpreted them in one way, and another in another, and we are still far from having a clear understanding of the laws of the arc.

A certain amount of knowledge has, however, been gained, and it is proposed in the present work to deal with some of the facts that have been acquired concerning direct-current arcs, maintained between carbon rods, the arc being not longer than the diameter of the positive carbon, and the potential difference between the rods being not greater than, say, 100 volts. It is proposed, in fact, to deal only with such direct-current arcs as are used in the lighting of our streets, and to leave on one side alternate-current arcs, very long arcs maintained with a large potential difference between the carbons, and arcs maintained between metals.

The arc is so bright that, if looked at with the naked eye, it appears to be simply a dazzlingly bright spot with needle-like rays diverging from it in all directions, but by projecting its image on to a screen its real shape and colour may be easily observed—if the image is magnified, so much the better.

In arcs maintained between vertical carbon rods, with the positive carbon uppermost, both shape and colour vary according to the length of the arc and the current flowing, but certain characteristics are common to all. In all, the end of

A

the positive carbon is more or less pointed, with a depression at the tip called the crater. This depression is shallower, the longer the arc, and is practically non-existent with arcs of lengths approaching the diameter of the positive carbon.

The end of the negative carbon is also pointed, but instead of a depression it often has a sort of little hillock on the tip. The tips of both carbons are white hot, and in the space between them there is a faint purple light, outlined by a deep shadow.

At a very early stage in the experiments made by the students of the Central Technical College, under Prof. Ayrton's direction, it was found that altering either the current or the length of the arc caused a change in the shapes of the carbons and visible arc, which in some cases was very considerable. It was therefore thought advisable to obtain a record of these changes under all circumstances, and diagrams of the carbons and arc were taken, when the potential difference had acquired its steady value, for all the currents and lengths of arc observed.

The diagrams were obtained by placing a piece of squared paper over the screen on which the enlarged image of the arc was projected, and drawing the complete outlines of the carbons and image. The carbons are always easy enough to draw for they are very definite, but the exact curve which outlines the purple image of the arc is much more difficult to obtain, for this image melts off very gradually into the surrounding darkness.

Fig. 1 is a reproduction of one of the diagrams half of the original size, and since the original diagram enlarged the carbons *ten* times, this reproduction shows them *five* times full size It may be observed that there is a white-hot crater at the end of the positive carbon, and a white-hot tip to the negative, and that the area of the crater in the positive is much larger than the area of the white-hot tip of the negative. That this glowing tip of the negative carbon gives out a fair amount of light may be easily seen by observing the beam of light from an arc after it has passed through a lens. This beam divides itself into two distinct parts, separated by a dark space, so that it looks like two beams, one coming from the crater of the positive, the other from the bright spot on the negative carbon. If a piece of paper be placed in the dark space between the two beams, it will have a faint violet

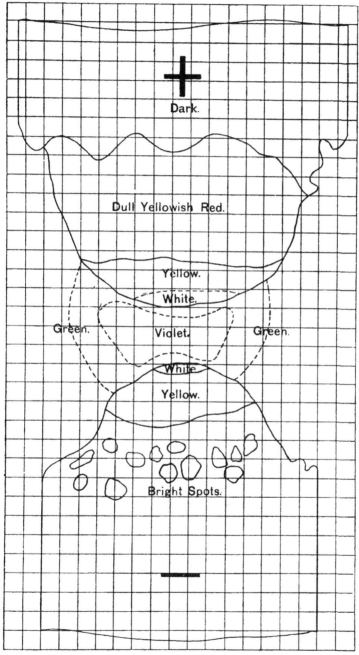

FIG. 1.—Image of Carbons and Arc 5 times full size.   Current, 10 amperes.
Length of Arc, 3mm.   P.D. steady at 46·5 volts.

light on it, but this light is evidently not sufficient to make the dust particles in the air visible.

The area of the bright spot on the tip of the negative carbon increases with the current, but at a much slower rate than the area of the crater in the positive, so that the ratio of the area of the crater to the area of the negative bright spot increases rapidly as the current is increased, with silent arcs.

The part marked " bright spots " on the negative carbon represents a circlet of seething balls, which, whatever they may be, always appear at the junction of the light and dark parts of the negative carbon. Above them, as far as the line which is marked " yellow," the carbon presents a granulated appearance, being covered with very small boiling balls, and the whole being of a reddish yellow colour. Then comes the yellow-hot part, marked " yellow," which is quite smooth, and finally the white-hot tip.

The positive carbon has also its smooth yellow-hot part marked " yellow," and its band of granulated darker yellow part above that, and higher still its circlet of seething balls—larger than those on the negative. The outlines of these balls are indicated (Fig. 1) in the highest wavy line, but they cannot generally be seen very distinctly, because no light is thrown on to them to be reflected back again, in the same way as the light from the crater is cast on to the negative carbon.

Looking at the arc itself through smoked glass, instead of at the image, it is seen that these balls are really the frayed ends of an outer crust of the carbon which is peeling off. It is as if the inner part of the carbon, being much hotter than this outer crust, caused it to expand and split, forming a sort of fringe hanging down over the inner hotter part from which it has broken away. Between this crust—which crumbles at a touch when cold—and the body of the carbon there is a space of from $\frac{1}{2}$mm. to 1mm. to the height of 5mm. or more, from which when the arc is burning sparks fly out, drop down to the edge of the crust, and then fly outwards and upwards, probably carried along by the strong upward draught of the column of hot carbon and air. It is possible that they finally settle on the positive carbon, for after the arc is extinguished this carbon is found to have numerous small particles of carbon on it arranged fairly

symmetrically. The tips of the strips of carbon that form
the outer crust apparently get burnt by the hot volatile carbon
into a semi-globular shape, and they boil and bubble under the
action of this heat just as a lump of sugar does when held in a
candle flame, and probably the action is really very much the
same in both cases.

Fig. 2 is a drawing of a section of a positive carbon with
its outer crust, A, showing the way in which this outer crust
bulges out and leaves a space between itself and the inner part
of the carbon, B.

FIG. 2. —Section of Positive Carbon with the Outer Crust curling away
from it.

Between the part of the arc marked " violet " in Fig. 1 and
that marked "green" there is a dark space, which is scarcely
perceptible with small currents and short arcs, but becomes
very wide and well marked with large currents and long arcs.
The "green" line shows the extreme edge of the luminous part
of the arc, or, at least, of that part which is bright enough to
show light on the image.

Figs. 3 to 6 show clearly the outlines of the purple and green
parts of the arc, and of the shadow between them, under
different conditions.

These figures I obtained by tracing the enlarged outlines
of the carbons and arc in the way already described ; very

special attention being given to the outlines of the purple part of the arc, the shadow, and the green outside part. The outlines were then shaded, so as to give as nearly as possible the *values* of the light given out by each part. Thus the most light is given out by the crater of the positive carbon, and by the tip of the negative carbon; therefore these were left white. The shadows between the purple and green parts of the arc are somewhat more abrupt than they really were, but their *shapes* are, I believe, correct. Unless the axes of the two carbons are absolutely in line, the arc is always a little to one side or the other, and that is the reason that the arc in all these four figures is slightly out of the centre; it is almost impossible to get it perfectly central.

In Fig. 4 the outlines of the balls of boiling carbon on the positive as well as on the negative carbon are shown. In Figs. 3 and 5 they could only be seen on the negative carbon, and in Fig. 6 the arc was so long that the screen was not large enough to show the boiling balls on either of the carbons.

The current used for the whole four figures was 20 amperes, the carbons were 18mm. positive and 15mm. negative.

In Fig. 3 the length of arc was 4mm., and the carbons used were both solid. It may be seen that the central purple part of the arc is of the form of an oblate spheroid, broken in upon by the tips of the carbons. Another diagram, made at the same time, of an arc of the same length and with the same current, but with the positive carbon cored, showed a central part of much the same shape, but of smaller area, smaller, in fact, not only in width, but in length, because although the distance between the tip of the negative carbon and the plane through the edge of the crater was in each case 4mm., the central part surrounded a much greater length of the point of the negative with the solid positive carbon than with the cored. The green part of the arc also started much higher up the negative with the cored than with the solid positive carbon, and touched the positive carbon 4mm. from its tip, whereas the green part could be seen at a distance of 14·5mm. up the solid positive carbon (Fig. 3). Thus, the whole visible part of the arc is much larger with a solid than with a cored positive carbon with an arc of 4mm. and a current of 20 amperes, but the general *form* of the arc is very much the same.

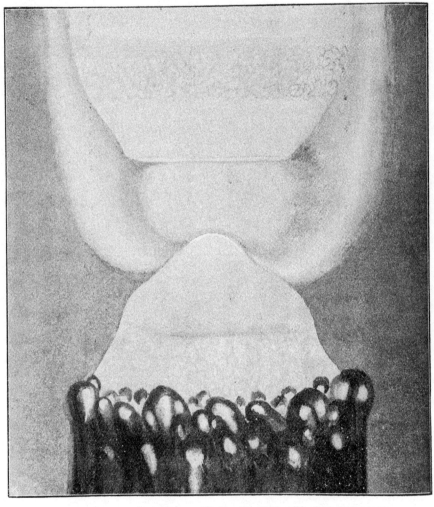

FIG. 3.—Carbons : Positive, 18mm. solid ; Negative, 15mm. solid.   Current, 20 amperes.
P.D. between Carbons, 48 volts.   Length of Arc, 4mm

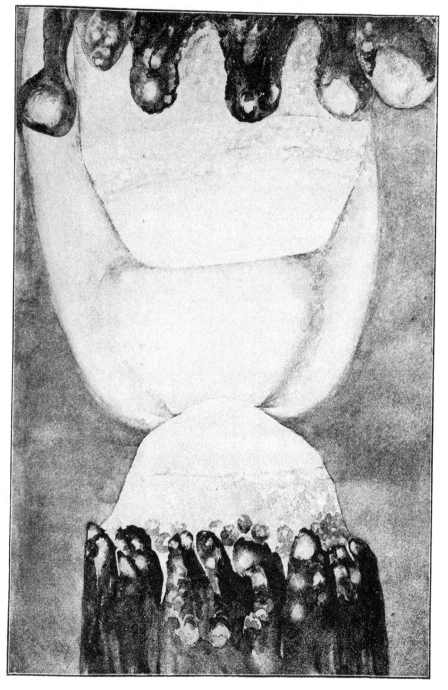

FIG. 4.—Carbons: Positive, 18mm. solid ; Negative, 15mm. solid. Current, 20 amperes. P.D. between Carbons, 56 volts. Length of Arc, 7mm.

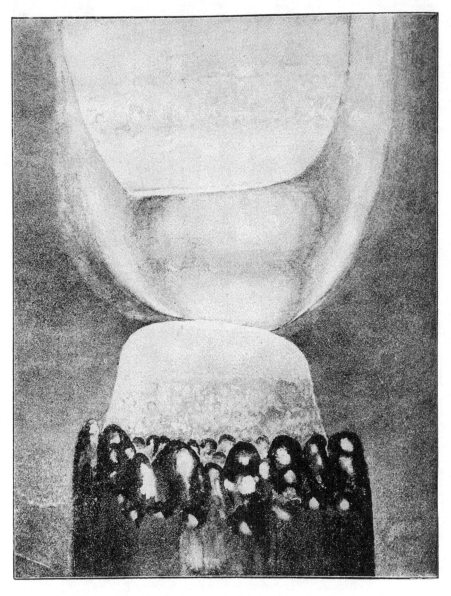

FIG. 5.—Carbons: Positive, 18mm. cored; Negative, 15mm. solid. Current, 20 amperes. P.D. between Carbons, 51 volts. Length of Arc, 7mm.

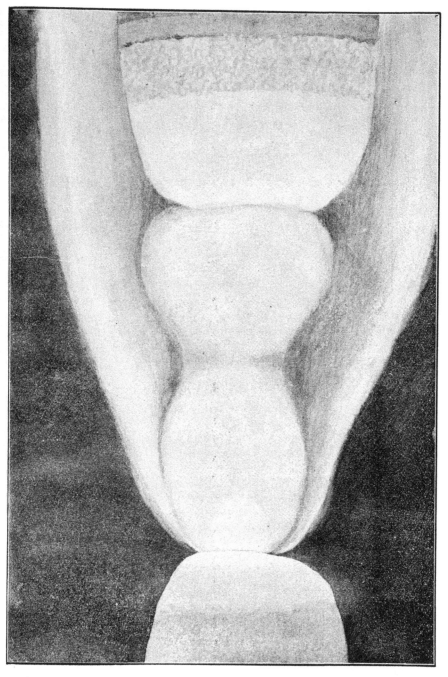

FIG. 6.—Carbons : Positive, 18mm. cored ; Negative, 15mm. solid.   Current 20 amperes.
P.D. between Carbons, 68 volts.   Length of Arc, 18mm.

In Figs. 4 and 5 the two arcs are of the same length, 7mm., the current 20 amperes; but for Fig. 4 the positive carbon was solid, and for Fig. 5 it was cored. Here, again, both the central purple part and the green portion surround the negative carbon to a greater distance when the positive carbon is solid than when it is cored ; again, also, both the central portion and the whole visible arc are larger with the solid than with the cored positive carbon. But in these two figures the *form* of the central part is also different. With the cored carbon it is gourd-shaped, with the solid, pear-shaped. With the cored carbon the arc has a dark shadow dividing it into two unequal parts, with the solid carbon this shadow is entirely absent. The tip of the positive carbon has a longer, and the tip of the negative a shorter point with the cored than with the solid carbon, which may be due to a lower temperature of the crater in the cored carbon, for, as will be shown later (page 14), greater heat is indicated by a more pointed negative and a less pointed positive carbon.

The balls on the positive carbon in Fig. 4 were not really luminous the whole time I was drawing that figure, but every now and then there was a little hiss caused by some imperfection in the carbon, which lighted them up, and during one of those periods I drew them.

In Fig. 6 the positive carbon was cored, and the arc was 18mm. in length, the current still 20 amperes. In this arc the gourd shape is much accentuated—the central part looks almost like two air balls, the one next the positive carbon placed horizontally, the other placed vertically below it, and touching the negative carbon. The vertical ball has inside it a small ball touching the negative. All three balls were of different shades of purple, the large one near the negative carbon was palest, the small one was darker, and the one near the positive was darkest of all.

The shade of the purple part of the arc was quite different according as solid or cored positive carbons were used, being much bluer with solid than with cored carbons.

I tried to obtain an arc of 18mm. with both carbons solid, in order to compare the two diagrams, but found it impossible to maintain an arc of more than 14mm. with two 18mm. and 15mm. solid carbons. Every time the length of the arc was increased beyond this the arc went out, because, as will be shown later, the E.M.F. of the dynamo was insufficient to maintain a longer arc with a current of 20 amperes flowing when both carbons were uncored. The shape of the 14mm. arc showed no tendency towards the double ball form observable with the cored carbon; it retained the pear shape noticed in Fig. 4. I have, however, found a tendency to assume the double ball form with arcs maintained between solid carbons when the current was very small; but, if the length of the arc is the same in both cases, the current has to be much smaller to produce this form when the positive carbon is solid than when it is cored.

Thus it is clear that when the current is kept constant, altering the length of the arc, alters both its size and shape ; and the use of a cored positive carbon, instead of a solid, changes the *size*, the *colour*, and, in long arcs, the *form* of the visible part of the arc.

The diagrams in Figs. 7 to 12 are reproductions of some of those made by Prof. Ayrton's students. They show very accurately the shapes of the ends of the carbons under the given conditions, but, as in Fig. 1, the dotted outlines of the arc must be taken to be only approximately correct. They show, for instance, that, with a given length of arc, the diameter of the visible part of the arc is smaller, the smaller the current, but not with absolute accuracy exactly how much smaller.

These diagrams have been reduced from ten times the full size of the carbons to two-thirds full size, and arranged in order of the sizes of the carbons, the lengths of the arc, and the currents. Figures in the same horizontal row are for the same length of arc with different currents, and figures in the same vertical column are for the same current with different lengths of arc.

We will first examine what is the effect on the shape of the negative carbon of changing (1) the current strength, and (2) the length of the arc ; and then we will see what effect these same changes have on the shape of the positive carbon. In all the figures with a short arc, say 0·5mm. (Fig. 7), the negative carbon is quite pointed, even with a small current, and it becomes more and more pointed as the current becomes larger and larger. With a 1mm. arc the negative is less pointed,

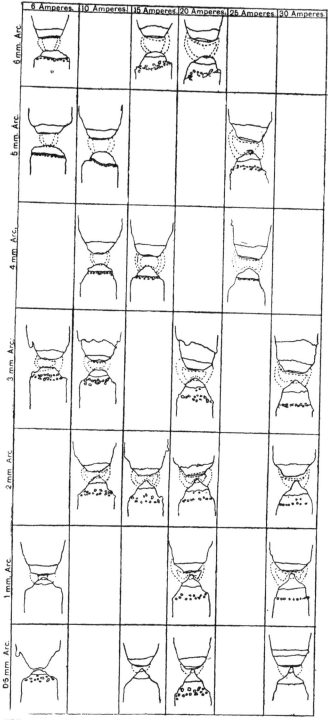

FIG. 7.—Carbons, Positive (upper) 18mm. cored. Negative (lower) 15mm. solid.

B

both with small and large currents, than with a 0·5mm. arc, and
as the arc gets longer the negative becomes blunter and blunter
although in every case it is more pointed with a large current
than with a small one for the same length of arc. At last,
when the arc is 6mm. long, the negative is quite blunt, even

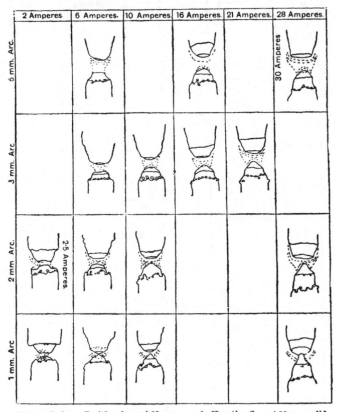

FIG. 8.—Carbons, Positive (upper) 13mm. cored. Negative (lower) 11mm solid.

when a current as great as 20 amperes is flowing. Thus, the
tip of the negative is more pointed—

        (1) The shorter the arc,
        (2) The larger the current.

In order to understand the causes of these phenomena, we
must examine the shapes of the negative tips a little more closely.

Comparing the 6-ampere 0·5mm. arc with the 6-ampere 3mm. arc in Fig. 7, one sees that the former is like the latter, with a sort of extra point added; comparing the 30-ampere arcs for the same two lengths, the same thing is observable, only it is much more pronounced. This sort of extra point is, in fact, found on all the negative carbons when the arc is short, whatever the current, and it is not found on any when the arc is long. This point must, therefore, depend upon the carbons being near together, and is caused, I think, solely by carbon deposited on the negative tip from the crater of the positive carbon. As the arc is lengthened, less and less of the carbon shot out by the positive reaches the tip of the negative carbon, and this little extra tip becomes smaller and smaller, and finally disappears.

When the distance between the carbons is great and the current small, the arc often plays about the edges of the carbons, sometimes seeming to travel round and round them, sometimes remaining stationary in one place for a short time, and then going to another, and so on, but always trying to make itself as long as possible. But in some cases, however the positive end of the arc may move, the negative end appears to remain in the same place, so that the negative bright spot remains fixed in position, and when this is so, on extinguishing the arc, it is found that there is a small crater at the end of the negative carbon.

It is possible that this crater would always exist in the negative carbon were it not that when the arc is shorter the crater is constantly filled up by the carbon deposited on it from the positive carbon.

When the arc is very short, and the current large enough to cause hissing, the deposition takes place so rapidly as to cause a " mushroom " to form, such as may be seen in the 25-ampere 1mm. arc in Fig. 9. When the arc is short, even when it is silent, the deposition is still often rapid enough to cause the negative carbon to grow longer instead of shorter, and this fact alone would seem to prove that the extra tip noticed on the negative carbon with short arcs is caused by carbon deposited on it from the crater.

It is true that with large currents the tip of the negative carbon is somewhat pointed, even when the arc is not very

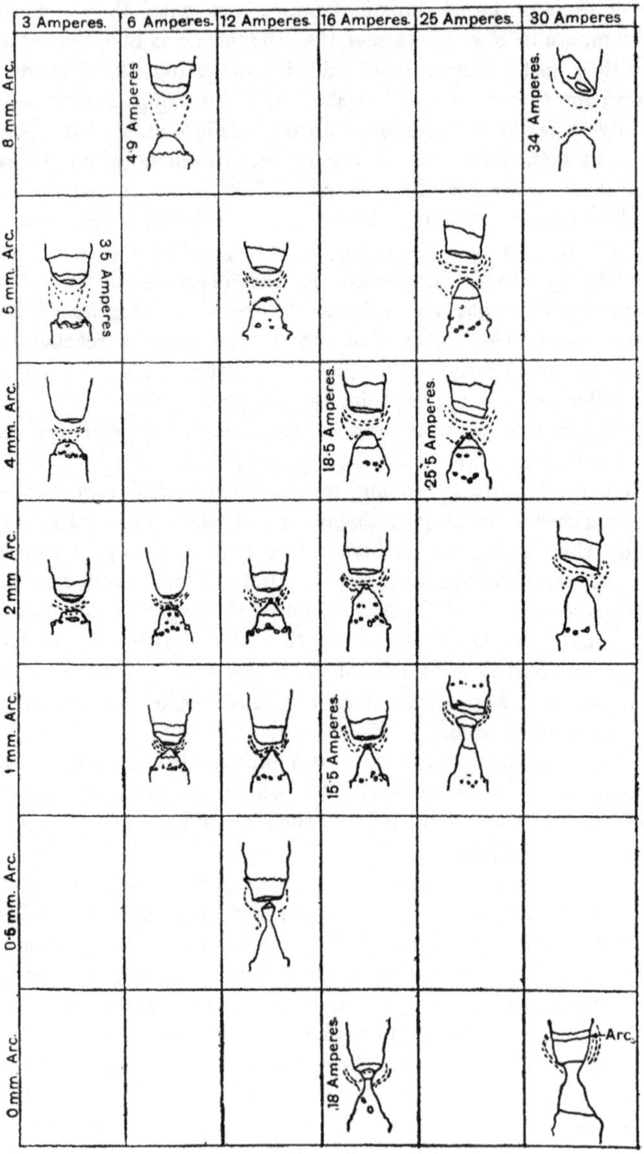

FIG. 9.—Carbons, Positive (upper) 9mm. cored. Negative (lower) 8mm. solid.

short, as in the 30-ampere 3mm. arc in Fig. 9, but it seems
to be pointed in a different way. The end of the carbon has
no little extra tip on it, but becomes more nearly cone shaped,
and this sort of pointing is attributable, I think, to quite a
different cause—namely, the burning away of the carbon. Most
of the heating of the negative carbon takes place from the outside, and is caused partly by radiation from the crater and
partly by the volatile carbon giving up its heat to it. When
the current is small the heat is comparatively small; therefore only a thin layer and a short length of the negative carbon
is made hot enough to burn away, and consequently the
point, which is the result of the burning away, is short and
blunt.

With large currents, however, the heat, being greater,
reaches farther, both through and down the negative carbon,
and the resulting tip is longer and more slender. It is more
or less conical, because the parts farther away from the positive
carbon get less heat, and, therefore, a thinner layer of them is
burnt away.

Thus with short arcs there are two distinct causes tending
to give the tip of the negative carbon its special shape, whereas
with long arcs one of these causes is nearly or quite inoperative,
hence the difference of character of the shapes of the negative
carbons in the two cases.

It is much more difficult to find out what happens to the positive carbon than to the negative when the arc is lengthened or the
current increased, because that carbon is very rarely luminous
over the whole length of its tapering part. As has already been
pointed out (page 4), the negative carbon is always bathed in
the light from the positive, and, therefore, its shape is very easily
discerned; but the positive sends all its own light away from it,
and can receive little, if any, from the negative. But even so,
by comparing the carbons in those cases in which it has been
possible to trace the positive carbon to its unburnt part, it
may be perceived that it also tapers more with a short arc than
with a long one when the same current is flowing, and has
a longer, but less pointed, tapering part with a large current
than with a small one.

For instance, take the 6-ampere 0·5mm. arc and the
6-ampere 5mm. arc in Fig. 7, evidently the positive carbon

tapers more in the former than in the latter, and in the 5mm. arcs in the same figure the positive carbon has a longer tapering part with 25 amperes than with 5 amperes. Also it will be found that wherever the craters show, the diameter of the crater is greater with a larger current than with a smaller one for the same length of arc and greater with a longer arc than with a shorter one when the same current is flowing.

Apparently what happens is this. The positive carbon is consumed in two ways—(1) by being shot out from the crater either in the form of vapour or of small particles, (2) by burning in combination with air. It is not probable that volatilisation, which requires an enormous temperature, can take place, if it takes place at all, anywhere except at the surface of the crater, and perhaps at the parts of the tip immediately surrounding that surface; therefore all the shaping of the positive carbon, except the formation of the crater, must depend upon its burning in combination with air. At the extreme tip the burning will take place most rapidly, because, owing to the immediate vicinity of the crater, the surface carbon will be hotter there than anywhere else, and the larger the crater the more rapidly the tip will burn away. Thus, instead of becoming more pointed as the current increases, the positive carbon actually becomes less pointed, because the tip is consumed so much faster than the sides. Although it is less slender, the point is longer, however, because the increased amount of volatile carbon extends farther up the sides, and thus burns away a longer portion of the carbon.

When the arc is short, the volatile carbon, not finding room between the tips, spreads out farther, and so also spreads farther along the surfaces of the carbons and causes them to burn away to a greater distance than when the arc is long. In fact, a given amount of volatile carbon must take up a given space with a given pressure all round it. If it can get this space between the carbons, it takes it; if not, it extends itself sideways and longways, and hence the ends of both the carbons have longer points with short arcs than with long ones, when the current is the same for each length of arc.

Fig. 10 shows very well the alteration that takes place in the shapes of the carbons when the current is suddenly changed from a higher to a lower, and from a lower to a higher value. The left-hand figure shows particularly well the alteration in the negative carbon. In this case, after the carbons had been formed by a current of 30 amperes, the current was suddenly changed to one of 4 amperes, and three minutes afterwards the dotted line diagram was taken, showing the negative carbon with a long tapering end. But with the smaller current the volatile carbon did not extend down nearly as far as before; consequently, as the tip of the carbon was gradually burnt away, the tapering of new parts of this

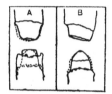

FIG. 10.—Carbons: Positive (upper) 13mm. cored. Negative (lower) 11mm. solid. Length of Arc 4mm. A, Current suddenly changed from 30 amperes to 4 amperes. Dotted lines show the shapes of the Carbons 3 minutes after the change. Continuous lines show their shapes 25 minutes after. B, Current suddenly changed from 10 amperes to 30 amperes.

negative carbon was not kept up. Hence, in 25 minutes after the current had been reduced from 30 to 4 amperes, the end of the negative carbon had become blunt, as shown by the continuous outside line. The small piece at the top of the

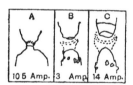

FIG. 11.—A, Carbons: Positive (upper) 13mm. solid. Negative (lower) 11mm. solid. Length of Arc, 0mm. B and C, Carbons: Positive (upper) 9mm. solid. Negative (lower) 8mm. solid. Length of Arc, 4mm.

negative carbon, which is of much smaller diameter than the remainder, indicates how far down the burning action of the amount of volatile carbon given off by the smaller current extended.

Fig. 11 shows the arc with solid carbons. In B a current of 3 amperes was flowing through an arc of 4mm. Comparing this with the arc of the same length, and with the same current in Fig. 9, for which the carbons were of the same size, the only thing to notice is that, as in Figs. 4 and 5, the visible arc, as shown by the dotted lines, is larger with a solid than with a cored positive carbon. Experience has shown that this is quite correct. In every case in which I have been able to compare the sizes of the arcs obtained with solid and cored positive carbons under similar conditions, I have found the area of the visible part to be larger when the carbons were both solid.

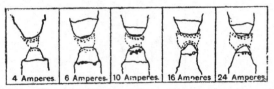

FIG. 12.—Carbons : Positive (upper) 13mm. solid.   Negative (lower) 11mm. cored.   Length of Arc, 4mm.

In Fig. 12 the positive carbons were solid and the negatives cored, with the result that the latter were burnt away farther down than they would have been if the cases had been reversed, and the negative carbons had craters in them.

SUMMARY.

When a direct-current silent arc is maintained between vertical carbon rods, the positive carbon being uppermost—

I. The tip of the positive carbon is white hot, and the tip of the negative has a white-hot spot on it.

II. A white-hot crater forms in the end of the positive carbon, and a more or less blunt point forms on the end of the negative.

III. The space between the two is filled by a violet light, the shape of which is defined by a shadow, which in its turn is bounded at its sides by a green light.

IV. The ends of both carbons are tapered, and the lengths of the tapering parts are increased both by increasing the current and by shortening the arc.

V. The diameter of the crater increases as the current increases, and also as the length of the arc increases.

VI. With uncored carbons the violet part of the arc is bluer, and all parts of the arc are larger than with cored carbons.

VII. With uncored carbons the violet part of the arc is of the form of an oblate spheroid when the arc is short, pear-shaped when it is long, and gourd-shaped when it is long and the current is very small.

VIII. With cored carbons the violet part is of the form of an oblate spheroid when the arc is short, gourd-shaped when it is long, and sometimes almost of the shape of a figure of 8 when the arc is very long for the current flowing.

IX. When the negative carbon is cored, a crater is formed in its tip exactly as if it were a positive carbon.

CHAPTER II.

A SHORT HISTORY OF THE ARC.

On March 20, 1800, Volta wrote his first letter announcing the discovery of his pile. The news was received by the scientific world with an enthusiasm only to be paralleled by that which was aroused at the end of 1895 by the discovery of the X-rays by Röntgen. New possibilities were opened up, and none could tell whither they might lead. A sort of experimental fever seized upon mankind, or at least upon the scientific part of it, and Paper after Paper was written describing new and interesting results obtained with the pile. So numerous were these Papers in the course of the next year that in the middle of 1801 a certain Dr. Benzenberg wrote to the editor of Gilbert's *Annalen*: "Could not the *Annalen*, in consideration of its object, be a little more varied? Galvanism, interesting as it is, is still only a very small part of physics. We can apparently only expect any real advance in knowledge from such work as is carried out on a large scale, and not from each experimenter, whose slight knowledge and small apparatus allow him to discover only what ten others have already found out before him."

The first question to which an answer was sought by all these numerous observers was, What is the nature of the new current? Is it a "galvanic" current? is it "common electricity"? or is it neither? Odd as it may now seem, many Papers were written to prove that the voltaic current had nothing in common with either galvanism or common—*i.e.*, frictional—electricity.

The early experiments may be divided into three classes, viz.:—(1) Those which dealt with the effect of the current on living things. (2) Those which produced chemical decomposition of inorganic matter, particularly of water. (3) Those which dealt with the heating power of the current, more par ticularly with the sparks produced by making or breaking a circuit. These last experiments led directly to the discovery of the arc, and are, therefore, the only ones with which we are immediately concerned.

One of the most ordinary ways of using frictional electricity was to produce sparks, and therefore one of the most obvious methods of showing that the voltaic current was of the same nature as "common electricity" was to make a spark by bringing together two conductors attached to the terminals of a battery. Most of the early observers were able to do this; but Sir Humphry Davy, towards the end of October, 1800, was the first to try the effect of using as conductors two pieces of well-burned charcoal, a substance which Priestley had already shown to be a good conductor of electricity.* In speaking of the result of using charcoal Davy said :—

"I have found that this substance possesses the same properties as metallic bodies in producing the shock and spark when made a medium of communication between the ends of the galvanic pile of Signor Volta." [1]

Later, in a lecture before the Royal Institution, given in 1801, Sir Humphry mentioned that the spark passing between two pieces of well-burned charcoal was larger than that passing between brass knobs, "and of a vivid whiteness; an evident combustion was produced, the charcoal remained red hot for some time after the contact, and threw off bright coruscations." [8]

This is evidently the description, not of an arc, but of a spark. For the essence of an arc is that it should be continuous, and that the poles should not be in contact after it has once started. The spark produced by Sir Humphry Davy was plainly not continuous; and although the carbons remained red hot for some time after contact, there can have been no arc joining them, or so close an observer would have mentioned it.

---

* Priestley's " History of Electricity," p. 598.

[1] To avoid the continual use of footnotes, the titles of the Communications referred to in the text, and of others of interest on the subject, are arranged in chronological order at the end of the Chapter. The numbers in the text refer to the numbers in this list.

In another lecture, delivered at the Royal Institution in 1802, in which he spoke of trying the effect of the electrical ignition of dry charcoal upon muriatic acid gas confined over mercury, Davy said, "The charcoal was made white hot by successive contacts made for nearly two hours." [9]

Hence it is quite certain, not only that he knew nothing about the arc at that time, but that the battery he used was incapable of maintaining an arc, otherwise the successive contacts would have been unnecessary.

In reading the accounts of the first experiments made upon the sparks produced by batteries, it seems as if the arc could hardly fail to be discovered very soon; as if in each case the next experiment *must* be the one that will produce a veritable arc. But this leaves out of account the resistance of the batteries used. The first batteries were mostly made of coins, such as the half-crown in England, and the double louis d'or in France and Italy, divided from pieces of zinc of the same size and shape by discs of cardboard soaked in dilute acid. The resistance of such a battery would be very great compared with what it should be in order to maintain an arc, and the passing of a spark would so lower the P.D. between the terminals that no other spark could pass till the battery had somewhat recovered. In fact, these batteries, having a high E.M.F. and great resistance, especially when they consisted of many pairs of plates, were exactly adapted to imitate the action of a frictional machine, and therefore to show that the voltaic current *was* an electric current—which was all that their devisers attempted at first to prove.

Cruickshanks very early discarded the cardboard discs, and arranged his pairs of plates in troughs containing dilute acid; and English experimenters immediately recognised the advantage of this arrangement. Most foreigners, however, continued for a long time to use the more primitive form of battery, and hence, although they made very numerous experiments upon the sparks produced by batteries both with charcoal and with metal terminals, their sparks probably remained sparks, and did not develop into arcs.

The diminution in the resistance of the battery caused by the use of *larger plates* was discovered early in 1801 by Fourcroy, Vauquelin and Thénard, who tried the effect of plates of different sizes. With eight pairs of plates eight inches in diameter they found that they could produce sparks brighter than with 120 pairs of smaller plates. Pfaff of Kiel says, in describing these experiments, " The rays streamed on all sides several lines wide, the crackling was very sharp, and in oxygen the wire burnt with a vivid flame." [5] Hence, the experimenters deduced the fact that batteries with large plates were the best to use for producing sparks and observing the heating effect of the current.

For some time after 1801 Davy and the other English observers confined themselves principally to experimenting upon the chemical effects of the current in decomposing substances which had proved refractory until this powerful agent was discovered. On the Continent, however, it was otherwise; the spark had its full share of attention for its own sake. In France Fourcroy, Vauquelin and Thénard, and in Germany and Austria Ritter, Tromsdorff, Gilbert and Pfaff, all experimented with it, and melted and burnt gold and silver leaf and thin wires by means of it, causing flames to arise between the two poles. Hence it is impossible to say when and by whom the arc was really discovered. For the arc, after all, is but a spark, which continues after the poles are separated, and which melts and burns or volatilises the substance of the poles. These experimenters probably did not see any great distinction between a continuous and rapidly following shower of sparks and a single spark which continued. They never mentioned the time of *duration* of their sparks, and they were so much accustomed to sparks passing between the two poles of a frictional machine without actual contact taking place, that it is very possible that a spark continuing to exist *after* the poles had been separated would appear quite natural to them.

The following abstracts and extracts will suffice to show the impossibility of judging when and by whom the arc was really discovered.

In 1801 Fourcroy, Vauquelin and Thénard, with a battery of plates a foot square, " ignited wires immediately, and in oxygen they burnt with a very vivid light." [10]

In the same year Ritter, of Jena, wrote to Gilbert, the Editor of the *Annalen,* that in trying to observe which end of a zinc-silver battery evolved the greater heat, he found that, when a silver leaf was attached to the zinc end and well burnt clean charcoal to the silver end, the silver could be completely burnt away by making contact between it and the charcoal ; while, if the position of the silver and charcoal were reversed, the silver did not burn, but there appeared on the charcoal "yellow, more than instantaneous sparks, which were not seen in the other experiment, quite sharp edges of the charcoal appeared to become blunt, in short, everything pointed to a *combustion of the charcoal.*" [6]   (The italics are Ritter's.)

In the same letter Ritter mentioned that when he had iron wires on each side of the battery, and a spark passed, they sometimes became melted together, and he had to use some force to separate them.   He also said, " I have spoken above of the big spark that the battery of 224 plates gave at the *closing* of the silver wire with the zinc plate above it.   But also on *breaking* the circuit it gave sparks. . . .   On quickly drawing away the iron wire attached to the silver plate in a vertical direction from the surface of the zinc plate, a small red spark appeared, which seemed to come with more certainty when the circuit had been closed for some time before it was opened."

In 1802 was published the following account of experiments made in April of the same year by Prof. Tromsdorff, at Erfurth.   A leaf of fine gold, after having been fixed to the zinc end of the pile, ignited and burnt with a crackling noise when the wire of the copper side was brought in contact with it.   " Other metals burnt with flames of different colours. . . . .   To prove that the ascension of the metals is a true oxidation, the experiments may be performed in a hollow glass sphere ; the oxide will adhere to the sides of the glass, and may be collected." [11]

The next account, also in 1802, is anonymous : " Two carbon rods, which were attached as conductors to a battery of 26 pairs, were brought into contact in a receptacle full of oxygen.   They caught fire and burnt." [12]

In 1803 Mr. Pepys, an Englishman, with a battery of 60 pairs of zinc and copper plates, disposed in two troughs, after the plan suggested by Cruickshanks, found that " carbons of boxwood not only ignited at their point of contact, but glowed red for a distance of quite two inches, and continued to do so for some time." [13]

In 1804 Cuthbertson, another Englishman, wrote "charcoal was deflagrated and ignited for about one inch by a battery." [14]

From 1804 till 1807 very little that was fresh was done. The old experiments were repeated by many observers, but no advance was made.   In 1807 Cuthbertson wrote more fully on the subject in his book on " Practical Electricity and Galvanism," quoted by Prof. Silvanus Thompson in his Cantor Lectures of 1895 :—

" *Experiment 209, Deflagration of Charcoal by Galvanic Action.*—The charcoal for this experiment must be made of some very close grained wood, such as boxwood or lignum vitæ, well charred, cut into pieces about an inch long, one end being scraped to a point, and the other so that it can be held by a port-crayon fixed to the end of one of the directors ; then, approaching the point of charcoal to the end of the other director, light will either appear or the charcoal will be set on fire.   The particular management required should be obtained by trials.   The light, when properly managed, exceeds any other artificial light ever yet produced." [15]

In his Bakerian Lecture delivered on November 16th, 1809, Davy said that when a current was sent by 1,000 double plates each four inches square through potassium vapour between platinum electrodes, over nitrogen gas, a vivid white flame arose.   " It was a most brilliant flame, of from half an inch to one and a quarter inches in length." [17]

In the library of the Royal Institution are two large thick volumes of manuscript notes, bound in leather, and carefully paged.   These are Sir Humphry Davy's laboratory notes for the years 1805 to 1812.   Faraday, who paged them, wrote a short note at the beginning of each volume, saying that Davy had a way, before he went to live at the Royal Institution, of tearing out pages from his note books, and taking them home with him to think over ; but that these two volumes being complete, he, Faraday, had paged them.   Finding that there was no more direct mention of the arc than the above in any of Sir Humphry Davy's published works, nor in the *Philosophical Transactions* of the Royal Society, nor in the *Philosophical Magazine* before 1812, I searched through these two volumes for any record of the first discovery of the arc, and found the following two passages which I am kindly permitted to publish.

" April 20, 1808.

"A given quantity of muriatic acid gas was acted upon by dry charcoal; there was a continued vivid light in the galvanic circuit."

" August 23, 1809.

" AN EXPERIMENT TO ASCERTAIN WHETHER ANY HEAT SENSIBLE TO THE THERMOMETER IS PRODUCED BY THE ELECTRIC FLAME IN VACUO.

"The jar which contained the apparatus consisted of a concave-plated mirror, so situated as to collect the light radiating from the charcoal, and to concentrate them (*sic*) on the bulb of a mercurial thermometer, which, together with the wires holding the two pieces of charcoal, passed through a collar of leather. No heat was apparently produced by the light excited in vacuo. The air being introduced, immediately the column of mercury rose. The light in vacuo was in part of a beautiful blue colour, and attended with bright red scintillations." [16]

The "vivid light" referred to in the first of these extracts is plainly an arc; but in the second, the words "electric flame" leave no room for doubt, not only that Davy was using an arc, but that it was no new phenomenon to him. When was the arc discovered then, and by whom? Was it not simply evolved through experiments on sparks and on the burning of metals, so gradually that no one realised it as a separate phenomenon until, with the large battery subscribed for by the members of the Royal Institution, Davy made a very long horizontal arc which formed a true arch, and therefore appealed to the imagination as something new? Even then, however, it was chiefly considered interesting as showing the immense power of the battery, as will be seen from the following accounts, the first of which, taken from the *Monthly Magazine*, a sort of popular journal of art, science, and literature, is, I believe, the first definite published account of the arc. The second is from Davy's "Elements of Chemical Philosophy," published in 1812.

"At the concluding lecture for the season at the Royal Institution the large voltaic apparatus, consisting of 2,000 double plates, four inches square, was put in action for the first time. The effects of this combination, the largest that has been constructed, were of a very brilliant kind. The spark, the light of which was so intense as to resemble that of the sun, struck through some lines of air, and produced a discharge through heated air nearly three inches in length, and of a dazzling splendour. Several bodies which had not been fused before were fused by this flame. . . . . Charcoal was made to evaporate, and plumbago appeared to fuse in vacuo. Charcoal was ignited to intense whiteness by it in oxymuriatic acid, and volatilised by it, but without being decomposed." [19]

Here is Davy's own account :—

"The most powerful combination that exists in which number of alternations is combined with extent of surface, is that constructed by the subscriptions of a few zealous cultivators and patrons of science, in the laboratory of the Royal Institution. It consists of two hundred instruments, connected together in regular order, each composed of ten double plates arranged in cells of porcelain, and containing in each plate thirty-two square inches, so that the whole number of double plates is 2,000, and the whole surface 128,000 square inches. This battery, when the cells were filled with one part of nitric acid and one part of sulphuric acid, afforded a series of brilliant and impressive effects. When pieces of charcoal, about an inch long and one-sixth of an inch in diameter, were brought near each other (within the thirtieth or fortieth part of an inch), a bright spark was produced, and more than half the volume of the charcoal became ignited to whiteness, and by withdrawing the points from each other a constant discharge took place through the heated air, in a space at least equal to four inches, producing a most brilliant ascending arch of light, broad and conical in form in the middle.

"When any substance was introduced into this arch, it instantly became ignited; platina melted as readily in it as wax in the flame of a common candle; quartz, the sapphire, magnesia, lime, all entered into fusion; fragments of diamond, and points of charcoal and plumbago, rapidly disappeared and seemed to evaporate in it, even when the connection was made in a receiver exhausted by the air pump; but there was no evidence of their having previously undergone fusion.

"When the communication between the points positively and negatively electrified was made in air rarefied in the receiver of the air pump, the distance at which the discharge took place increased as the exhaustion was made, and when the atmosphere in the vessel supported only one-fourth of an inch of mercury in the barometrical gauge, the sparks passed through a space of nearly half an inch ; and, by withdrawing the points from each other, the discharge was made through six or seven inches, producing a most beautiful coruscation of purple light, the charcoal became intensely ignited, and some platinum wire attached to it fused with brilliant scintillations, and fell in large globules upon the plate of the pumps. All the phenomena of chemical decomposition were produced with intense rapidity by this combination. When the points of charcoal were brought near each other in non-conducting fluids, such as oils, ether, and oxymuriatic compounds, brilliant sparks occurred, and elastic matter was rapidly generated ; and such was the intensity of the electricity that sparks were produced even in good imperfect conductors, such as the nitric and sulphuric acids." [20]

FIG. 13.—Horizontal Arc, copied from the figure in " Davy's Elements of Chemical Philosophy."

This very definite and beautiful description of the arc leaves no doubt that Sir Humphry Davy was the first to show the long horizontal arch of flame that gives the arc its name ; although the question whether or not he was the first person to obtain an arc of any shape and size will probably remain for ever a mystery.

After 1812 no important work on the arc was done till 1820, when Arago suggested that it would probably behave like a flexible conductor, and both attract a magnet and be attracted by it. He thought a very powerful battery was needed to produce such an arc as would show the deflection; and not having one himself, he suggested that someone who had should try the experiment.[21] Meanwhile Davy, working on the same lines, had come to the same conclusion; and without having seen Arago's suggestion, which, however, had been published before

he made the experiment, he tried the effect of an arc and a
magnet on one another, and found that they deflected one
another, just as Arago and he had predicted. It was on this
occasion that he gave the name of *arc* to the electric flame.[22]

In 1821 Dr. Robert Hare, a Professor in the University
of Pennsylvania, published an account of his "Galvanic
Deflagrator," an improved form of battery, with which he
found that he could get much finer heating effects than with
the older forms. His notions were peculiar, for he thought
that "the fluid extricated by Volta's pile " was " a compound
of caloric and electricity," both of which were material fluids.
He remarked that "the igneous fluid appeared to proceed from
the positive side," which later observers construed into his having
been the first to notice that material was carried from the positive
to the negative pole. He also said that, when the positive pole
was of charcoal and the negative of steel, the light was the most
vivid that he had ever seen, and the charcoal assumed a pasty
consistence as if in a state approaching to fusion. He was the
first to suggest that the charcoal could retain this state without combining with the air, and burning away, " because of
the volatilisation of the carbon forming about it a circumam-
bient air." [23]

Silliman himself, the editor of the Journal, next obtained a
deflagrator and made some very notable discoveries. It is a
little difficult at first to follow the course of his observations,
for the poles of his deflagrator appear to have got mixed up ;
but he explained this in a subsequent number of the Journal,
and even if he had not done so, in the light of our present
knowledge there would have been no real possibility of mistake.

He first observed what must have been a hissing arc, for he
described very clearly the formation of a mushroom at the end
of the negative charcoal, and pointed out that the negative
charcoal *grew* in length during the process, and that, therefore,
particles must be shot out from the positive charcoal on to it.
To confirm this observation he described and named the crater
in the positive pole, and observed that, as he moved the negative over the surface of the positive pole, it produced a crater-
shaped cavity over every place where it rested. The first
notice of the peculiar smell of the arc is also due to him. " I
should observe that during the ignition of the charcoal points,

there is a peculiar odour somewhat resembling electricity."
He examined the negative charcoal after the arc was extinguished, and found that it appeared to have been fused by the
heat.[24] He described the boiling bubbles that appear on both
carbons while the arc is burning, examined some of these when
cold, and came to the conclusion that they consisted of melted
carbon. He examined under the microscrope a mushroom formed
between charcoal as the positive pole and plumbago as the
negative, and found "a congeries of aggregated spheres with
every mark of perfect fusion and with a perfect metallic lustre."
These spheres were of many colours, and some were black and
some white. The coloured ones were attracted by a magnet,
proving that they contained iron. The black ones he considered
to be melted carbon.[25] Later he fused two carbons together by
allowing them to touch while an arc was burning between
them; but his experiments were, unfortunately, stopped by ill
health, and were never resumed.[26]

A great gap now appears in the history of the arc, and
there is nothing noteworthy to record till 1838, when Gassiot
showed that the temperature of the positive electrode was
much greater than that of the negative. Ritter had already
shown this for a spark, unless, perchance, he had an arc, which
is possible ; but Gassiot showed that, of two wires of the same
substance and diameter, that which formed the positive pole
of a horizontal arc was melted so far along as to bend down,
while the negative remained perfectly stiff.[27] In the same year,

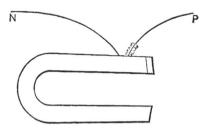

FIG. 14.—Rotation of the Arc at the Pole of a Magnet. (Copied from the
Transactions of the London Electrical Society.)

in conjunction with Walker, Sturgeon, and Mason, he also
first observed the rotation of the arc at the pole of a magnet.

They found that if they completed the circuit of a powerful battery through the pole of a magnet, so that an arc was maintained between a wire from the positive terminal of the battery and this pole, then the arc would rotate—clockwise if the pole were north-seeking, and counter-clockwise if it were south-seeking.[28]

Daniell, the inventor of the cell which bears his name, made many experiments with large batteries. Using charcoal for the positive and platinum for the negative pole of an arc, he found that the platinum became coated with carbon, which was beautifully moulded to its shape; while, if he used platinum for the positive pole and charcoal for the negative, the latter became covered with little globules of platinum after the arc was extinguished. This confirmed Silliman's discovery that the material of the positive pole was shot out on to the negative. Finding that with his battery the arc would not start without actual contact of the poles, Daniell tried sending a spark from a Leyden jar between the poles when they were apart, and succeeded in igniting the arc by this means.[29]

In 1840 Grove made some very notable experiments to determine whether the amount of matter separated from the poles by a given quantity of electricity was constant for a given material. He came to the conclusion that it was, and that "the all-important law of Faraday is capable of much extension." The law he alluded to was, of course, the law of electrolysis, and shows that he considered the action of the arc to be purely electrolytic. Indeed, in the Paper in which he described these experiments he said, "The passage of the current is, as proved in these experiments, materially modified by the nature of the elastic medium through which it passes, and is greatly aided when such medium is capable of uniting chemically with the electrodes. In pure hydrogen I have never yet been able to maintain a continuous arc, except with charcoal, which forms carburetted hydrogen." Grove tried using the carbon from gas retorts, but found that charcoal gave a larger and more diffuse flame. He described the three requisites for a brilliant discharge in an oxydating medium as being oxydability, volatility, and looseness of aggregation of the particles.[30]

To Becquerel is due the honour of having discovered that the electric light had the same chemical effect on the salts of silver as sunlight,[31] and De la Rive first used this power of the arc to obtain a daguerrotype, a faint one it is true, of a bust.[32]

In the same year Mackrell obtained an arc between two fine iron wires in dilute sulphuric acid, and he found that, if he used a positive pole of iron and a negative one of charcoal, if the iron were put into the acid first the charcoal became brilliant, but if the charcoal were immersed first, no such result took place.[34]

Casselmann, in 1844, first described the shape of a long arc, as two cones with their bases in contact with the carbons, and their points touching one another. He allowed the arc to burn away till it became extinguished, under different conditions, and found that with the same battery a longer arc could be maintained between charcoal electrodes than between electrodes of carbon prepared as the plates of a Bunsen battery are prepared. Between these carbons also the arc hissed, but if they were heated red hot beforehand, and steeped in solutions of such volatile substances as sodium or potassium, the arc burnt quietly and steadily, and would grow to a greater length before it became extinguished.[35]

In 1845 a very curious discovery was made by Neef, who wished to find out at which part of the arc the light first appeared. In order to eliminate any secondary heating effect produced by the combination of the hot positive pole with the oxygen of the air, he used platinum poles—a plane for the positive and the point of a cone for the negative. Between these he struck an arc with a very small current by means of a spark from a Leyden jar. He then observed the effect with a microscope, and found that the light started at the negative pole, with no perceptible heat. The results he considered he obtained were the following :—

(1) The light always appears first at the negative pole, and this first light is independent of combustion.

(2) The source of the heat is the positive pole, and this heat is originally dark heat,

(3) The light and heat do not, at first, mingle, but only when they have attained a certain intensity ; from this fusion the phenomena of combustion and the flame are produced.

He was the first to suggest that the carbon which was fused and shot off from the positive pole was condensed at the negative pole to the specific gravity of crystallisable graphite.[37]

De la Rive, after having made the first daguerreotype with the arc, made a series of experiments as to the longest arc that could be maintained by a given battery when the poles were of various substances. He used a voltameter to measure the current, and found that the *current* that was flowing when the arc became extinguished was the same for all substances, but that the *length* of the arc varied with the substance. He noticed that when he had a slab of carbon for the negative pole and a pointed carbon rod for the positive, the deposit from the positive took a regular form. Also the longest arc was only half as long with this arrangement as it was when the positions of the slab and the rod were reversed. When the poles were of magnetised iron the longest arc was much shorter than when the iron poles were unmagnetised.[38]

In the same year Van Breda, experimenting on the arc in vacuo, found that with copper poles, when he placed a slip of iron in the arc between them, the copper of *both* poles became covered with iron, and there were traces of copper on the iron slip. On weighing both electrodes and the iron slip, he found that the electrodes had each gained in weight, and the iron lost, but that taking all three together there had been a loss of weight. With one electrode of iron and one of coke, the iron lost more weight than the coke, whether it was at the positive or negative pole.[39]

Matteucci made some very accurate and important experiments on the amount of matter lost by each electrode in a given time. Like Van Breda, he came to the conclusion that matter was shot out by both poles, and he pointed out that the two sets of particles, being in opposite electric states, must attract one another. He found that the difference of temperature between the poles was greater the smaller the conductivity of the electrodes, and that the amount of matter lost by each depended upon its temperature, its oxydability, and the volatility and fusibility of the products of oxydation. Both poles lost more in air than in a vacuum in a given time. With coke electrodes he estimated that the proportion of the loss at the positive pole to that at the negative varied between 2 to 1 and 5 to 1, according to the length of the arc.[40]

Quet, placing a vertical carbon arc perpendicular to the common axis of the coils of an electromagnet, found that the arc shot out horizontally, like the flame of a blow-pipe, and that, unless the carbons were very close together, the arc became extinguished with a loud noise.[41] Several electric blow-pipes on this principle have since been invented.

In 1852 Grove made a most interesting experiment, which showed that a liquid could act as one pole of an arc. He attached fine platinum wires to the terminals of a battery of 500 cells, dipped the ends of both wires into distilled water, and then gradually withdrew the negative wire till it was a quarter of an inch above the surface of the water. " A cone of blue flame was now perceptible, the water forming its base, and the point of the wire its apex. The wire rapidly fused, and became so brilliant that the cone of flame could no longer be perceived, and the globule of fused platinum was apparently suspended in air, and hanging from the wire ; it appeared sustained by a repulsive action, like a cork ball on a *jet d'eau*, and threw out scintillations in a direction away from the water. The surface of the water at the base of the cone was depressed and divided into little concave cups, which were in a continual agitation." When the conditions were reversed, so that the negative wire was immersed and the positive out of the water, the effect was the same, but not so marked. The cone was smaller, and its base was much narrower in proportion to its height.[42]

The first accurate quantitative experiments on the arc were made by Edlund in 1867. In spite of the disadvantages under which he laboured—for he had no dynamo, and at that time there were no recognised units of current, P.D., or resistance— Edlund discovered one of the fundamental conditions of the arc, namely, that with a *constant current* the apparent resistance is equal to a constant resistance, plus a resistance which varies directly with the length of the arc.

Edlund's method of experimenting was as follows : He used a battery to send a current through an arc of definite length, then, putting out the arc and pressing the carbons tightly together, he measured the distance by which he had to separate the plates in a copper voltameter so as to bring the current to

the same value as before. Doing this for various lengths of arc
he found that, *as long as the current was kept constant,* the length
of the copper voltameter which represented the arc could be
expressed by a constant length plus a length which varied
directly with that of the arc.

Hence, putting his result in the form of an equation, he
found out that

$$r = a + b\,l,$$

where r is the apparent resistance of the arc, l its length, and a
and b constants *for a constant current.*

When, however, the current was varied, he found that a and b
both diminished as the current increased, so that the apparent
resistance of the arc for a given length was smaller the greater
the current. The numbers which he obtained for a justified
him, he considered, in concluding that a varied inversely as the
current, but those for b were too small to enable him to arrive
at the law connecting b with the current.

Edlund started his experiments with the idea that there
was a back E.M.F. in the arc caused by the disintegration of
the carbon particles, and therefore, having obtained several
values of a and b with different currents, he calculated the
back E.M.F. in the arc when each of those currents was
flowing.

From these calculations he concluded that the *back E.M.F.
in the arc had a constant value equal to that of about 23 Bunsen's
cells, for all the currents he used,* and that *the true resistance of the
arc was directly proportional to its length, increasing, however, as
the current decreased.*

He next considered the question theoretically, and gave
what he regarded as a proof that the back E.M.F. in the arc
must be independent of the current, and also that the work
performed in the arc by the current was proportional to the
current as long as the E.M.F. of the battery remained
constant.

He then made a new series of experiments to determine
whether the back E.M.F. depended upon the E.M.F. of the
battery used to produce the current, and decided that it did
not.[44]

Edlund's next series of experiments was undertaken to
find out whether the back E.M.F. was constant with smaller

currents than had been employed in the first series. The deflections produced by the currents in the tangent galvanometer used were not published in this case, as they had been before; but Edlund considered that his results showed that, in the case of small currents, the back E.M.F. *diminished* as the current was decreased. He used copper, brass, and silver pole points for these experiments, as well as carbon.[45]

Edlund's third series of experiments was very striking. He found that, when an arc was produced with a somewhat large current between *carbon* poles, the arc continued for a short time after the circuit was broken, so that, if the circuit were closed again fairly quickly after the break had been made, the arc was not put out. But, on the other hand, when *silver* poles were used, the arc did not continue for even one-eightieth of a second after the circuit was broken.

From this he concluded that if carbon poles were used, and if immediately after breaking the circuit the carbons were switched on to a galvanometer, a momentary current would be sent by the arc through the galvanometer, and the existence of the back E.M.F. in the arc would thus be made certain. A series of experiments carried out in this way, with and without a battery being switched with the arc into the galvanometer circuit, led Edlund to conclude that the back E.M.F. existing in the arc *after* the main circuit had been broken could not be less than that of from 10 to 15 Bunsen's cells. Further, from the momentary current being increased when the negative carbon was warmed by a Bunsen's burner, and not diminished as might have been expected had the back E.M.F. been a thermo-electric one (set up by the positive carbon being hotter than the negative), he concluded that the back E.M.F. in the arc was *not* due to thermo-electric action.[46]

In 1876 the Jablochkoff candle came into use. It consisted of two upright parallel carbon rods, separated from one another by a layer of solid insulating material, which burnt away at about the same rate as the carbons. This solid material was supposed to be necessary to keep the arc from running down the carbons and burning anywhere but at the tips.

When, later, it was discovered that the arc still remained at the ends of the carbon, even when there was nothing but air between them, this was attributed to the action of the upward current of hot air and vapour. In 1878, however, Prof. Ayrton gave the true explanation, and proved that the hot-air theory could not be correct, by showing that the arc remained at the ends of the carbons, even when they were held upside down. He pointed out that in remaining always at the ends of the carbons the arc was simply following Ampère's law concerning the tendency of a circuit in which a current is flowing to enlarge itself, on account of the repulsive action of the current in one part of the circuit on the current in another part at right angles to it. Prof. Ayrton's own explanation will make this clearer.

"The figure below, in which the continuous arrows indicate the directions of the currents, and the dotted arrows the line of action of the repulsive forces, show this action clearly. The carbons are placed further apart in the figure than they are in reality, merely for convenience of drawing. The two forces are

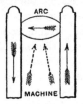

FIG. 15.

oblique to the carbons ; but the resultant is parallel to them, and will always be away from those ends of the carbons which are connected with the magneto-electric machine, no matter how often the whole current be reversed." [49]

This explanation also applies to the form taken by a horizontal arc, and shows that the upward direction of Sir Humphry Davy's long arc (*see* Fig. 13, p. 27) depended principally on the position of the poles relatively to the remainder of the circuit. The same cause makes the vertical arc, when it is long and the current is small, take up a position between the points of the carbon which are farthest from each other, and thus make itself as long as possible.

Mr. Schwendler, during the course of an investigation on "The Electric Light," carried out in 1878 for the Board of

Directors of the East Indian Railway Company, gave the following conclusions in a *précis* of his report :—
" There appears to be no doubt that an appreciable E.M.F. in the arc is established, which acts in opposition to the E.M.F. of the dynamo machine. This E.M.F. of the arc *increases* with the current passing through the arc. The resistance of the arc for constant length is also a function of the current passing through it, *i.e.*, the resistance of the arc *decreases* with the current (*see* the following table)."

Table I. (*Schwendler*).

Current in webers.	Resistance of the arc in Siemens units.	E.M.F. of the arc in volts.
28·81	0·91	2·02
23·87	1·72	1·91
16·27	1·97	1 86

Mr. Schwendler said, further, that he considered it highly probable that the resistance of an arc of constant length was inversely proportional to the current. No details were given in this *précis* regarding the way in which the E.M.F. of the arc was measured.[50]

In 1879 Messrs. De la Rue and Hugo Müller, while experimenting on the discharge in vacuum tubes, came to the conclusion that "the stratified discharge in a vacuum tube is simply a magnified form of arc." In order to test this theory they made a series of experiments on discharges in various gases at various pressures, with the poles at various distances from one another, and of different shapes. The poles were fixed in a bell jar that could be filled with the different gases and exhausted ; a suitable arrangement was made for altering the distance between the poles, and the gases used were air, hydrogen and carbonic acid.

In air the pressures varied from 2·6mm. to 761mm., the distances varied between 0·54in. and 6 4in., the current ranged from 0·01390 to 0·04474 webers, and the number of chloride of silver cells used varied from 10,940 to 11,000. The substance of the electrodes was only definitely mentioned in one case, when it was brass. Many beautiful engravings, showing the appearance of the discharge under different circumstances, were published; from these it appeared that in air the light usually divided itself into at least two and sometimes more parts, with dark spaces between them.

In hydrogen, in some cases, the discharge showed a very definite stratification. In carbonic acid, when the pressure was very small, there was very little evidence of stratification; but in both gases the discharge was divided, as in air, into light and dark parts.

Mr. Seaton, one of the assistants of Messrs. De la Rue and Müller, first noticed a very interesting circumstance connected with the discharge. Whenever contact was first made the pressure in the bell jar increased far more than could be accounted for by the rise of temperature of the enclosed gas. The moment contact was broken the pressure fell almost to what it had been before contact was made, the slight increase being due to rise of temperature. It was found by experiment that the increase of pressure took place at both terminals equally. The experimenters considered it to be accounted for by "the projection of the gas-molecules by electrification against the walls of the glass vessel, producing thereby effects of pressure, which, however, are distinct from the molecular motion induced by heat."

The parts of the Paper which concern the arc are summed up in the following manner:—

"When the discharge takes place there is a sudden dilatation of the medium, in addition to and distinct from that caused by heat. This dilatation ceases instantaneously when the discharge ceases.

"The electric arc and the stratified discharge in vacuum tubes are modifications of the same phenomenon." [53]

In the same year Rossetti found, from a large number of experiments, that the maximum temperature of the positive pole was 3,900°C., and that of the negative about 3,150°C. The temperature of the arc itself he found to be about 4,800°C. whatever the intensity of the current flowing. The method of experimenting was not given.[54]

In the following year, 1880, the first measurements of the diameter of the crater of the positive carbon were made by Mr. J. D. F. Andrews, who was also the first to remark that, in order that a definite result might be obtained with each current and length of arc employed, the arc should be allowed to burn steadily and quietly for at least half an hour before any measurements were taken.

He measured the diameter of the crater, when cold, with a rule divided into 100ths of an inch, and, neglecting its depth, which was very small with the length of arc he used ($\frac{3}{16}$in.), he found the area from the diameter. He considered that his experiments showed that the area of the crater was " directly proportional to the quantity of current producing it." In proof of this he gave the following table :—

Table II. (*Andrews*).—*Comparison of Observed Currents and Currents calculated on the Assumption that Area of Crater is directly proportional to Current.*

Diameters of craters, in inches.	Areas of craters, in inches.	Quantities of cur. in webers measured.	Quantities of current in webers cal. from craters.
0·140	0·0196	9	9·9
0·156	0·0243	12	12·4
0·186	0·0272	...	13·8
0·203	0·0324	...	16·5
0·266	0·0556	29	28·3
0·326	0·0825	42	42·0
0·453	0·1602	81	81·6

The diameters of the carbons used were not given.

Mr. Andrews remarked that when the arc hissed the end of the positive carbon was covered with a number of small craters, showing that it moved about, and that a number of very small arcs appeared to try to spread over the end of the positive carbon, each detonating the air, and thus causing a hissing noise.[55]

Le Roux considered that the great fall of potential at the positive carbon was caused by a back E.M.F., the result of a thermo-electric effect. His idea was that the carbon of the positive electrode was electro-positive to the vapour of the arc in a degree which increased as the temperature increased. He found that with a high resistance galvanometer in the arc circuit he could detect the back E.M.F. 0·2sec. after he had stopped the arc by hand.[57]

Niaudet in 1881 first noticed that the hissing of the arc was accompanied by a sudden fall of P.D., and gave the following table of observations that he had made :—

Table III. (*Niaudet*).—*Currents and P.D. with Silent and Hissing Arcs.*[58]

Current in webers.	P.D. (presumably in volts).	——
34	54·3	Silent
36	43	Hissing
34	49	Silent
43	41·4	Hissing
38·1	49	Silent

In a Paper on "The Resistance of the Electric Arc" Profs. Ayrton and Perry in 1882 described experiments made with Grove's cells to test the accuracy of Mr. Schwendler's conclusions concerning the back E.M.F. and the resistance of the arc, and they found that, when carbons 0·24in. in diameter were used and the length of the arc kept constant, the current could be varied from about 5 to 15 amperes without much change being produced in the P.D. between the carbons. They also used a Brush machine to produce an arc up to 1·25in. in length, and gave as the equation connecting V, the P.D. between the carbons in volts, and *a*, the length of the arc *in inches*

$$V = 63 + 55\,a - 63 \times 10^{-10a},$$

and this equation they regarded as being true for currents between 5·5 and 10·4 amperes with the carbons employed.[59]

In 1882 Prof. Dewar made some very interesting experiments on the internal pressure of the arc. He used a horizontal arc maintained between carbon tubes, each of which was attached to a manometer. With a steady silent arc the manometer attached to the positive carbon showed a fixed increase of pressure of about 1mm. to 2mm. of vertical water pressure, while at the negative carbon there was a slight diminution of pressure. The same results were obtained when the arc and ends of the carbons were enclosed in a block of magnesia to equalise the temperature of the poles, and also when the manometers were filled with carbonic oxide or nitrogen instead of with air.

When the arc hissed, it often no longer covered the ends of the carbons, so that the manometers showed no pressure ; but when this did not happen, the positive carbon showed a *diminution* and the negative an *increase* of pressure with a hissing arc.

From these experiments Prof. Dewar concluded that the arc acted as if it had a surface tension.[60]

In 1883 Frölich, desiring to find out whether Edlund's equation $r = a + b\,l$ were true for the larger currents that could be obtained with dynamos, used the results of experiments made for the purpose of testing the dynamos constructed by Messrs. Siemens and Halske, by various people at various times, to obtain an equation connecting the P.D. between the carbons with the length of the arc. In the table given in his Paper the P.D.s corresponding with different currents for the same length of arc showed wide variations, yet he took it for granted that the P.D. was really independent of the current, and so he deduced the equation

$$V = a + b\,l$$

as the relation that would connect the P.D. between the carbons with the length of the arc, where a and b were the same, *whatever the current*, if there were no errors of observation in the experiments.

Frölich found what he considered to be the numerical values of a and b, and so he put his equation connecting V. the P.D. in volts between the carbons with l the length of the arc in millimetres in the form

$$V = 39 + 1 \cdot 8\,l.$$

From this formula he calculated what he considered ought to have been the P.D.s for the various currents and lengths of arc with *any current from 1 to 100 amperes*, in the 47 measurements the results of which he quoted; and the differences between the results given by his formula and by experiment he put down to errors of observation.

By dividing by A, the current in amperes, Frölich's formula became

$$r = \frac{39}{A} + \frac{1 \cdot 8\,l}{A},$$

and this he considered was the formula which gave r the apparent resistance in ohms of an arc l millimetres long produced by a current of A amperes. He used the latter formula to calculate a table of 273 apparent resistances of arcs, varying from 0mm. to 20mm. in length, and produced by currents of 1, 5, 10, 15, &c.,

up to 100 amperes; but no experiments were adduced giving results agreeing with his calculated values.

He then proceeded to inquire if his "newly-discovered facts" threw any light upon the question whether there was a back E.M.F. in the arc, or merely a contact resistance, and he concluded that either explanation was consistent with the facts, provided that the cross-section of the arc was directly proportional to the current, which he considered likely.

Multiplying both sides of his equation by A, he obtained the following equation, giving W, the power in watts expended in the arc :—

$$W = (a + b\,l)\,A\;;$$

and he concluded from this equation that the power expended in an arc of fixed length was directly proportional to the current.[61]

In 1885 Peukert published an account of some experiments he had made with constant currents and varying lengths of arc, for the purpose, as he said, of "eliminating one of the three possible variables, P.D., current, and length of arc." In order to avoid having a crater, and thus complicating the measurement of the length of the arc, which was made by projecting the distance between the carbons on to a millimetre scale, he had the carbons filed flat before each experiment, and the determination made before the carbons had had time to burn away much.

Having obtained the P.D.s needed to send currents of 10, 15, 20, 25 and 30 amperes through various lengths of arc, he plotted curves connecting the apparent resistance with the length of the arc for each current, and found that they were all straight lines, so that Edlund's equation,

$$r = a + b\,l,$$

was proved true for large as well as for small currents.

Peukert, like Edlund, found that a varied inversely as the current, but he pointed out that b diminished more quickly than the current increased. This latter fact, he thought, was partly explained by the air surrounding the arc becoming very hot when the current was large, and so acting itself as a conductor.

He considered the question of a back E.M.F. in the arc, and showed that the mean value of $A\,a$ in his experiments was about

35 volts. He thought, however, that an E.M.F. of this large value could not be set up by the disintegration of the carbon, as Edlund had supposed, and that a back E.M.F., if so produced, ought to increase with the current. He also considered the possibility of the back E.M.F. being produced thermo-electrically, or by the combination of the carbon points with the hydrogen gas liberated by decomposition of vapour in the air, but both these suggestions he rejected also.

Peukert finally concluded that most probably the resistance of the arc was really composed of two parts, one that of the arc proper, while the other, and much the larger part, was a mechanical, or contact, resistance between the arc and the carbon. And he added that a certain minimum P.D. being

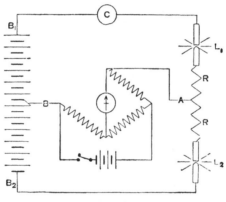

FIG. 16.

necessary to maintain an arc, which fact had been adduced as a proof of the existence of a back E M.F., might arise from a certain P.D. being necessary to tear off the carbon which formed the conducting medium between the two carbons.[62]

In 1885 Von Lang endeavoured to measure the *true* resistance of the arc, as distinguished from the *apparent* resistance, which alone had been determined by previous experimenters. To do this he used a battery, $B_1 B_2$ (Fig. 16), to produce two arcs L_1 and L_2, in series, each $\frac{1}{3}$mm. long, formed with carbons 5mm. in diameter; and between the arcs there was placed a resistance, R R. A point, A, in the resistance was then found, having the same potential as B, the middle point of the battery. Then, since these two points of the quadrilateral were at the same potential, the parallel resistance of the two halves of the quadrilateral between A and B could be measured with a Wheatstone's bridge, exactly as if there were no E.M.F.s in the quadrilateral.

This resistance Von Lang found to be 1·82 ohms, and therefore the resistance right round the quadrilateral was $4 \times 1·82$, or 7·28 ohms. Next the arcs were replaced by equal resistances of such a value that the current flowing round the quadrilateral, as measured by the ammeter, C, was the same as before, viz., 4·32 amperes. The parallel resistance between the points A and B was now found to be 6·29 ohms, and consequently the resistance right round the quadrilateral was $4 \times 6·29$, or 25·16 ohms. Hence the E.M.F. of the battery. $B_1 B_2$ was $4·32 \times 25·16$, or 108·7 volts. And, as the quadrilateral had a resistance of 7·28 ohms when the two arcs were in the circuit, a P.D. of $4·32 \times 7·28$, or 31·4 volts, must have been employed in sending the current of 4·32 amperes round this quadrilateral as far as its resistance was concerned. Consequently a P.D. of $108·7 - 31·4$, or 77·3 volts, must have been employed in overcoming the back E.M.F. of the two arcs, or the back E.M.F. of each arc, was 38·6 volts.[63]

Stenger in 1885 made some experiments in order to confirm the conclusion come to by De la Rue and Müller (*see* p. 37), that there is no sharp distinction between an arc and the discharge in a vacuum tube, and also to find out what are the conditions that tend to make the discharge take one form rather than the other, when the current is passed through electrodes in a vacuum, and under what conditions the two sorts of discharge may merge into one another.

He considered that they were distinguished from one another in four ways :—

(1). The gaseous portion has less resistance in the arc than in the tube discharge.

(2). In the arc the anode is hotter than the cathode, in the tube discharge the reverse is the case.

(3). In the spectrum of the arc the light of the substance of the electrodes predominates over that of the vapour between them, while in the tube discharge the spectrum only gives the lines of the gas and is independent of the nature of the electrodes.

(4). In the arc both electrodes waste away, in the tube discharge only the cathode does so.

As regarded the first two distinctions, he quoted Hittorf to show that the resistance in a vacuum tube diminished as the pressure increased. When contact was made with a tube of nitrogen at a pressure of 17mm., the current was extremely small, and the cathode was hotter than the anode. When the pressure was raised to 53mm. and all the other conditions were the same, the resistance diminished so much that the current increased to two amperes, "a strength that might be used for an arc" as Stenger said, and the anode became hotter than the cathode.

He next tried whether with carbon electrodes, in a good vacuum of either hydrogen or nitrogen, the poles of the arc were of equal temperature, as Grove had said. With the vacuum he could at first obtain, the positive pole was the hotter in both gases, though the difference of temperature was less than in air. He then tried a pressure of 0.1mm., but found that as soon as the arc was struck the pressure increased immensely, owing to the hot vapour sent off by the carbon poles. He continued to strike the arc and improve the vacuum till at last the pressure hardly increased at all when the arc was struck, "*at the same time the difference of temperature between the carbons disappeared.*" It was thus evident that Grove was right, and that Stenger had only been unsuccessful before because his vacuum was not good enough. On account of the carbon vapour he could never get a vacuum of less than from 1mm. to 2mm., otherwise he thought it probable that he might have succeeded in making the negative pole hotter than the positive, in which case he would have passed from an arc to a tube discharge by simply improving the vacuum.

The increase of pressure when the arc was struck was in some instances very striking, and was evidently of the same kind as that obtained by De la Rue and Müller. Also the moment contact was broken, the pressure fell to what it had been before contact was made, as it had done in their experiments.

Stenger considered that the small resistance of the gaseous part of the arc was due to the presence of metallic vapour, of which, he said, there was a considerable amount, even when the poles were of carbon. [65]

Shortly afterwards Edlund summed up the results obtained by himself, Frölich, Peukert, and Von Lang, and proceeded to consider Peukert's view that the term a in the equation $r = a + b\,l$ represented a resistance of transition at the surface of contact of the electrodes and the arc. Edlund concluded that, since the value of the back E.M.F. of the arc—39 volts —obtained by Von Lang's method, which depended on the measurement of the true resistance of the arc, did not differ much from either the values obtained by himself, 41·97 volts, or by Frölich, 39 volts, or by Peukert, 35 volts, it followed that *there was no resistance of transition in the voltaic arc, and that, therefore, the entire diminution of the strength of the current which resulted from the production of the arc-light was caused by the resistance of the arc $b\,l$ and by the E.M.F. contained in it.* [66]

In 1886 Messrs. Cross and Shepard made experiments with both silent and hissing arcs to determine whether the apparent resistance of the "whistling" arc was a linear function of the length for a constant current, as it had been shown to be by Edlund and others for a silent arc. They used currents varying from 3·27 to 10·04 amperes, and lengths of arc from 0·25/32 to 16/32 of an inch, and found that for a given current the resistance of the whistling arc *could* be represented by a straight line in terms of the current, but that this line was steeper than that representing the resistance of a silent arc for the same current.

Without expressing any opinion as to the existence of an inverse E.M.F. in the arc, they gave this name, for convenience sake, to the product A a, where A was the current and a was given by the resistance equation

$$r = a + b\,l,$$

and they found by experiment that both with silent and with whistling arcs this product A a diminished as the current increased, having a mean value of 39·33 volts for a silent and 14·98 volts for a whistling arc. With very long arcs, however, and with long arcs in which metallic salts were volatilised, they found that the apparent resistance tended to become abnormally small.

Next they experimented with an inverted arc, and although it was much less steady than the upright arc, the results obtained were practically the same.

The arc having been restored to its original position, they surrounded the positive carbon with a deep cup-shaped shield of fire-clay in order to retain the heat, and found that this increased the inverse E.M.F. both for silent and for whistling arcs, and that it also increased the length of the arc at which whistling occurred for each particular current. On the other hand, b, the coefficient of l, was diminished.

Cooling the positive carbon with a water jacket caused the E.M.F. to fall to 11·7 volts for a current of 7 amperes and to 5·6 volts for one of 8 amperes.

Finally, a few experiments were made on an arc in a partial vacuum under a pressure of only 4in. of mercury. The equation $r = a + bl$ still held, but the arc hissed for all the lengths and currents tried, and the lines connecting resistance and length were less steep than those for hissing arcs under normal pressure for the same currents respectively.[67]

In the same year, 1886, Nebel used a method somewhat similar to Von Lang's for the determination of the back E.M.F. of the arc, but instead of the two arcs employed by Von Lang his method required but one. He did not publish the results of his experiments, but contented himself with making a mathematical determination of the conditions that should exist between such results when obtained.

He afterwards made experiments for the purpose of determining the P.D. that existed between carbon electrodes when arcs of various lengths were burning with various currents flowing. He used two carbons of equal diameter, of which the positive was cored and the negative uncored. Pairs were employed having diameters of 10, 12, 14 and 16 mm. He found the depths of the craters by filling them with half-melted sealing wax, which he afterwards measured with a spherometer.

Nebel drew curves, the first of the kind ever published, connecting the P.D. between the carbons with the current flowing for various constant lengths of arc; and he pointed out that, with the carbons he used the P.D. diminished, fell to a minimum, and then rose again as the current increased from its lowest to its highest value. The minimum P.D., as he remarked, corresponded with a larger current the longer the arc.

Table IV.—(*Nebel*).

Positive Carbon Cored, Negative Uncored.

Amperes.	10 mm. Carbons.						12 mm. Carbons.					
	1 cm. 1·8	2 cm. 3·1	3 cm. 4·4	4 cm. 5·6	5 cm. 6·9	6 cm. 8·2	1 cm. 2·1	2 cm. 3·4	3 cm. 4·7	4 cm. 5·9	5 cm. 7·2	6 cm. 8·5
	P. D. between the Carbons in Volts.						P. D. between Carbons in Volts.					
4	46·28	55·88	...	...	...	...	...	...	...	...	...	...
6	43·18	49·32	51·94	...	...	...	42·66	49·00	...	...	...	...
8	42·08	46·16	50·20	55·18	...	...	40·90	45·36	51·68	...	...	...
10	43·00	45·06	49·24	51·82	57·26	...	40·16	44·12	47·40	50·74	55·04	...
12	43·40	45·30	48·38	51·22	55·88	57·06	40·82	44·34	47·32	50·54	54·10	57·62
14	43·46	46·08	47·68	50·84	54·64	56·20	...	...	47·72	50·06	53·66	56·44
16	43·52	46·26	47·02	50·14	53·36	56·02	40·90	44·52	47·82	49·42	53·23	56·36
18	43·72	46·58	47·18	49·66	51·06	55·62	...	...	...	...	52·04	...
20	37·6*	46·74	47·32	49·96	50·98	54·76	41·82	45·92	48·14	50·00	52·10	56·08
22	...	46·78	47·69	49·96	51·14	53·10	...	...	...	...	52·22	...
24	...	47 00	47·90	50·66	51·38	53·92	41·98	46·16	49·72	50·06	52·62	56·52
26	...	44·14*	...	50·84	51·84	...	...	...	49·98	51·58	53·22	...
28	...	...	...	...	...	...	...	...	...	51·80	53·22	57·10

Amperes.	14 mm. Carbons.						16 mm. Carbons.					
	1 cm. 2·7	2 cm. 4·0	3 cm. 5·2	4 cm. 6·7	5 cm. 7·8	—	1 cm. 2·7	2 cm. 4·0	3 cm. 5·2	4 cm. 6·5	—	—
10	40·84	44·42	51·16	55·50	58·66	...	...	...	...	...	...	...
12	40·00	43·34	48·80	51·76	53·56	...	40·76	42·98	48·80	...	...	...
14	39·80	42·46	46·70	49·12	51·20	...	...	...	...	...	...	...
16	40·10	42·07	45·78	48·68	50·70	...	40·36	42·44	45·96	55·46	...	...
20	40·34	42·16	45·08	48·42	50·54	...	40·30	42·32	45·74	50·42	...	...
24	40·39	42·56	45·32	47·92	50·76	...	40·52	42·08	45·64	48·88	...	...
26	40·78	42·80	46·04	48·22	50·78	...	...	...	...	...	...	...
28	41·23	42·96	46·50	48·78	50·82	...	40·58	41·94	45·72	47·44	...	...
30	37·68*	43·40	46·68	48·92	50·90	...	...	...	...	...	...	...
32	...	..	...	...	...	...	41·12	42·12	45·72	47·20	...	...
34	...	...	...	..	...	...	...	...	...	47·74	...	...
36	...	...	...	..	...	...	...	42·22	45·78	47·90	...	...
38	...	...	...	...	...	...	...	...	...	48·06	...	...

He also drew curves connecting the P.D. between the carbons with the length of the arc for constant currents, but did not publish them. These curves were not all straight

---

* Hissing.

lines, as he pointed out, but dipped down towards the axis of length for small currents. He remarked that they made a smaller angle with the axis of length the larger the current, and applied Frölich's formula

$$V = a + b\,l$$

to them, thus treating them as if they *had* been straight lines. He found what he considered to be the values of a and b in this equation for each of his sets of results, and from these values it appeared that with his carbons a *increased* as the current increased. He came to the conclusion, however, that it was still uncertain whether a depended upon the current or not, but showed that it diminished as the diameter of the carbons increased. He considered his experiments to prove that b depended on the current.

He calculated the apparent resistances of the arc from the P.D.'s and currents and plotted curves, which he published, connecting these with the lengths of the arc for various constant currents.

In the table of Nebel's results on p. 48, the top line gives the magnified length of the arc as seen on the screen, 7·89 times its true length ; the second line gives this length with the depth of the crater added. Each P.D. is the mean of five observations.[68]

Arons tried to find the *true* resistance of the arc by placing it, with the battery producing it, in one arm of a Wheatstone's bridge. One of the diagonals of the bridge contained the stationary coil of a dynamometer, while the other contained the secondary coil of an induction apparatus. The three arms of the bridge were adjusted until this inductor sent no current through the stationary coil of the dynamometer, as proved by there being no deflection of the movable coil of the dynamometer, which was traversed by an independent alternating current produced by an auxiliary induction apparatus.

Next the arc was short-circuited and a resistance put in its place of such a value that, without making any alteration in the resistances of the other three arms of the bridge, no current was sent through the dynamometer-diagonal of the bridge by the secondary coil of the first inductor which formed the other diagonal. Under these circumstances he considered that this added resistance was equal to that of the

E

arc. With a current of 3·4 amperes, Arons found this resistance to be 2·1 ohms, and the back E.M.F. 40·6 volts, while with a current of 4·1 amperes he found the resistance to be 1·6 ohms, and the back E.M.F. 39·6 volts. The length of the arc is not stated in his Paper.[69]

In the same year Von Lang repeated his method, illustrated in Fig. 13, and now found 37 volts as the value of the back E.M.F. between the carbons. This method he also employed with electrodes of other materials, and he found that when they were both composed of one of the following substances, arranged in the order indicated—platinum, nickel, iron, copper, zinc, silver, cadmium—the back E.M.F. of the arc fell steadily from about 27 volts for the platinum electrodes to about 10 for those of cadmium.

Von Lang expressed V, the P.D. between the carbons, as

$$V = a + b \, l \, A \, ;$$

but a, the back E.M.F., was, for no one of the pairs of electrodes used, given with a probable error of less than + or − 3 per cent., and in some cases it was as high as + or − 34 per cent.[70]

Uppenborn, in 1888, found from experiments with a constant current of 7·7 amperes, producing an arc with 10mm. carbons, that

$$V = a + b \, l,$$

where l was the length of the arc. a, however, in this equation varied from 35·4 to 45·4, and b from 1·74 to 3·2.

When the current was increased he found that a *increased,* while b *diminished.* a, he considered might be replaced by $x + y$ A, where A was the current and x and y depended on the carbons. He concluded that, since a increased both with increase of current and with increase in the cross section of the arc, the effects were probably due to surface resistance rather than to a back E.M.F.[73]

Dr. Feussner, in the same year, pointed out that by the method used by V. Lang, Arons and others, only part of the resistance of the arc was measured. The method depended, he considered, upon Ohm's law and Kirchhoff's two laws, for which it was essential that both the E.M.F. and the resistance should be independent of the current flowing. In the arc, the second of these conditions was not fulfilled, for the conductivity of the arc, and hence its resistance, depended on the current flowing.

His idea was that the resistance of the arc should be obtained simply by dividing the P.D. between the carbons by the current, and hence that it could be measured by the Wheatstone's bridge used in the ordinary way. He considered that there was no direct proof of the non-existence of a back E.M.F. in the arc, but as it apparently existed in alternate current as well as in direct current arcs, it could not be a thermo-electric effect, since the temperature of the electrodes must be very nearly equal in alternate current arcs.

He pointed out that gases, even at a high temperature, are non-conductors, and that therefore the arc itself cannot consist of gases, but must be volatilised carbon. He considered, there-fore, that in order that an arc should exist, at least one of the electrodes must be at its temperature of volatilisation. This temperature, he said, if there were no other source of heat than the current, could only be given by the current, *i.e.* the current must "encounter a transition resistance so great that the consequent loss of P.D. multiplied by the current strength must be equivalent to the energy given out at the place in question, either as heat or light, at the temperature of the beginning of volatilisation. Hence the transition resistance must be the greater the higher the temperature of volatilisa-tion."

Dr. Feussner gave reasons for considering that the full equation representing the relations between the P.D., current, and length of the arc should be

$$-E = W J = (w_0 J + w_1 J_1) (a + \beta) + (w_0 J + w_1 J_1) \gamma l,$$

where E is the back E.M.F. of the arc,

    J is the current flowing,

    J_1 is the unit of current,

    W is the total true resistance of the arc,

    w_0 is that part of the resistance of the arc that is independent of the current flowing,

    w_1 is the part of the resistance that depends on the current when unit current is flowing.

E 2

He finally gave the three qualities necessary in electrodes for
the arc between them to give the most light with the least
power expended.   They were

(1) High temperature of volatilisation.

(2) Great power of radiation.

(3) Small heat conductivity.[74]

Fig. 17 shows the apparatus used by Lecher to try to find a
back E.M.F. in the arc between carbon electrodes.

D was the gramme dynamo, from which the conductor a led
to the arc L, and thence through a' through a commutator $c\,c'$
to the galvanometer G.   From the other terminal of G the
conductor went through $c\,c'$ again and through b' and b back
to the machine.   The galvanometer had a stop so that the
needle could only move in one direction, and as the full
current from the dynamo would have carried the needle far
beyond the scale, a shunt d was inserted in the galvanometer
circuit.   The commutator $c\,c'$ was first arranged so that when

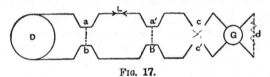

FIG. 17.

the arc was burning the needle was deflected.   Next the
commutator was reversed, so that, but for the galvanometer
stop, the needle would have given the same deflection as
before, only in the opposite direction.   The shunt d was then
cut out, so that had the needle been free to move the deflection
would have been from five to seven times as great as the whole
length of the scale.   The machine was now short circuited at
$a\,b$ for a very short time, and would therefore not act over the
remainder of the circuit, which might thus be considered as a
closed circuit.   Lecher considered that if there were a back
E.M F. in the arc, the back current created by it should now
have caused a deflection of the needle of the galvanometer,
which was free to move in the direction caused by such a
current.   No such deflection was observed, but as both stop
and galvanometer needle were a little springy a slight deflection followed on the short circuiting, such as might have

been caused by a back E.M.F. of 2 volts or so. That this deflection was really caused by spring and not by a back E.M.F. was shown by the fact that the deflection of the needle was the same, whether the dynamo was short-circuited at *a b* or at *a′ b′*.

Lecher measured the P.D. between the carbons when the arc was burning under various conditions by means of a Thomson electrometer. The carbons were 5mm. in diameter, the current about 5 amperes, and the length of arc 2mm. The following table gives the various conditions and positions of the carbons and the results of the experiments :—

Table **V.**—(*Lecher.*)

State of Carbons.	Position of Carbons.	P.D. between Carbons in Volts.
Ordinary	Horizontal	42
Negative, heated by gas burner	,,	52
Positive ,, ,,	,,	48
Both cooled by being wound with thick copper wire up to points	,,	less than 35
Ordinary	Vertical, positive carbon above	47
,,	,, negative ,,	46
Lower cooled to tip by mercury bath in water jacket	,, positive ,,	43
	,, negative ,,	41

The carbons enveloped in mercury developed mushrooms, and the arc was very unsteady during the last experiment. Lecher considered the results to prove that the P.D. between the carbons depended on their temperatures. He thought hissing was caused by the discharge springing to and fro to cooler places as the previous places became too hot, and thus setting up a vibration which produced sound.

Lecher used an exploring carbon of 1·2mm. diameter to find the P.D. between each of the carbons and the arc. By placing the exploring carbon vertically midway between the carbons of a horizontal arc, he obtained 35 volts as the P.D. between the positive carbon and the exploring carbon, and 10 volts as the P.D. between the exploring carbon and the negative. He found that the exploring carbon could be some little distance out of the visible arc without the P.D. between it and either of the other carbons being materially altered.

Using a Ruhmkorff coil and a condenser, he found that the silent arc was not discontinuous, but that the hissing arc was.[75]

Uppenborn also explored the arc with a carbon rod, having tried copper and platinum wires embedded in clay, steatite, and glass tubes, to no purpose. He found that with arcs of from 6mm. to 16mm. in length, the P.D. between the positive carbon and the exploring carbon placed near it, varied from 38 volts to 32·5 volts, while he found only 5 volts P.D. between the negative carbon and the exploring carbon placed near it. He gave a very good description of the shape of the arc.[76]

Luggin, in a long and very interesting paper, discussed the Wheatstone's bridge methods previously used for measuring the resistance of the arc, all of which he considered were open to criticism. He pointed out that, whereas when a small increase of E.M.F. was applied to an ordinary conductor through which a current was flowing, the P.D. at its ends increased, and therefore the variation of P.D. had a positive sign; in the arc, on the contrary, an increase of current was attended by a diminution of the P.D. between the ends of the carbons, and consequently the variation of the P.D., caused by an increase of the E.M.F., must have a negative sign.

Having shown the connection between the resistances in an ordinary Wheatstone's bridge, and having mentioned that the relation remained the same even if there were constant E.M.F.'s in all the arms, he continued thus: "If now an arc were started in the arm a, while in all the other resistances the bridge remained unaltered, then the P.D. at the ends of a would be less than before, and we should have to conclude from the diminished P.D. that the resistance w_1 had become less, *and that the arc had a negative resistance.*"

Luggin himself used a method somewhat similar to Von Lang's for finding the resistance of the arc. In one of the arms of a Wheatstone's bridge he placed an arc, in the second a comparing resistance of 2 ohms, and in the third and fourth liquid resistances amounting together to 300 ohms. In one of the diagonals was a battery of accumulators and a rheostat to regulate the current, which was shunted by a liquid resistance of 600 ohms and an electrically-driven tuning-fork. In the second diagonal, a condenser and telephone were in series,

the former being used to protect the telephone from the direct
current, and to take such a quantity of electricity from the
alternating current as to allow the telephone to be sufficiently
sensitive.

Luggin attributed the fall of P.D. between the carbons that
took place when the current was increased to a diminution in
the resistance of the gaseous medium caused by its rise of
temperature. He pointed out that this change could not be
an instantaneous one, and that therefore if the current were
alternated with sufficient rapidity the P.D. would rise and
fall with the current, "and the arc would show a positive
resistance." He considered that the reason that Arons found
a positive resistance in the arc was that the alternating current
he impressed on the direct current in his arc alternated very
rapidly, and he was quite unable to understand why Frölich
did not find a negative resistance.

To find the variation in the P.D. between the carbons when
the length of the arc was varied and the current kept constant,
Luggin used two Siemens carbons 10mm. in diameter, the
positive cored and the negative uncored. For each figure he
took the mean of three or four sets of observations. The following table gives the results of his experiments with a current
of 7 amperes. The length of the arc is given in scale divisions, each of which was 0·434mm. He found that the formula

$$V = 40·04 + 1·774 \, l,$$

where V was the P.D. between the carbons in volts and l the
length of the arc in scale divisions, fitted his observations
extremely well, as will be seen from the following table.

Table VI.—(*Luggin.*)   Current 7 amperes.
Positive Carbon Cored, Negative Uncored.

l	V (volts) observed.	V (volts) calculated.	Difference.
1	41·6	41·81	− 0·21
2	43·3	43·59	− 0·29
3	45·4	45·36	+ 0·04
4	47·6	47·14	+ 0·46
5	49·3	48·91	+ 0·39
6	50·8	50·68	+ 0·12
7	52·3	52·46	− 0·16
8	53·9	54·23	− 0·33

Luggin tried using a horizontal disc of carbon 15cm. in diameter as one of the electrodes of an upright arc, and a carbon rod for the other. When the disc was placed under the rod and used as the negative electrode, it was found that even a very slow rotation of the disc extinguished the arc, however strong the current. When, however, the disc was placed above the rod and used as the positive electrode, it could be rotated fairly quickly even with a current of only 7 amperes, and with currents of from 27 to 33 amperes it could be rotated very fast indeed without the arc becoming extinguished. With slow rotations the disc was pitted with small craters, but with the faster ones these craters ran into one another. When the arc hissed, while the disc was still, the P.D. between the carbons sank as usual 10 or 11 volts, but if, while the hissing continued the disc was set in rotation, the P.D. sank from 3 to 5 volts lower still.

He next experimented with iron electrodes, and finally he used an exploring carbon to find the difference in volts between the fall of potential between the positive carbon and any point in the arc, and the fall of potential between that same point and the negative carbon. This difference is denoted by E in the following table, while V denotes the P.D. between the carbons in volts, and l the length of the arc in mm. Siemens carbons were used, and a current of 6·8 amperes for the first set of results, and one of 8·9 amperes for the second.

Results (I.) were obtained with ordinary Siemens carbons. Results (II.) with carbons sprinkled with soda.

Table VII.—(*Luggin.*)

(I.)		(II.)		
V	E	V	E	l
39·8	25·9	17·9	0·43	2·9
42·5	27·1	19·4	1·76	3·0
46·3	26·9	20·0	0·69	3·9
48·7	32·2	21·4	3·77	4·0
49·3	31·0	22·6	2·89	5·0
51·7	33·3	28·0	7·00	6·8
52·7	32·9			
57·7	34·6			

From the immense diminution in E in (II.), caused by sprinkling the carbons with soda, so much greater than the corresponding diminution in V, Luggin concluded that there was an enormous leap of potential at the positive pole, which was almost neutralised by sprinking the carbons with soda. This onesidedness diminished as the length of the arc diminished, and he mentioned that once, with the ordinary Siemens carbons, when the point of the negative carbon was entirely surrounded by the crater, V was 35·5 volts and E was as low as 18·8 volts.[77]

Dubs, in 1888, found that if two carbon plates were placed one above the other with a distance of 1mm. between them, and a blowpipe flame was directed on to the lower, so as to carry carbon particles to the upper, a galvanometer connected with the plates showed a weak current flowing against the blast. Also that if a carbon plate were heated on an oxyhydrogen flame and laid on a cold carbon plate, a current flowed from the cold to the hot plate. With copper plates the effect was less, and was *nil* with iron. Dubs regarded the effect as analogous with the back E.M.F. in the arc, and considered that that depended, at any rate in part, on the mechanical action of the current.[79]

In 1889, Luggin followed up his previous researches by a long and close investigation of the arc with exploring carbons. Some parts of the Paper in which he gave the results of these investigations it is impossible to follow; for not only is the German very involved, but there is a complete dearth of Figures, and the explanations given are too scanty to enable the reader to construct these for himself. This is all the greater pity, as Luggin's work, where it can be understood, is both original and suggestive.

The first experiments were undertaken to find out how the potential varied in different parts of the same cross-section of an arc 8mm. in length. Siemens carbons, 12mm. in diameter, were used for the electrodes, and Carré carbons, 1mm. to 2mm. in diameter, for exploring. These always burnt to a point in the arc. The results differed according as the positive carbon was solid or cored. When it was solid, it was found that when one exploring carbon just touched the purple core of the arc and another just dipped into the outer flame, both in the lowest cross-section of the arc, the potential of the inner carbon was three volts lower than that of the outer one.

For higher cross-sections of the arc the case was reversed, the inner carbon having a potential two volts higher than the outer in the cross-section midway between the electrodes. When the point of one carbon was placed at the centre of the purple core, and that of the other touched the outer flame in the same cross-section, the potential of the inner carbon was higher than that of the outer. When the positive carbon was cored, very little difference was found between the potentials of the inner and outer portions of the same cross-section of the arc, which led Luggin to doubt if the differences found with solid carbons had not been caused by the exploring carbon itself.

Luggin found that a thick exploring carbon disturbed the arc considerably, and that it seemed to repel it. Its insertion also raised the P.D. between the carbons, and sometimes made the arc hiss and sing. Putting a thick exploring carbon about 2mm. from the positive electrode caused two arcs to form, one between the positive and the exploring carbon, and the other between the exploring carbon and the negative.

An exploring carbon of 1·3mm. diameter was placed with its point immediately under the crater (carbons both solid, 12mm. in diameter), with a current of 15·5 amperes flowing. The P.D. between the positive and exploring carbons was found, from five measurements, to be 33·7 ± 0·46 volts. Changing the length of the arc appeared to have no influence on this P.D. The exploring carbon was next placed in the arc, with its point as near as possible to the bright spot on the negative carbon, and the P.D. between exploring and negative carbons was found, from six measurements, to be 8·78 ± 0·17 volts.

This one-sidedness in the potential of the arc, which Luggin called E, was also measured in another way, on the supposition that the fall of potential in the arc itself was perfectly uniform. E is, of course, the difference between the fall of potential between the positive electrode and the exploring carbon and the fall of potential between the exploring carbon and the negative electrode. If, then, the exploring carbon is placed in the middle section of the arc and v, the P.D. between the positive and exploring carbons, is measured simultaneously with V, the P.D. between the two electrodes, then

$$E = v - (V - v) = -V + 2v.$$

If the potential of the exploring carbon differs by an amount dv from the surrounding gases, the above value of E is incorrect by $2dv$, as Luggin showed. He made measurements of V and v in the manner described, only allowing the exploring carbon to touch the outer edge of the flame surrounding the arc, however, so as to cause the least possible disturbance to the arc. This probably introduced some error, but the value of E thus obtained was the same as with the earlier experiments, namely, 24·9 volts. Other measurements made with a current of 16·8 amperes and with increasing length of arc gave

V = 49·9 volts.	E = 24·5 volts.
V = 53·8 „	E = 25·8 „
V = 65·2 „	E = 25·6 „

With cored positive and solid negative carbons 12mm. in diameter the following values were obtained with increasing length of arc :—

V = 44·9 volts.	E = 28·3 ± 0·24
V = 51·4 „	E = 29·6 ± 0·61
V = 55·2 „	E = 30·3 ± 0·29
V = 58·7 „	E = 31·4 ± 0·40
V = 64·8 „	E = 34·4 ± 0·54

For these last experiments the exploring carbon was placed by eye only. Luggin pointed out that the two sets of experiments show that while with solid carbons the P.D.s between the positive carbon and the middle cross-section of the arc and between the same cross-section and the negative carbon increase at about the same rate with increasing length of arc, with a cored positive carbon the P.D. between the positive and the middle cross-section increases more rapidly than the P D. between that cross-section and the negative carbon.

From another experiment it appears that, at least with a cored positive carbon, the potential of the arc itself does not fall at a perfectly uniform rate, as was supposed in making the experiment. He fastened two carbons, one on either side of a plank 3mm. thick, so that they were held at a constant vertical distance from one another. He then placed the carbons horizontally, so that the point of the upper was vertically over that of the under, in an arc burning between positive cored and solid negative carbons. According as the pair of carbons was near the positive or negative electrode, the P.D. between them was

12·4 or 8·05 volts. The same experiment was not tried with two solid electrodes.

Placing an exploring carbon in the outside aureole of the arc, near the positive carbon, the P.D. between the positive and exploring carbons was found to be about 1 volt greater than when the latter was placed as near as possible to the centre of the crater. Luggin considered these experiments showed that the P.D. between the positive carbon and the surrounding gases was constant as a first approximation.

He gave a careful description of the different parts of the arc and carbons, and mentioned that the outer green part, or aureole, as he called it, started higher up the positive carbon than the edge of the crater, and that there was a space between it and that carbon wide enough for him to insert a thin slip of carbon into it. He mentioned that carbon pencils glowed brightly directly they were placed within the aureole, and that the latter widened out where it touched these pencils. He also noticed that the arc proper and the aureole varied in form according to the material, cross-section and position of the electrodes. He described an upright arc between solid carbons as being like an inverted bell standing on the point of the negative carbon, and said that with long arcs the middle cross-section of the arc gave less light than the ends, and that this was particularly the case with cored carbons, or with carbons in which volatile salts were mixed. He suggested also that the appearance of dividing itself into two unequal parts, which he observed in the arc with cored carbons, was connected with the irregularity in the fall of potential near the positive and negative electrodes that he had found to exist.

Some experiments with hissing arcs between solid carbons led Luggin to observe that the hissing took place when the current was so strong that the crater filled the whole of the end of the positive carbon. This is a very important point that he was the first to observe. He noticed also that the longer the arc the larger the current that could be used before the arc hissed ; but that with longer arcs the end of the carbon, and therefore the crater, was larger. With long *silent* arcs he found that the end of the positive carbon was convex, instead of its having a crater.

In the following table he called the ratio of the current strength to the size of the surface of the end of the positive carbon, D ; hence, when the current is measured in amperes and the surface of the carbon in millimetres, D is the current strength in amperes per millimetre of end of carbon.

Table VIII.—(*Luggin.*)

l (mm.).	A (Amperes).	Diameter of End-surface (sq. mm.)	D (Ampere).
3·7	19·2	6·87	0·51
8·5	26·3	8·25	0·49

Luggin found that the quantity he called E, *i.e.* the difference between the fall of potential between the positive carbon and the arc, and the fall of potential between the arc and the negative carbon, was somewhat smaller with hissing than with silent arcs, showing that the fall of potential that took place when the arc began to hiss was chiefly between the positive carbon and the arc.

He tried to find evidences of polarisation and hence a back E.M.F. in the arc immediately after the current was cut off, but was unsuccessful, and he considered that his experiments showed that there is no important back E.M.F. in the arc 0·005 second after the current is turned off.[80]

In 1890 Prof. Ayrton, with some of his students at the Central Technical College, began the series of experiments that were described in the Paper he read at the Electrical Congress in Chicago in 1893. The Paper was unfortunately burnt before it could be published, but the experiments and their results, of which the laboratory notes were retained, will be discussed later on.

In 1891, in an account that included a revision of all that was known of the physics of the arc up to date, Prof. Elihu Thomson mentioned that he had found that, while both short and long arcs could be made to burn steadily, there was an intermediate stage of flickering and unsteadiness.

He considered that the hollowness of the crater was due to the evaporation of carbon from its surface, and that the temperature of the crater was that of the boiling-point of carbon, or, " more correctly, of sublimation at atmospheric pressure." He found that the carbon of the crater was in a plastic state, and proved it by pressing the carbons together

with arcs of from 150 to 200 amperes, and finding that they
would fit each other perfectly afterwards. He was also able to
bend sticks of carbon a quarter of an inch thick, by passing a
big enough current through them almost to vaporise them,
and causing them to emit a light nearly as intense as that of
the arc. Hence he argued that it was probable that carbon
might be liquefied if subjected to the temperature of the arc
under high pressure in an inert gas.

He considered that the formation and maintenance of the
arc might be due to electrolytic action, the hot vapour taking
the part of the bath, and acting by a molecular interchange of
carbon atoms. He mentioned that constant potential arcs

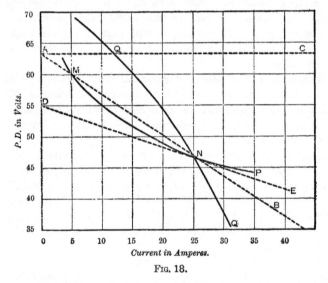

FIG. 18.

were impossible, and that "for stability the resistance should
not be dependent wholly on the current passing."[81]

In a series of articles on the alternate-current arc, published
in 1890, M. Blondel showed how, with a direct-current arc, to
determine graphically the conditions of dynamo and of resistance outside the arc, so that a steady arc of given length
might be maintained between carbons of given diameter.

Having drawn experimentally the curve M N P (Fig. 18),
which showed the connection between the P.D. at the

terminals of the lamp and the current flowing while the arc
was maintained at the given constant length, he drew Q Q,
the characteristic of the circuit taken at the terminals of the
lamp. He then pointed out that if this curve cut M N P at N,
the necessary and sufficient condition for stable equilibrium
was that Q Q should cut M P from above downwards in the
direction of increase of current. If the potential of the
dynamo were a constant V, represented by the horizontal
straight line A C, a resistance R would have to be added
to the lamps, such that the new characteristic of the feeding
circuit A B (E = V − R A) should satisfy the preceding condition.

Suppose, for instance, that M N P were the curve for an arc
of 4mm., and that E D, the tangent to M N P at N, cut the
axis of volts at the height of 55 volts. Then, as M. Blondel
pointed out, it is impossible to supply an arc of 4mm. with
a current of 25 amperes with a constant potential of less than
55 volts, with the given carbons. And in order that the arc
should be perfectly steady, the constant potential of the
dynamo would have to be 60, or even 65 volts.[82]

Cravath made many experiments in 1892 to discover the
causes of hissing in the arc. He considered that the principal
of these were air currents, impurities in the carbons, and short-
ness of the arc. As he noticed that during hissing the stream
of carbon vapour appeared to issue from a *part* only of the
crater, and that this part was constantly changing, it occurred
to him that perhaps hissing was due to the heating and cooling
of the carbons caused by this change of position. He, therefore, tried the effect of moving one carbon horizontally over the
other while a steady silent arc was burning. When the positive
carbon was pointed and the negative flat the arc burned silently
as before, but when the negative carbon was pointed and the
positive flat, each change of position caused a hiss.

Cravath measured the diameter and depth of the crater, the
length of the arc and the diameter of the knob of the negative
carbon under varying circumstances, and he found that "the
arc burns away the carbons so as to keep all points at an equal
distance from each other."[83] He considered that the sudden
diminution of the P.D. between the carbons when hissing began
was due to the dampening effect of the cool carbon preventing
the consumption of energy.[84]

In 1892 Stenger sought for evidence of a back E.M.F. in the arc thus : A shunt-dynamo sent a current through an arc, 5 accumulators, and a tangent galvanometer, in series. The accumulators were joined up so as to oppose the current, and the needle of the galvanometer was pressed against a stop so that it was not deflected by the dynamo current, but could move freely in the opposite direction.   On short-circuiting the shunt-dynamo the conduction of the arc lasted long enough for the accumulators to send a current and produce a deflection of over 90deg. in the galvanometer ; but, when the experiment was repeated with the accumulators removed from the circuit, a deflection of only about $\frac{3}{4}$deg. occurred, produced by the spring of the stop.   Hence any back E.M.F. in the arc either ceased with the main current, or was very small compared with 10 volts.[86]

In a Paper read before the British Association in 1892, Prof. Silvanus Thompson gave an approximate formula connecting V, the P.D. in volts between the carbons, l, the length of the arc in millimetres, and A, the current in amperes, viz. :—

$$V = a + \frac{b\,l}{A},$$

where the constant a, however, varied between 35 and 39 volts and the constant b between 8 and 18.   The constant part of the P.D. which is independent of the current for an invariable length is sometimes called the apparent back E.M.F., and, although Dr. Thompson did not affirm that there was an actual back E.M.F., he considered that the arc acted as though it were the seat of a back E.M.F.

He described his experiments, made with an auxiliary exploring carbon, which showed that the drop of potential in the arc itself was small, and that the main drop was at the positive carbon.   The latter, he considered, was accounted for by the volatilisation of the carbon at the crater, which, he suggested, was always at the temperature of boiling carbon, and this idea was confirmed, he thought, by Capt. Abney's discovery that the brilliancy of the same quality of carbon per square centimetre was a constant.   Mr. Crookes's experiments, which showed that the flaming discharges produced by very high-pressure very short period alterna-

ting currents were endothermic flames of nitrogen
and oxygen, had led him to try whether the combination of
nitrogen and oxygen produced by the high temperature of the
arc had anything to do with the E.M.F. observed there.
To test this he surrounded the arc with a glass tube, and
introduced successively oxygen, nitrogen, carbon dioxide,
hydrogen, &c., but with a normal arc taking 10 amperes not
one volt difference in the P.D. was observed.   When chlorine
or carbon monoxide surrounded the arc the positive carbon was
flattened over the end, and the end of the negative became a
very obtuse cone, while with hydro-carbon gas the crater was
very deep; and, lastly, when the arc was formed in oxygen
the carbons burnt away very rapidly.   The gases had to be
introduced quietly, since blowing on an arc in ordinary air
Dr. Thompson found raised the P.D. to 75 volts.[87]

Later in the same year M. Violle published an account of
some experiments he had made in order to determine the
temperature of the positive carbon and the arc.

He cut a deep trench all round the positive carbon a little
way from the end, so that a small piece was left, only connected with the rest of the carbon by a narrow neck.   When
most of this piece was burnt away, and the remainder was all
of the same brightness, it was shaken off, and fell into a little
cup, after which the amount of heat given off by it was
measured in the usual way.   Assuming the ordinarily received
value of the specific heat of carbon, M. Violle found by this
method that the temperature of the crater was about
3,500°C., whatever power was spent in the arc—whether 10
amperes were flowing at 50 volts, or 400 amperes at 85 volts.
He concluded, therefore, that the candle-power per square
centimetre of the crater was the same for all arcs with the
same kind of carbons.[88]

In 1893 M. Violle made some further experiments on the
temperature and brightness of the crater.   The carbons were
placed horizontally, and inclosed, in order to reduce the cooling
effect of the surrounding air.   The temperature was measured
in the same way as in his previous experiments, and led to the
same result, namely, that the temperature of the crater was
about 3,500°C.   He considered that the arc itself was at the
same temperature, and that this was the temperature of boiling

F

carbon. He laid great stress on this, as may be seen from the following quotation : "The most important result of my researches is to establish the fact that the voltaic arc is the seat of a perfectly definite physical phenomenon, the ebullition of carbon."

To determine the amount of light given out per square milli-metre of crater he employed two methods, (1) the use of the spectrophotometer, (2) photography. They both led him to the same conclusion, that the amount of light per square mm. was constant, and independent of the current used.

Finally M. Violle gave reasons for believing that the arc is an electrolytic phenomenon, in which there is a continual stream of carbon vapour passing from the positive to the negative electrode.[89]

In August, 1893 Messrs. Duncan, Rowland and Todd pub-lished an account of the effects they had obtained on producing an arc under pressure and in a vacuum. They began their Paper by stating that the two causes alleged for the back E.M.F. in the arc—the vaporisation of the positive pole and the thermo-electric effect produced by the carbon vapour in contact with the unequally heated carbons—must both be operative ; but that the first must cease to exist directly the main current ceased to flow, and could not therefore be detected even immediately after the circuit was broken, while they con-sidered it must be possible to detect evidence of the latter for a short time after the main current had been stopped.

The first part of the back E.M.F., viz., that due to the volatilisation of the carbon, they considered must be a constant a, while the second part must be a function of the current and the length of the arc. From these considerations they sug-gested that the complete equation connecting P.D., current, and length of arc should be

$$V = a + f(l) f(A) + b f'(l) f'(A),$$

where V was the P.D. between the carbons, l the length of the arc, and A the current.

They next called attention to the fact that the P.D. between the carbons diminished as the current increased, and gave the results of experiments supporting this. The following they suggested as an explanation of the phenomenon. If V be the

P.D. between the carbons, a the *constant* back E.M.F. due to volatilisation of the carbon, and a' the back E.M.F. due to thermo-electric action,

$$A = \frac{V - a - a'}{\dfrac{b\,l}{A}};$$

$$\therefore \quad V = a + a' + b\,l.$$

Now, a' diminishes with increase of current, since the temperature of the negative carbon increases while that of the positive remains constant. Hence V diminishes as A increases.

Keeping the current constant at six amperes, and the arc of a fixed length, they found that, starting with the atmospheric pressure, V increased as the pressure was increased, but at a slower rate. When, on the other hand, the pressure of the surrounding air was reduced below that of the atmosphere, V also increased except in the case of the $\frac{1}{16}$th-inch arc when V diminished on reducing the pressure. This except tional result, however, arose probably from the arc hissing. Hence, for a given current and length of silent arc, V has a minimum value for atmospheric pressure. The increase of V as the pressure was reduced below that of the atmosphere they considered was caused by an increase of a', for they said that a, the constant counter E.M.F., was probably lower in a vacuum, and the positive carbon not so hot, but the negative carbon seemed to cool proportionately faster than the positive.

The counter E.M.F., they concluded, increased apparently with the pressure above one atmosphere, while the ohmic resistance of the arc did not greatly change. As regards the formula, their experiments showed that a varied with l for each pressure employed, and they failed to arrive at any exact law connecting V, l and A, even for one pressure, the equation that most nearly fitted their results being, they thought,

$$V = a + \frac{b\,l}{A^{0.7}},$$

but this, they remarked, was only approximately correct.[90]

In some articles published in 1893 M. Blondel gave it as his experience that, although the *maximum* brilliancy of the crater was independent of the current flowing in the arc, yet that the *average* brilliancy of the incandescent portions increased both

F 2

with the intensity and with the density of the current, until the crater was well saturated. If, he said, the value of the current be suddenly varied, the intrinsic brilliancy undergoes a temporary and very appreciable variation which may reach ten per cent., and which diminishes gradually until the dimensions of the crater are so altered as to restore the surface of emission to the value that it ought to have for the new current. He considered that the heating of the crater only took place at the surface, and that the temperature of volatilisation was only reached by a very thin superficial film.

In order that the light of the crater should not vary with the carbons employed, he was of opinion that it was necessary that the carbons should contain a *very* small admixture of foreign substances, and that the molecular condition of the light giving surface should be always the same. He thinks that the surface of the crater is always turned into graphite with carbon electrodes, so that they always fulfil the second condition. He found the intrinsic brilliancy of the crater to vary between 152 and 163 " bougies decimales."

With the hissing arc M. Blondel observed that the current flowed jerkily, the violet part of the arc became blue-green, and the arc lost its transparency, being transformed into an incandescent mist that hid the crater. When hissing ceased, and the arc became violet again, he noticed that the crater appeared to be covered with black specks which gradually disappeared, a fact which he thought pointed to a lowering of the temperature during hissing. The arc while hissing gave from 10 to 20 per cent. less light than when silent, and M. Blondel concluded therefore that while with the silent arc most if not all of the carbon that is transferred from one pole to the other is volatilised, with the hissing arc a large part of it is transferred by disruptive discharge.[91]

In 1894 M. Violle renewed his inquiries into the temperature of the arc with a still larger range of current than before, namely, 10 to 1,200 amperes. He found that in all cases the temperature and intrinsic brilliancy of the positive carbon remained constant. The arc he employed was enclosed, and he found by spectroscopic methods that the brilliancy of the arc itself increased as the current increased. While doubting that the brilliancy of luminous rays contributing to the spectra of gases was connected with the temperature by the same relation as the brilliancy of the corresponding regions in spectra of solid bodies, he still considered that his experiments led to the conclusion that the temperature of the arc proper increased with the current, and was in general higher than that of the positive carbon.[92]

In the same year, while experimenting with a view to using a fixed portion of the supposed uniform light of the crater of an arc as a practical standard of light, Mr. Trotter made his notable discovery of the rotation of the arc. By the use of a double Rumford photometer, giving alternating fields, as in a Vernon Harcourt photometer, his attention was called to a bright spot at or near the middle of the crater. The use of rotating sectors accidentally revealed that a periodic phenomenon accompanied the appearance of this bright spot, and though it is more marked with a short humming arc, the author believes that it is always present.

An image of the crater was thrown on to a screen by a photographic lens ; and a disc having 60 arms and 60 openings of 3°, and rotating at from 100 to 400 revolutions per minute, was placed near the screen. Curious stroboscopic images were observed, indicating a continually varying periodicity, seldom higher than 450 per second, most frequently about 100, difficult to distinguish below 50 per second, and becoming with a long arc a mere flicker. The period seemed to correspond with a musical hum of the arc, which generally broke into a hiss at a note a little beyond 450 per second. The hum was audible in a telephone in the circuit, or in shunt with it. The current was taken from the mains of the Kensington and Knightsbridge Electric Light Company, often late at night, after all the dynamos had been shut down. The carbons were *not cored ;* six kinds were used.

A rotating disc was arranged near the lens, to allow the beam to pass for about $\frac{1}{100}$th of a second, and to be cut off for about $\frac{1}{100}$th of a second. It was then found that a bright patch, occupying about one-quarter of the crater, appeared to be rapidly revolving. Examination of the shape of this patch showed that it consisted of the bright spot already mentioned, and of a curved appendage which swept round, sometimes changing the direction of its rotation. This appendage seemed to be approximately equivalent to a quadrant sheared concentrically through 90°.

The author inclined to the theory of constant temperature of the arc, and attributed this phenomenon, not to any actual change in the luminosity of the crater, or to any wandering of the luminous area, such as is seen with a long unsteady arc, but to the refraction of the light by the heated vapour. All experiments, such as enclosing the arc in a small chamber of transparent mica, or the use of magnets, or an air blast, failed to produce any effect in altering the phenomenon. A distortion of the image of the crater while the patch revolved was looked for, but nothing distinguishable from changes of luminosity was seen.[93]

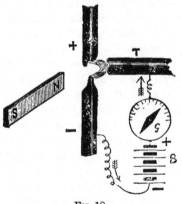

Fig. 19.

Prof. Fleming, in the course of experiments on the arc in 1894, was led to believe that the carbon boiled at the crater, and that the violet core of the arc consisted of a torrent of carbon vapour passing towards the negative pole. This violet core may be heated to a higher temperature than that of the crater by the passage of the current through it. He thought that at the cooler negative pole some of the carbon vapour was condensed, but that some of it was deflected back again on to the positive carbon, " causing the golden aureole or flame and creating thus a double carbon current in the arc." The negative carbon, he thought, gave evidence after use of having been worn away by a kind of sand blast action.

Joining the positive carbon, an electric bell, and a third carbon which dipped into the arc, in series, he found that sufficient current passed in the circuit to make the bell ring, but if the negative carbon were joined up instead of the positive, the bell gave no sound.   This led him to conclude that there was no perceptible P.D. between the arc and the negative carbon.

Fig. 19 shows the arrangement used by Prof. Fleming in experimenting on the conductivity of the arc.

The third carbon T, upon which the arc was made to play steadily by means of the magnet, was joined up in series with the galvanometer G, the battery of 15 secondary cells B, and the negative carbon.   When the *negative* pole of the battery was joined to the negative carbon the galvanometer needle was deflected, showing that a current was passing, but when the poles of the battery were reversed there was no deflection of the galvanometer.   Hence Prof. Fleming concluded that the arc possessed a unilateral conductivity, allowing a *negative* current to flow through it from the negative carbon to the positive, but not allowing a positive current to flow in the same direction. After performing this experiment it was found that the third carbon T was cratered, and that its tip was converted into graphite.[94]

A measurement of what was considered to be the *true* resistance of the arc was made in 1895 by Mr. Julius Frith with the Wheatstone's bridge seen in Fig. 20.   Two of its arms, P, Q, consisted of the two halves of a stretched platinoid wire, each having a resistance of 5·35 ohms.   The third arm was composed of a battery of 26 accumulators, E, which, together with a shunt dynamo D_1, sent a current through a resistance, R_1, an ammeter C and the arc X in series, the arc being 2mm. in length, formed with carbons 11mm. in diameter.   The fourth arm consisted of a shunt dynamo, D_2, exactly similar to D_1, only at rest, together with resistances R_2 and R_3 and a coil L, whose self-induction could be varied by moving an iron core.

The resistances of D_1 and D_2 were each 0·04 ohm ; of R_1 and R_2, each 8 ohms ; of E, 0·25 ohm ; and of R_3 0·3 ohm.

Before closing the switch S, the speed of the dynamo D_1 and the length of the arc were varied until the potential of the points A and B were equal, as tested by the voltmeter V, which, by means of the switch K, could be connected either across the accumulators E or across the arc X.

An alternator D_3 sent an alternating current through a wire, any two points of which could be tapped to supply the alternating P.D. to the bridge, and, after closing the switch S, the resistance of R_3 and the self-induction of L were varied until the sound in the telephone T became a minimum, the condenser M, inserted in the telephone circuit, cutting off from the telephone any direct current effect that might be caused by want of perfect equalisation of the potentials of the points A and B.

The best results were obtained with R_3 having a resistance of about 0·6 ohm, and an alternating P.D. of 5 volts supplied to the bridge. This makes the resistance of the arc about 0·6 ohm, which, together with the readings of the ammeter C and the voltmeter V, give the back E.M.F. in the arc as 39 volts.

Other methods of testing were employed by Mr. Frith, and results were obtained agreeing with the above.[95]

Mr. Wilson has experimented on the electric arc under considerable atmospheric pressures. As the pressure was increased, and the current kept constant, he found that the apparent resistance of the arc increased. When the pressure had been raised to five atmospheres the temperature of the crater had fallen, while at 20 atmospheres the brilliancy of the crater fell to a dull red colour. Diminishing the pressure, on the contrary, increased the brightness.

Mr. Wilson therefore concluded that "the temperature of the crater, like that of the filament in an incandescent lamp, depends on how much it is cooled by the surrounding atmosphere, and not on its being the temperature at which the vapour of carbon has the same pressure as the surrounding atmosphere." [97]

Mr. Freedman made some experiments on "The Counter Electromotive Force in the Electric Arc," using small currents up to two amperes, with electrodes of different substances. His conclusions were as follows :—

"1st. There is a counter E.M.F. present in an electric arc depending simply upon the material and temperature of volatilisation of the electrodes, and this counter E.M.F. has a constant definite value for that material.

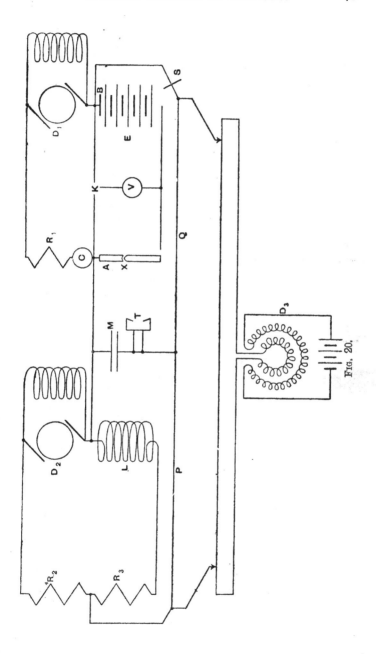

Fig. 20.

"2nd. On account of the different temperatures there must be a thermo-electric effect. The counter E.M.F. due to this phenomenon must depend on the difference of temperature between the electrodes. Consequently, it must increase with the length of the arc as the temperature of the negative electrode falls; and it must decrease with the current as the temperature of the negative electrode rises.

"3rd. It is fair to assume that two amperes of current will tear off twice as many molecules, in the same length of time, from the positive electrode as one ampere; in strict analogy to electro-deposition. If the quantity of matter is doubled the resistance is most likely halved; so that c r would remain a constant quantity. This would be so, provided the temperature remained constant. But since the temperature rises with increase of current, c r must actually decrease with increase of current."

From these considerations Mr. Freedman formed the following "complete formula for the difference of potential between the electrodes."

$$v = x + y + c\,r,$$

in which

$x =$ the constant counter E.M.F., depending upon the material of the electrodes and its temperature of volatilisation.

$y =$ the counter E.M.F. due to the thermo-electric effect, being a function of the material of the electrodes and the difference of the temperature; the higher temperature being that of volatilisation of the electrode and the lower depending upon the material and size of the electrode, the current and the length of the arc.

$c =$ the current strength.

$r =$ the ohmic resistance of the arc, depending upon the material of the electrodes, the length of the arc, the temperature and the current.

For any given material of electrodes analytically expressed in terms of length and current, said the author,

$$v = x + f\,(l,\,c) + c\,f'\,(l,\,c)$$

or in terms of temperature and currents,

$$v = x + f\,(t_2,\,t_1) + c\,f'\,(t_2,\,t_1).[99]$$

At the meeting of the British Association at Ipswich in 1895,

Prof. Ayrton read a short Paper on "The Resistance of the Arc."
He had been led, by the study of the various curves connecting
the P.D. between the carbons with the current flowing for constant lengths of arc, to the conclusion that if there were, as most
observers seemed to think, a back E.M.F. and a true resistance
in the arc, then the resistance must be negative. Some experiments made by Mr. Mather at his suggestion strengthened the
idea.*

"In one of these experiments two points of equal potential
were found in a circuit consisting of an arc, a battery, and a
resistance. Another battery, consisting of a few cells of known
E.M.F. and resistance was applied between these two equi-
potential points, and the current flowing through the battery
was noted. The resistances of the two parallel halves of the
circuit, excluding the arc, were known, so that the current
which, taking the arc resistance as zero, should flow through
this battery, could be calculated. Now the value of this calculated current was found to be less than the observed value, no
matter in which direction the P.D. was applied, and this result
was also obtained when an alternating P.D. was used. Hence
the resistance of the arc was apparently less than zero."

"The other experiment consisted in running the arc at a
steady P.D. and current, suddenly altering the resistance in
circuit by a small amount, and noting the changes in the
ammeter and voltmeter-readings so produced. The new conditions were maintained only long enough to allow of these
readings being taken. The arc was then brought back to its
former condition before taking another reading. It was found
that a change of P.D. in *one* direction was always accompanied
by a change of current in the *opposite* direction. The results
of both experiments were however only qualitative."

It may be mentioned, that although the idea of a negative
resistance in the arc occurred to Prof. Ayrton quite independently, before he had ever heard of Luggin's work on the
subject, yet that able experimenter made the same suggestion
as long ago as 1888 (*see* p. 54).

In their Paper read before the Physical Society, in May,

---

* Prof. Ayrton's Paper was not published, but the account of it given
here is taken from the Paper on the same subject read by Messrs. Frith
and Rodgers, before the Physical Society, in May, 1896.

1896, Messrs. Frith and Rodgers gave the results of a long and very complete series of "Experiments on the Resistance of the Arc," undertaken with a view to throwing some light on the discrepancy between the negative resistance obtained by Prof. Ayrton and the positive resistance found by all other experimenters. They tried several methods of experimenting, the most successful of which is represented diagrammatically in Fig. 21.

D is the armature of an alternator, the current from which passes round two circuits in parallel, one of which contains the arc X, and the other an adjustable resistance R. By adjusting R the alternating currents in the two halves can be made equal.

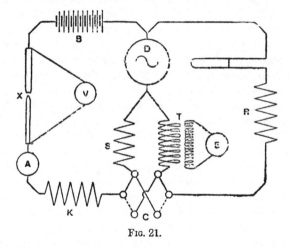

Fig. 21.

When this is the case the impedances of the two halves to alternating currents must be equal.

The continuous current circuit shown to the left consists of a battery of accumulators B, the hand-adjusted arc lamp X, the resistance K, the ammeter A, and (with the commutator C as shown) the resistance S and the alternator D. The alternator D carries the continuous current, but this does not prevent its acting as an alternator.

The air transformer T was used to measure the small alternating current independently of the continuous current flowing. For this purpose its thick wire coil was placed in series with the alternator D, and its thin wire coil was connected with an electrostatic voltmeter E. By means of the commutator C, the air-transformer T could be thrown into either circuit, the resistance S being thrown by the same operation into the other circuit. The resistance S was equal to that of the thick wire coil of T, so that when S replaced T the continuous current was unaffected by the change.

When experimenting, the arc was run at the required current and P.D. by altering the number of cells in B, K being always kept the same. R was then adjusted till the deflection of E was the same when T was in either circuit. If the value of R when balance was obtained were R_1, then

$$R_1 = k + b_1 + l + x \quad \cdots \cdots \quad \text{(i)}$$

where k was the constant resistance at K, b_1 was the resistance to alternating currents of the battery B, l the resistance of the arc lamp and connections, and x the resistance of the arc.

The carbons were next firmly screwed together and the number of cells in B reduced, till the continuous current was the same as before. R was again adjusted till the deflections of E were equal, then if R_2 were the new value of the resistance, and b_2 the resistance of the portion of the battery now used

$$R_2 = k + b_2 + l \quad \cdots \cdots \quad \text{(ii)}.$$

Next the cells were cut out and the mains leading to them were short circuited, so that the third value of R obtained was

$$R_3 = k + l \quad \cdots \cdots \quad \text{(iii)}.$$

From (ii) and (iii) the resistance of b_2 was obtained, and, by proportion, of any number of cells. Putting these values in (i) the value of x in ohms was found.

Messrs. Frith and Rodgers defined the resistance of the arc as the ratio of a small increment of P.D. applied, to the small increment of current produced; that is, they were measuring the value of $\dfrac{dV}{dA}$ when an alternating current was applied to the arc which they considered to be too small to produce any visible effect on it. They called this the "instantaneous" $\dfrac{dV}{dA}$ to distinguish it from the steady $\dfrac{dV}{dA}$, the tangent of the inclination of the tangent line to the curve representing the steady values of V and A with a constant length of arc.

In order to show the difference between these two values, and also to show that in an analogous case, where the resistance could be measured apart from the back E.M.F., the instantaneous $\frac{dV}{dA}$ found by superimposing a small alternating current on a continuous one did really give the value of the resistance, a glow lamp taking a current of 10 amperes at a P.D. of about 8 volts was joined in series with three accumulators. Thus, they had a resistance which they could measure separately in series with a back E.M.F. They then sent a small alternating current through the circuit *against* the E.M.F. of the accumulators, and plotting the curve obtained for the instantaneous $\frac{dV}{dA}$ and current, they found it very nearly coincided with the curve representing the observed resistance and the current, while the values found for the steady $\frac{dV}{dA}$ were all smaller than the corresponding resistances.

This experiment, the authors considered, justified them in concluding that if the arc consists of a back E.M.F. and a resistance, the actual value of the resistance was given by their method.

They varied the conditions of their experiments as much as possible. They studied the effect on the resistance of the arc of variations in the amount, frequency, and wave form of the alternating current; the effect of different kinds of carbons and different P.Ds and currents; the effect of using different combinations of cored and solid carbons, of carbons cored with substances other than carbon, and the effect of the relative sizes of the carbons.

The largest alternating current used had a root mean square value equal to about 10 per cent. of the continuous current. Frequencies between the limits of 250 and 7 complete alternations per second had no effect on the resistance of the arc. Complete information respecting the carbons is given in the figures. It will be seen from these that the ordinates of + solid − solid are all negative, those of + cored − cored are all positive, and that the other curves all lie between these two extremes. The curves in Fig. 22 connect resistance with current for a constant P.D.; those in Fig. 23 connect resistance and P.D. for a constant current. The curves for solid carbons

are in each case all very close together, while those for which
cored carbons were used show much greater divergence owing

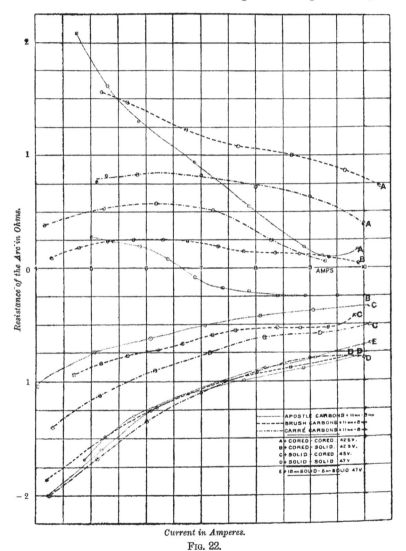

Current in Amperes.

FIG. 22.

to variations in the formation and diameter of the cores with
different makes of carbons.

From the curves the authors concluded that with both
carbons solid the resistance of the arc was always negative;
with both cored it was always positive, and with one cored and

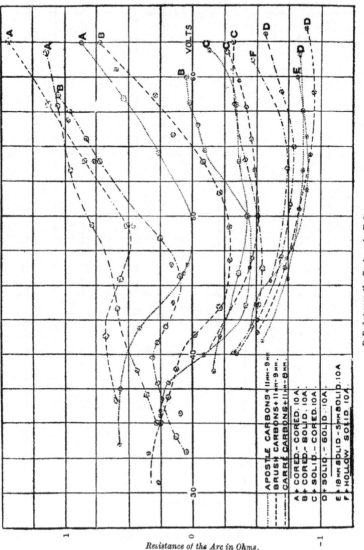

FIG. 23.

Resistance of the Arc in Ohms.

the other solid it was sometimes positive and sometimes negative.

They pointed out that with a constant current the resistance of the arc appeared always to reach a minimum as the arc was lengthened out, and then to increase again. This minimum is more strongly marked and occurs with a smaller P.D. with cored than with solid carbons. It is reached with cored carbons with the shortest arc in which the dark central space appears (*see* p. 7).

With inverted arcs the resistance with solid carbons was practically unchanged, but with cored carbons no dark space was seen, and the resistance was much less than for ordinary arcs. The authors considered that the degree of contact between the purple glow and the negative carbon had great effect on the resistance of the arc, which was most negative when that contact was most perfect.

When both carbons were cored it was found that above a certain frequency the instantaneous $\frac{dV}{dA}$ was positive, and below that frequency it was negative. The critical frequency was about 1·8. With the positive carbon cored and the negative solid, at 35 volts the resistance was positive with all frequencies, at 45 volts it was negative with all frequencies, and at 55 volts it was positive with frequencies above 1·8, and negative with frequencies below that.

The authors found that the current flowing through a hissing arc was oscillatory, the oscillatory current amounting in one case to 3 per cent. of the continuous current.[103]

Arons made some fresh experiments in 1896 to prove the existence of a back E.M.F. in the arc, and to determine its value. He worked on the same lines as Stenger (*see* p. 64), using, however, the town mains, which gave 105 to 110 volts, instead of a dynamo. His arrangement was as follows.

The mains (Fig. 24) were joined in series with two variable resistances, R, at least 3 ohms, and r, about 0·4 ohm, the tangent galvanometer T, the arc A and a battery of accumulators, B. Between the point of connection of R and r and the negative main, a Dubois key K was inserted. Hence, when K was open, the arc was fed by the mains, but when it was closed, a current flowed through the arc from the accumulators

G

in the direction *opposite* to that which flowed from the mains.

The carbons used were 15mm. in diameter, and both cored. The arc was from 1·5 to 2mm. in length.

The points which Arons wished to determine were :—

(1) What was the least E.M.F. in the accumulators with which a current could be made to flow through the hot vapour of the arc immediately after it was extinguished. (This current would, of course, with his arrangement, flow in the reverse direction, thus *helping* the back E.M.F. if there were any which continued after the arc was extinguished.)

(2) What E.M.F. was necessary in the battery to enable it to maintain an arc in the reverse direction, for even only a short time after the original arc was extinguished.

It is evident, from the arrangement, that closing the key K both stopped the current from the mains (and therefore extinguished the arc) and turned on the reverse current from the accumulators. Hence no time was lost between the two operations.

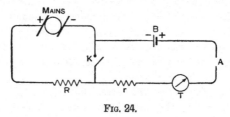

FIG. 24.

As regards his first point, Arons concluded that with the carbons he employed the smallest E.M.F. of the accumulators that would send a reverse current through the arc immediately after it was extinguished was 18 volts. He attributed the fact of Stenger's having been able to send such a current with an E.M.F. of 10 volts to his having used different carbons. He considered, therefore, that both his and Stenger's experiments showed that the condition of the carbon electrodes and the vapour after the extinction of the arc was of such a nature that it required a definite outside E.M.F. to send a current through the gaseous space.

Regarding the second point, Arons found that the accumulators could produce an arc in the reverse direction, after the extinction of the original arc, with a very small P.D. at the first moment, but that the P.D. necessary to maintain this arc then rose rapidly till it reached its normal value. This, he thought, was because the E.M.F. of the accumulators was assisted at the first moment by the still active back E.M.F. of the original arc, which, however, very rapidly died away. From his experiments he calculated this back E.M.F. to be from 10 to 14 volts.[104]

Some experiments made in 1896 by Mr. W. E. Wilson and Prof. G. F. Fitzgerald, "to determine, if possible, whether the temperature of the crater in the positive carbon varies when the pressure in the surrounding gas is changed," led to the conclusion that there was not sufficient evidence to affirm that the temperature of the crater was either raised or lowered by pressure.

The experimenters first used compressed air, and found that with any pressure greater than that of the atmosphere, some of the radiation was cut off by the formation of red fumes of NO_2, which became very plentiful when the pressure was as great as 100lb. per square inch. In compressed oxygen the arc burnt very steadily, but there was sufficient nitrogen present in the oxygen for enormous quantities of NO_2 to be formed at high pressures.

These results caused the experimenters to conclude that the red hot appearance of the crater and the reduction of radiation in their former experiments (*see* p. 72), were due to the formation of large quantities of NO_2 when the arc was under pressure.

In hydrogen contaminated with hydrocarbons, at atmospheric pressure, the arc burnt very far along both carbons, especially the negative. Trees of soot and a deposit of hard graphitic carbon formed all round the crater, as if there were electrolysis of the hydrocarbon, and carbon were electro-negative compared with hydrogen. The arc was very unsteady, both current and P D. varying continually, and the soot trees hid the crater, so that the attempt to get any measures of radiation under pressure with hydrogen was abandoned.

Finally, carbon dioxide was tried, a cylinder of CO_2 being connected with the arc box. At pressures above 150lb. the arc could not be maintained long enough for radiation measures to be obtained, but at lower pressures some good measurements were made. With from 1 to 6 or 7 atmospheres very little change of radiation appeared to take place.

Messrs. Wilson and Fitzgerald consider that the results of their experiments render it very improbable that it is the boiling point of carbon that determines the temperature of the crater, and this opinion is strengthened by the fact that the carbon is so slowly evaporated. "The crater of mercury" they say, "is dark, but then it volatilises with immense rapidity, and the supply of energy by the current being more than 100 times that required merely for evaporation, there seems very little reason why even a considerable difference in latent heat should make any sensible difference in the rate of evaporation of mercury and carbon, especially as, at the same temperature, the diffusion of carbon vapour is nearly three times as fast as that of mercury vapour, and the temperature immensely higher."[105]

M. Guillaume, in a Paper read before the French Physical Society, said that he believed the reduction of brilliancy found by Mr. Wilson in his experiments on the arc under pressure (see p. 72) was due to carbon being dissolved in the surrounding atmosphere. He considered his theory was proved by the fact of Messrs. Wilson and Fitzgerald having found in their later experiments above that, with an arc burning in CO_2 at a pressure of 150lb., a fog formed when the pressure was suddenly reduced.[106]

The much vexed question of a back E.M.F. in the arc was attacked by Herzfeld, who, like Stenger and unlike Arons, came to the conclusion that it did not exist. He used modifications of Edlund's method ; extinguishing the arc, and switching into the circuit immediately after, some instrument for measuring the P.D. between the carbons. In his final experiment the time that elapsed between the two operations was only $\frac{1}{270}$sec., yet he could detect no P.D. between the carbons even so short a time after the arc had been extinguished.

His next experiment was made to determine whether the P.D. between the carbons depended in any way on the amount of carbon transferred from one pole to the other. The arc was placed between two plates 6cm. high and 8cm. broad, the distance between them being varied from 2cm. to 10cm. The P.D. between them was 1,800 volts. The plates could be connected with a Leyden jar, and either both insulated, or one insulated and the other, with the part of the Leyden jar to which it was attached, earthed. As it was found that the ultra-violet rays of the arc quickly discharged the condenser, the charge was kept up by means of a Holz machine.

It was found that although, when the electric field was excited, the particles sent out from the positive towards the negative carbon were continually attracted to the insulated plate, yet that neither the current nor the P.D. between the carbons was changed within the limits of sensibility of a Schuckert's ammeter and voltmeter by this withdrawal of carbon particles from the arc.

The particles were always attracted to the *insulated* plate and arranged themselves radially on it (Fig. 25) whether it

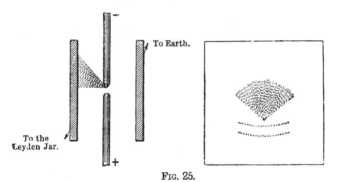

To Earth.

To the
Leyden Jar.

FIG. 25.

was charged positively or negatively, even when it was 8cm. from the carbons, while the earthed plate was only 0·6cm. from them. If both plates were insulated the particles went to both equally. Herzfeld thought that perhaps the narrow double ring of particles that ranged themselves below the radial lines (Fig. 25) were those sent out from the negative to the positive carbon. The flow of particles could be plainly seen in an image of the arc projected through a lens.

The author considered that his experiment showed that the supposed back E.M.F. in the arc cannot be due to a polarisation of the electrodes taking place through separated solid particles, since the P.D. did not vary with the number of solid particles that really reached the one electrode from the other.

To see if a polarisation took place through the gaseous portion of the arc, and if, therefore, the vapour of the arc was affected by the electric field, an enlarged image of the arc was made through a spectroscope, the slit of which was perpendicular or parallel to the electric lines of force, but no change could be detected when the electric field was excited.

Next, it was sought to discover the cause of the formation of a mushroom with hissing arcs, and of the growth of the negative carbon that takes place, even with silent arcs, under certain circumstances. It was shown, in the following way, that an absence of sufficient oxygen to burn the carbon will cause these growths. The carbons were enclosed in glass tubes 18mm. in diameter, to which the access of fresh air was restricted by a cork in one end of each. The carbon particles thus flew unburnt from the positive crater to the negative point and rested there. Thus the length of the arc remained unchanged for from five to ten minutes, the crater becoming deeper and the point of the negative carbon becoming longer all the time. The growth on the negative was in the shape of a corkscrew.

To test whether this form was due to magnetic forces induced by the regulating magnet in the base of the lamp, the negative carbon was surrounded by an electromagnet to within 5cm. of its point, and it was found that changing the polarity of this magnet changed the direction of the corkscrew, which was from 1·2 to 2mm. in height, while the maximum number of screw turns was $2\frac{1}{2}$. If the spirals were formed from positively electrified particles on their way from the positive to the negative carbon, then those particles moved in the opposite direction from Ampère's molecular streams of electromagnets.

To try what effect cooling each carbon and the arc itself separately had on the P.D. between the carbons, Herzfeld directed a thin jet of carbonic acid gas against each in turn, using 7mm. cored carbons placed in a Dubosq lamp from which the regulating mechanism had been removed. He found that whether the positive carbon was above and the negative below, or *vice versâ*, or if the arc was horizontal; whether the stream was directed against the positive or the negative carbon, in all cases the P.D. between the carbons *increased* when the cooling

jet was applied, and, although it diminished after the cooling
was discontinued, it remained, in most cases, above its normal
value. At the same time that the P.D. increased, the current
in all cases diminished, under the cooling action of the carbonic
acid.

When the carbons were surrounded by an atmosphere of
carbonic acid gas by being enclosed in a glass tube closed
beneath and filled with the gas, it was found that directing a
stream of the same gas against either carbon increased the
P.D. between the carbons much less than when they were in
air, unenclosed.

A current of air from a foot-bellows directed against either
of the carbons produced no effect on the P.D. between them
when the arc was burning silently.

Herzfeld considered that the effect of the carbonic acid was
two-fold; it both cooled the tips of the carbons, and increased
the resistance of the arc itself by lowering its temperature.
That the effect was not thermo-electric was, he thought, proved
by the following experiment. A rod of graphite 1mm. thick
was placed in the arc midway between the two carbons, and
the P.D. between the graphite and each of the carbons was
measured with a d'Arsonval galvanometer with 170,000 ohms in
circuit. The P.D. of about 35 volts between the positive
carbon and the graphite was *increased* by 2·1 volts when the
positive carbon was cooled by the carbonic acid, and the P.D.
of 6 volts between the graphite and the negative carbon was
also *increased* by 2·8 volts when the latter was cooled in the
same way.

In his concluding experiments Herzfeld made a comparison
between open arcs and those enclosed in glass tubes, the
E.M.F. of the accumulators being constant, and the arc being
allowed to burn away unregulated till it went out. The
following results were obtained :—

(1) The time between the P.D. between the carbons attain-
ing a given value and the arc going out was much greater
when the arc was enclosed.

(2) The P.D. between the carbons when the arc went out
was greater when it was enclosed.

(3) Greater lengths of the ends of the carbons glowed after
the arc went out when it was enclosed.

When the enclosing tube was filled with carbonic acid gas
the carbons burnt away very quickly, and after the arc was
extinguished a blue light was frequently seen, both of which
facts appeared to the author to show that the carbonic acid was
split up, in higher temperatures, into carbonic oxide and oxygen,
the latter enabling the carbons to burn away more quickly.

The conclusion arrived at by the author, as the result of
his experiments, was that the great heat produced at the crater
is not a Peltier effect, but that a substance of great resistance
is accumulated at the boundary between the positive carbon
and the air, which is heated by the passage of the current
through it, and which vaporises the positive carbon. This
vapour, he thinks, condenses into fluid and solid drops in the
cooler parts of the arc.[107]

The same question of the existence of a back E.M.F. in the
arc was attacked by M. Blondel in the following manner. The
circuit of a continuous current arc was periodically interrupted

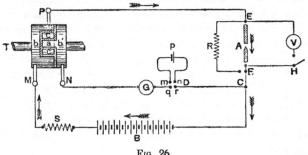

Fig. 26.

at very short intervals and for very short periods, and during
each interruption the two carbons were connected with a
galvanometer.

These operations were carried out by the revolving commutator in Fig. 26, which also shows the circuits and general
arrangements. The commutator T, driven by a continuous-
current constant speed motor, consisted of an ebonite core
on which there were two copper rings b and b', one of which
was broader than the other. The ring b had a piece cut
out of it, and in this indentation there was a tongue a
forming part of the ring b' and two copper insulated plates c

and c'. All these parts were separated by strips of mica, and the brushes themselves were also insulated from their holders by ebonite, so that the insulation resistance between any two of the brushes resting on the commutator and between each brush and earth always exceeded 5 megohms. This commutator revolved at a speed of about 40 revolutions a second. The piece cut out of the ring b was about one-fifth of the circumference. The arc lamp was fed by the battery B, giving about 70 volts. The current traversed successively the steadying resistance S, and the commutator between M and P across the ring b, then the lamp E F and the switch C. At every revolution the circuit was broken during $1/5 \times 1/40 = 1/200$ of a second by the passage of the indentation beneath the brush P, the spark being taken on the insulated plate c. These interruptions were very brief and followed very close on one another. The arc was perfectly stable, and could not be distinguished from an ordinary continuous-current arc.

The arc having become steady, q and r were connected so that the arc was short-circuited by the galvanometer G, which was fairly sensitive, during the passage of the tongue a under the brush P (about $\frac{1}{600}$th of a second). The author considered that with this arrangement there was no reason to fear the influence of cooling on the physical conditions of the arc during its extinction, nor, consequently, during the passage of the tongue a. If there existed, therefore, an E.M.F. or ordinary polarisation, it should betray itself, the author thought, by producing a permanent and easily-observed deflection of the galvanometer.

A battery, p, usually a single cell, interposed in the galvanometer circuit, first one way and then the other, enabled him to estimate the value of this E.M.F. and satisfy himself as to the sensitiveness of the method ; it was only necessary to take two readings of the deflections obtained with the battery plus the arc, and to compare them with that given by the arc alone.

Finally, he substituted the resistance R, taking the same current at the same voltage, for the arc itself, and then carried out the same series of measurements as on the lamp, and was thus able to discover in what the two phenomena differed.

These experiments were carried out under the most diverse conditions, with long arcs and short, silent ones and whistling,

with carbons far apart and sticking together, with solid carbons S, or cored carbons C, and the results presented no
other difference than such as would arise from experimental
errors. The speed of rotation of the commutator was also
varied within wide limits without causing any appreciable
difference. It was found, however, that a speed of the order
previously mentioned or even a higher one, was necessary in
order to obtain steady arcs and steady galvanometer deflections.

Table IX. is a summary of a few of the series of figures
obtained.

Table IX. (*Blondel*).

No of Experiment.	Nature of Carbons. Upper	Lower	Arc. Amp.	Terminal Volts.	Galv. Deflections Arc alone.	Arc plus cell.	Resistance. Amp.	Terminal Volts.	Galv. Deflections. Resist. alone.	Resist. plus cell.
						+ −				+ −
1	C	S	5	35	7	70 −78	5	34·5	0	71·5 −75
2	S	S	8	25	1	75 −72	8	27·7	−9·5	66 −83
3	S	S	10	18	0	75 −73	10	18	−4	73 −78
4	S	S	8	18	−3·5	73 −75	8	18	−8	67 −82
5	S	S	11	4	1·3	80 −73	11	4	0·5	76 −73
6	C	S	7	20	1	73 −73	7	20	1	76 −74
7	C	S	7·5	20	2	71 −74	7·5	20	−3	71 −75
8	C	S	8	18	−5	70 −78	8	17·7	−6	68 −79
9	S	S	8	19	−1	72·5 −77	8·25	17·5	1·2	75·5 −73
10	C	S	6	29	2·5	70 −75	6	29	2·5	77 −74

M. Blondel pointed out that the deflections produced by the
arc plus the cell of 2·25 volts E.M.F. were very large compared with those produced by the arc alone. He considered
that the above table showed that if there were a constant
counter E.M.F. in the arc it could not be greater than

$$\frac{5}{70} \times 2 \cdot 25 = 0 \cdot 16 \text{ volt.}$$

M. Blondel concluded from his experiments that, although
the arc may not be of the exact nature of an ordinary resistance, yet that it behaves sensibly like a resistance, and possesses no counter E.M.F., in the ordinary sense of the term,
comparable with the observed P.D. It is not due, therefore,
he thinks, to an electrolytic phenomenon, and if there be a
residual E.M.F. due to thermo-electric causes, this cannot
exceed a fraction of a volt.[109]

Granquist, in criticising Arons' latest experiments on the back E.M.F. of the arc (*see* p. 81), said that he thought the reason Arons could not apparently get a current to flow in the reverse direction through the carbons, immediately after the arc was extinguished, with a less P.D. than 18 to 22 volts, was that the tangent galvanometer he used was not sensitive enough to detect the current.

He pointed out that since the dying out of the vapour between the carbons after the arc was extinguished was a cooling effect, it was probable that the time taken by the operation depended to some extent on the amount of vapour existing while the arc was burning, and therefore on the current flowing through the arc. The very sending of a current through the carbons would tend to increase the heat of this vapour, and hence to retard the dying-away process. He had himself found this to be the case, for he could send a current through the arc for a much longer time after it was extinguished when a large current had been flowing, than when there had been a small one. Also, he could send a large current in either direction for a longer time than a smaller one, after the arc was extinguished.

Granquist himself had found that he could send a current in either direction through the carbons immediately after the arc was extinguished, with a single Daniell's cell. The experiments were made in 1894, but as the results were only published in Swedish, he gave a short account of them in a Paper published in German in 1897.

Fig. 27 shows the apparatus used. D was a Siemens shunt dynamo, G an ammeter, A the arc, B a Daniell's cell, M a mercury contact, and R a reversing switch by means of which the current from B could be reversed; G' was a galvanometer constructed by Granquist himself on the principle of unipolar induction, which, as it contained no coils of wire, might be considered to be free from self-induction; a, b and c, were three metal brushes which rubbed against the wheel W and the two wheels attached to it, each 235·5mm. in circumference, the one of ebonite and the other of metal. All three wheels were joined firmly together, so that they rotated on the same axis. In the periphery of the metal wheel a slot 34mm. in length was cut and filled in with ebonite. Similarly, in the periphery of the ebonite wheel a slot, 21mm. in length, was cut right down to the metal axis, and filled in with brass, so that when this part of the wheel came under the brush *b* there was electric connection between *b* and the wheel W. The wheels were so arranged that when the brush *a* was on the ebonite part of the metal wheel, the brush *b* was on the metal part of the ebonite wheel. Thus, when the dynamo circuit was broken, the circuit *c* A M R *b* was closed.

It was necessary that the dynamo circuit should be open longer than the galvanometer circuit was closed, in order to allow time for the spark which would pass at breaking to die away, and for the arc circuit to be really completely·broken before the galvanometer circuit was closed. Hence, the ebonite part of the metal wheel was 34mm. in length, while the metal

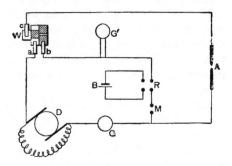

Fig. 27.

part of the ebonite wheel was only 21mm. Thus, the dynamo circuit was broken 1·5 times as long as the galvanometer circuit was closed, so that by moving the brush *b*, and changing the speed of rotation of the wheels, the time between the opening of the dynamo circuit and the closing of the galvanometer circuit could be altered at will.

The following was the method of experimenting. After the wheels had been set in motion and the carbons brought into contact, the dynamo. circuit was closed. As soon as the arc was well established, the circuit of the cell B was also closed by means of the mercury contact M. A deflection U_1 was thus obtained in the galvanometer G'. The current was then reversed by means of the switch R, and the deflection U_2 observed. Then, if E were an E.M.F. in the arc, and the E.M.F. of the cell were e,

$$E = \frac{U_1 + U_2}{U_1 - U_2} e.$$

In the following table U_1 and U_2 were the deflections of the galvanometer, and E the supposed back E.M.F. of the arc. The time which elapsed between the complete breaking of the current and the closing of the galvanometer circuit was 0·0009 second.

Table X. (*Granquist.*)

Current in arc.	U_1.	U_2	E.
6·2	+30·0	− 18·1	0·27
6·2	24·0	14·1	0·26
5·0	22·0	13·5	0·26
5·2	23·8	14·5	0·24
3·2	19·7	11·5	0·26
7·5	20·2	13·5	0·20
5·6	14·4	10·0	0·20
8·9	14·5	17·7	0·11
5·0	20·5	12·7	0·23
4·0	17·0	10·5	0·24
			mean 0·227 volt.

Hence Granquist found, as Lecher, Luggin, and Stenger had already done, that there was no back E.M.F. in the arc *after it was extinguished* greater than about 0·227 volt. Unlike Blondel, however, he did not think this precluded the possibility of a far greater back E.M.F. in the arc *while it was burning*, but he considered that the larger back E.M.F. and the current ceased to exist at the same moment.[110]

In commenting on M. Blondel's method, given above, of proving that the arc possesses no back E.M.F. in the ordinary acceptation of the term, Prof. Fleming suggested that there might be a back E.M.F. due to a "Thomson effect" along the hot vapour of the arc itself—that is to say, a back E.M.F. due to the temperature gradient of the hot vapour. Prof. Fleming mentioned that he and Prof. Dewar had shown that in carbon between the temperatures − 200° and + 200°, the E.M.F. caused by the "Thomson effect" acts from cool to hot as it does in copper. If in carbon vapour the "Thomson effect" keeps the same sign as in solid carbon, he thinks there might be a back

E.M.F. due to it along the column of vapour. This, he said, would account for the fact that when the negative carbon is heated the P.D. between the carbons for the same current is less than when it is unheated, and also for the smaller P.D. necessary for an alternating current arc, because the positive and negative carbons must be more nearly equal in temperature.[111]

CHRONOLOGICAL LIST OF ORIGINAL COMMUNICATIONS CONCERNING THE ARC.

[1] Nicholson's *Journal*, 4to, 1801, Vol. IV., p. 326 ... ... DAVY.
[2] Gilbert's *Annalen*, 1801, Vol. VII., p. 161 ... ... GILBERT.
[3] Gilbert's *Annalen*, 1801, Vol. VII., p. 248 ... ... PFAFF.
[4] Gilbert's *Annalen*, 1801, Vol. VII., p. 516 ... ... PFAFF.
[5] Gilbert's *Annalen*, 1801, Vol. VIII., p. 370 ... ... PFAFF.
[6] Gilbert's *Annalen*, 1801, Vol. IX., p. 341 ... ... RITTER.
[7] Nicholson's *Journal*, 8vo, 1801, Vol. III., p. 136 ... DAVY.
[8] *Journal* of the Royal Institution, 1802, Vol. I., p. 166 ... DAVY.
[9] *Journal* of the Royal Institution, 1802, Vol. I., p. 209 ... DAVY.
[10] *Annales de Chimie*, An. IX. (1801), Vol. XXXIX. ... { FOURCROY, VAUQUELIN, THÉNARD.
[11] Nicholson's *Journal*, 4to, 1802, Vol. V., p. 238 ... ... TROMSDORFF.
[12] Gilbert's *Annalen*, 1802, Vol. XI., p. 396 ... ... ANON.
[13] *The Monthly Magazine*, 1803, Vol. XV., p. 259 ... ... PEPYS.
[14] Nicholson's *Journal*, 1804, Vol. VIII., p. 97 ... ... CUTHBERTSON.
[15] " Practical Electricity and Galvanism," 1807, p. 260 ... CUTHBERTSON.
[16] MS. Note Book at the Royal Institution, 1808 ... ... DAVY.
[17] *Philosophical Transactions*, 1809, p. 46 ... ... DAVY.
[18] MS. Note Book at the Royal Institution, 1809 ... ... DAVY.
[19] *The Monthly Magazine*, August 1, 1810, Vol. XXX., p. 67 DAVY.
[20] "Elements of Chemical Philosophy," 1812,Vol. I., p.152... DAVY.
[21] *Annales de Chimie et de Physique*, 1820, Vol. XV., p. 101 ARAGO.
[22] *Philosophical Transactions*, 1821, p. 18 ... ... DAVY.
[23] Silliman's *Journal*, 1821, Vol. III., p. 105 ... ... HARE.
[24] Silliman's *Journal*, 1822, Vol. V., p. 108... ... ... SILLIMAN.
[25] Silliman's *Journal*, 1823, Vol. VI., p. 342 ... ... SILLIMAN.
[26] Silliman's *Journal*, 1826, Vol. X., p. 123... ... ... SILLIMAN.
[27] *Philosophical Magazine*, 1838, p. 436 ... ... GASSIOT.
[28] *Transactions* of the London Electrical Society, 1837 to 1840, p. 71 ... ... { GASSIOT, WALKER, STURGEON, MASON.
[29] *Philosophical Transactions*, 1839, p. 92 ... ... DANIELL.
[30] *Philosophical Magazine*, 1840, Vol. XVI., p. 478 ... GROVE.
[31] *Comptes Rendus*, 1840, Vol. XI., p. 702 ... ... BECQUEREL.

[32] *Comptes Rendus*, 1841, Vol. XII., p. 910 ... ... DE LA RIVE.

[33] *Comptes Rendus*, 1841, Vol. XIII., p. 198... ... ... BECQUEREL.

[34] *Archives de l'Electricité*, 1841, p. 575 ... ... MACKRELL.

[35] Poggendorff's *Annalen*, 1844, Vol. LXIII., p. 576 ... CASSELMANN.

[36] *Comptes Rendus*, 1844, Vol. XVIII., p. 746 .. ... { FIZEAU, FOUCAULT.

[37] Poggendorff's *Annalen*, 1845, Vol. LXVI., p. 414 ... NEEF.

[38] *Comptes Rendus*, 1846, Vol. XXII., p. 690 ... ... DE LA RIVE.

[39] *Comptes Rendus*, 1846, Vol. XXIII., p. 462 ... ... VAN BREDA.

[40] *Comptes Rendus*, 1850, Vol. XXX., p. 201 ... ... MATTEUCCI.

[41] *Comptes Rendus*, 1852, Vol. XXXIV., p. 805 ... ... QUET.

[42] *Philosophical Transactions*, 1852, p. 88 ... ... GROVE.

[43] *Comptes Rendus*, 1865, Vol., LX., p. 1,002 ... ... DE LA RIVE.

[44] Poggendorff's *Annalen*, 1867, Vol. CXXXI., p. 586. ... EDLUND.

[45] Poggendorff's *Annalen*, 1868, Vol. CXXXIII., p. 353 ... EDLUND.

[46] Poggendorff's *Annalen*, 1868, Vol. CXXXIV., pp. 250, 337 EDLUND.

[47] Poggendorff's *Annalen*, 1870, Vol. CXXXIX., p. 354 ... EDLUND.

[48] Poggendorff's *Annalen*, 1870, Vol. CXL., p. 552... ... BEZOLD.

[49] *The Electrician*, 1879, Vol. II., p. 76 ... ... AYRTON.

[50] *The Electrician*, Vol. II., 1879, Jan. 18th, p. 107 ; and 25th, p. 117 ... ... SCHWENDLER.

[51] Royal Engineering Committee Extracts for 1879, Appendix III.

[52] *La Lumière Electrique*, 1879, Vol. I., p. 41 ... ... DU MONCEL.

[53] *Philosophical Transactions*, 1879, p. 159 ... ... { DE LA RUE, MÜLLER.

[54] *La Lumière Electrique*, 1879, Vol. I., p. 235 ... ... ROSSETTI.

[55] *Journal* of the Society of Telegraph Engineers, 1880, Vol. IX., p. 201... ... ... ANDREWS.

[56] *La Lumière Electrique*, 1881, Vol. III., p. 220 ... ... ROSSETTI.

[57] *La Lumière Electrique*, 1881, Vol. III., p. 285 ... ... LE ROUX.

[58] *La Lumière Electrique*, 1881, Vol. III., p. 287 ... ... NIAUDET.

[59] *Proceedings* Physical Society, 1882, Vol. V., p. 197 ... { AYRTON, PERRY.

[60] *Proceedings* of the Royal Society, 1882, Vol. XXXIII., p. 262 ... ... DEWAR.

[61] *Elektrotechnische Zeitschrift*, 1883, Vol. IV., p. 150 ... FRÖLICH.

[62] *Zeitschrift für Elektrotechnik*, 1885, Vol. III., p. 111 ... PEUKERT.

[63] *Wiener Akad.*, 1885, Vol. XCI., § 844 ... ... VON LANG.

[64] *Centralblatt für Electrotechnik*, 1885, Vol. VII., p. 443... VON LANG.

[65] Wiedemann's *Annalen*, 1885, Vol. XXV., p. 31 ... ... STENGER.

[66] Wiedemann's *Annalen*, 1885, Vol. XXVI., p. 518 ... EDLUND.

[67] *Proceedings* of the American Academy of Sciences, 1886, p. 2 ... ... { CROSS, SHEPARD.

[68] *Centralblatt für Elektrotechnik*, 1886, Vol. VIII., pp. 517, 619 ... ... NEBEL.

[69] Wiedemann's *Annalen*, 1887, Vol. XXX., p. 93 ... ... ARONS.

[70] Wiedemann's *Annalen*, 1887, Vol. XXXI., p. 384 ... VON LANG.

[71] *Centralblatt für Elektrotechnik*, 1887, Vol. IX., p. 219 ... VOGEL.

[72] *Centralblatt für Elektrotechnik*, 1887, Vol. IX., p. 633 ... UPPENBORN.

[73] *Beiblätter*, 1888, Vol., XII., No. 1, p. 83 ... ... UPPENBORN.
[74] *Centralblatt für Elektrotechnik*, 1888, Vol X., p. 3 ... FEUSSNER.
[75] *Centralblatt für Elektrotechnik*, 1888, Vol. X., p. 48 ... LECHER.
[76] *Centralblatt für Elektrotechnik*, 1888, Vol. X., p. 102 ... UPPENBORN.
[77] *Centralblatt für Elektrotechnik*, 1888, Vol. X., p. 567 ... LUGGIN.
[78] *Centralblatt für Elektrotechnik*, 1888, Vol. X., p. 591 ... SCHREIHAGE.
[79] *Centralblatt für Elektrotechnik*, 1888, Vol. X., p. 749 ... DUBS.
[80] Wien *Sitzungsberichte*, 1889, Vol. XCVIII., p. 1,192 ... LUGGIN.
[80a] *Proceedings* of the Royal Society, 1889, Vol. XLVII.,
p. 118 ... ... FLEMING.
[81] *The Electrical World*, 1891, Vol. XVII., p. 166 ... ... { ELIHU THOMSON.
[82] *La Lumière Électrique*, 1891, Vol. XLII., p. 621 ... BLONDEL.
[83] *The Electrical World*, 1892, Vol. XIX., p. 195 ... ... CRAVATH.
[84] *The Electrical World*, 1892, Vol. XX., p. 227 ... ... CRAVATH.
[85] *The Electrician*, 1892, Vol. XXVIII., p. 687, Vol. XXIX.,
p. 11 ... ... TROTTER.
[86] Wiedemann's *Annalen*, 1892, Vol. XLV., p. 33 ... ... STENGER.
[87] *The Electrician*, Vol. XXIX., 1892, p. 460 ... ... S.P.THOMPSON
[88] *Comptes Rendus*, 1892, Vol. CXV., p. 1,273 ... ... VIOLLE.
[89] *Journal de Physique*, 1893, Vol. II., p. 545 ... ... VIOLLE.
[90] *Electrical Engineer* of New York, 1893, p. 90 ... ... { DUNCAN, ROWLAND and TODD.
[91] *The Electrician*, 1893, Vol. XXXII., pp. 117, 145, 169 ... BLONDEL.
[92] *Comptes Rendus*, 1894, Vol. CXIX., p. 949 ... ... VIOLLE.
[93] *The Electrician*, 1894, Vol. XXXIII., p. 297 ... ... TROTTER.
[94] " Electric Lamps and Electric Lighting," p. 153 ... FLEMING.
[95] *Memoirs* and *Proc.* of the Manchester Lit. and Phil. Soc.,
1895, Vol. IX., Series IV., p. 139 ... ... FRITH.
[96] Wiedemann's *Annalen*, 1895, Vol. LV., p. 361 ... ... LEHMANN.
[97] *Proc. Roy. Soc.*, 1895, Vol. LVIII., p. 174 ... ... WILSON.
[98] *Beiblätter*, 1895, Vol. XIX., p. 97 . . .. GRANQUIST.
[99] *The Electrical Review*, 1895, Vol. XXXVII., pp. 230,253,301 FREEDMAN.
[100] *The Electrical World*, 1895, Vol. XXV., p. 277... ... MARKS.
[101] *The Electrical Engineer* of New York, 1895, Vol. XIX.,
p. 198 ... ... MARKS.
[102] *The Electrical World*, 1896, Vol. XXVII., pp. 262, 378 { PUPIN, FREEDMAN.
[103] *The Philosophical Magazine*, 1896, p. 407 ... ... { FRITH, RODGERS.
[104] Wiedemann's *Annalen*, 1896, Vol. LVII., p. 185 ... ARONS.
[105] *Proceedings* of the Royal Society, 1897, Vol. LX., p. 377 { WILSON, FITZGERALD.
[106] *The Electrician*, 1897, Vol. XXXVIII., p. 642 ... ... GUILLEAUME.
[107] Wiedemann's *Annalen*, 1897, Vol. LXII., p. 435 ... HERZFELD.
[108] *L'Éclairage Électrique*, 1897, Vol. X., pp. 289, 496, 539 BLONDEL.
[109] *The Electrician*, 1897, Vol. XXXIX., p. 615 ... ... BLONDEL.
[110] *Öfversigt af Kongl. Vetenskaps-Akademiens Förhandlingar*, 1897. N : o 8, Stockholm, p. 451 ... ... GRANQUIST.
[111] *The Electrician*, 1898, Vol. XL., p. 363 ... ... FLEMING.

CHAPTER III.

PHENOMENA CONNECTED WITH THE "STRIKING" OF THE ARC
AND WITH SUDDEN VARIATIONS OF CURRENT.

AT the Electrical Congress held in Chicago in August, 1893,
Prof. Ayrton read a long Paper on the subject of the Electric
Arc, which gave the results of experiments that he had been
carrying out with his students during the three preceding
years. Neither the Paper, nor any abstract of it, was published
in the report of the Congress, for while it was in the hands of
the secretary of Section B of the Congress, it was unfortunately
burnt five months after it had been read.

The experiments to which Prof. Ayrton specially directed
his attention were briefly :—

1. Obtaining the time variations, after striking the arc, of
the P.D. between the carbons, with various constant currents,
various constant lengths of arc, and with the ends of the
carbons variously shaped.

2. Obtaining the time variation of the P.D. between the
carbons when the current was suddenly changed, and the
length of the arc was kept constant.

3. Obtaining curves connecting the steady final values of the
P.D. between the carbons with the current, for different
currents, lengths of arc, and sizes of carbons, cored and
uncored.

4. The influence of varying the current and the length of
the arc on the depth and width of the crater.

5. The distribution of potential throughout the arc.

6. The candle-power and efficiency of the arc with various
currents, P.Ds., and lengths of arc.

E

The lamp used in these experiments (Fig. 28) was hand-regulated, the adjustments being effected by turning pinion P_1

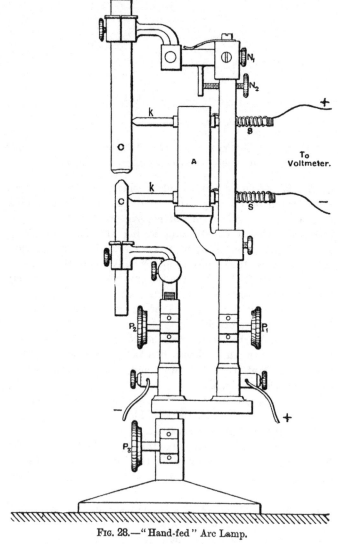

Fig. 28.—" Hand-fed " Arc Lamp.

to alter the height of the positive carbon, pinion P_2 to alter the height of the negative carbon, and pinion P_3 to raise both carbons together. By turning the nut N_1 the positive carbon could be turned about a horizontal axis in the plane of the figure, and by turning the nut N_2 the positive carbon was moved round a horizontal axis at right angles to the plane of the figure. To measure the P.D. between the tips of the carbons, the voltmeter was attached to two thin carbon rods $k\,k$, sliding in tubes in a block of asbestos, A, and pushed against the main carbon rods C C by spiral springs S S. Had the P.D. been measured between the lamp terminals, a variable error would have been introduced, from the drop of pressure in the carbons themselves, which would have been serious with large currents and long carbons. Since the voltmeter had a resistance of about 80,000 ohms in circuit

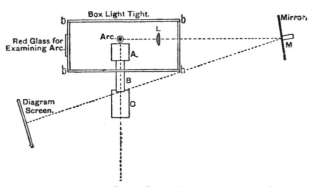

Fig. 29.—Plan of Arc Lamp, Lens, Mirror and Diagram Screen.

with it, the resistance between the ends of these auxiliary voltmeter carbons $k\,k$ and the main carbons C C introduced no practical error.

The length of the arc was always taken to be the vertical distance between the point of the negative carbon and the horizontal plane drawn through the edge of the crater of the positive carbon. Length of arc "0 millimetres" does not, therefore, mean that the carbons were in contact, but that the point of the negative carbon was just entering the crater at the end of the positive carbon. This distance was measured on an image formed on the diagram screen, as shown in Fig. 29 by the lens L and the plane mirror M. This image of the arc was exactly ten times full size.

When these experiments were first started, at the beginning of 1890, it was not known what were the conditions necessary for the P.D. between the carbons to remain constant when the current and length of arc were both kept constant, and consequently it was found, as had been found by all previous experimenters, that a given current could be sent through an arc of given length by many different potential differences, and that no set of experiments made one day could be repeated the next.

In the earlier experiments a reading was taken soon after the current had been brought to the desired value; hence the curves connecting P.D. with current for a constant length of arc were different for each set of experiments, and were always too steep.   For example, with both positive and negative carbons cored, and both 13mm. in diameter, the early curves show that when the arc had a constant length of 5mm. the P.D. fell from 59 volts for a current of 4 amperes to 24 volts for a current of 30 amperes; that is, it was diminished by 35 volts.   Whereas, in the later experiments, with a 13mm. cored positive carbon and an 11mm. solid negative, the P.D. fell from 60·5 to 48·8 volts, or only 11·7 volts for the same change in current with the same length of arc.

The first step in advance was made by keeping each new current flowing for a certain minimum time before taking the observations.   This resulted in making the curves connecting P.D. with current for a given length of arc much flatter, and not quite so widely different with different sets of experiments; but it was still found that the curves for values of current ascending and those for values of current descending were different.

Fig. 30 gives one of the sets of curves drawn during this series of experiments made in 1890, when it had been found that allowing a certain time to elapse before the reading was taken after the current had been altered made the readings for ascending and descending current more nearly equal, and also made the curve connecting P.D. with current for a given length of arc less steep.   It was also found that, if the whole series of readings could be taken without the arc going out, better results were obtained; therefore, during the five hours occupied by the experiments from which Fig. 30 was taken the arc was never extinguished.

A large number of experiments were now carried out, each of which occupied the greater part of a day, as the current, which was made to slowly vary backwards and forwards between two limits, was never stopped, nor the arc allowed to go out, for many hours at a time.

However, even with all these precautions, looped curves similar to those in Fig. 30 were obtained, and it is apparent from these curves that the P.D. needed to send a given current through the arc, which was kept at a constant length of 4mm., was never twice the same. For instance, a current of 10 amperes (Fig. 30) was sent through the arc by

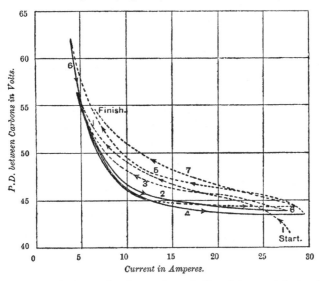

Fig. 30.—18mm. cored carbons. Arc not allowed to go out during the 5 hours' run. Termination of experiment due to wasting of carbons. Arc at a constant length of 4mm. Current increasing, ———→——— Current decreasing, - - -←- - - Latter half of curve 6 is dotted owing to the length of arc being indeterminate. This was due to inequality in the carbons.

P.Ds. of 51, 49·5, 49, 48, 46·5, 46·2, and 46 volts respectively, so that for this one current the P.Ds. ranged from 46 to 51 volts. Hence, from these curves it would be impossible to find any exact relation between P.D. and current for a given length of arc.

Consequently, when the work was taken up again, it was thought advisable to make a complete investigation of the variation of P.D. with the time that elapsed after the arc had been started or the current suddenly changed, the current and length of arc subsequently being kept constant during each experiment.

The questions to which answers were sought were the following :—

(1) Does the P.D. *ever* become a constant for a given current and length of arc ?

(2) If there is a final constant P.D. for each current and length of arc, is this P.D. the same, and is the time taken to reach it after starting the same, whether a cored or an uncored positive carbon is used ?

(3) What are the *causes* of this variation of P.D. ?

(4) How is the time before this P.D. is reached affected by the employment of (*a*) different lengths of arc, (*b*) different currents ?

(5) How is this period of time affected by the current that was flowing through the arc before the change was made ?

The experiments from which Figs. 31 to 35 were taken answered the first three questions. They showed that the P.D. *does* reach a final constant and steady value, that coring the positive carbon increases the period of time which elapses after starting the arc before the final value of the P.D. is reached, and they showed the causes to which the variation of the P.D. is due.

These experiments were all made with positive carbons 18mm. and negative carbons 15mm. in diameter. The positive carbon was in some cases cored as is indicated in the figures, and in other cases solid, and the negative carbons were solid in all cases. Also the current used was 10 amperes, and the length of the arc was 3mm. for all the experiments except those which answered question (4). The first point on each of the curves was taken the moment the carbons had been separated to a distance of 3mm. after striking the arc.

It is evident that question (1) is answered in the affirmative, for after a shorter or longer time the P.D. in all cases finally reached a constant steady value of from 44 to 46 volts, when the positive carbon was cored, and from 47 to 50 volts when it was solid (Figs. 31 to 35), the slight variation of P.D. finally arrived at with the same current and length of arc for the same kind of carbons being accounted for by the fact that every pair of carbons differs slightly from every other pair in hardness and structure.

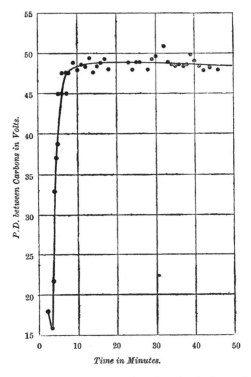

FIG. 31.--Current suddenly started and maintained at 10 amperes. Length of arc maintained at 3mm. Carbons : Positive, 18mm. cored ; negative, 15mm. solid. The positive carbon was shaped as it came from

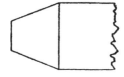

the makers, thus, shown full size. The negative was shaped by being previously used in a 3mm. arc with a current of 10 amperes.

Fig. 31 shows the connection between P.D. and time after
starting the arc, when the positive carbon is cored and shaped
as it came from the maker, and the negative has previously
been used with a current of 10 amperes and an arc of 3mm.
till the P.D. has become constant.   To save using this
expression again and again, I shall call a carbon 'normal' when
it has been burnt long enough, with a given current and an
arc of given length, for the P.D. to have reached its steady
value, and I shall call an arc 'normal' when it is burning
with 'normal carbons.

It will be observed in Fig. 31 that *after* the length of the
arc had been adjusted at 3mm. the P.D. between the carbons
fell to the low value of 16 volts.   Hence a genuine arc 3mm. in
length can be maintained silently, at any rate for a short time,
with a P.D. of only 16 volts.   It was thought probable
that this very low P.D. might have been caused in some
way by the soft core of the positive carbon, and experiments were, therefore, made with cored and uncored positive
carbons under precisely similar conditions.   The tips of the
positive carbons were filed flat, and they were used with
negatives that were 'normal' for 10 amperes and 3 millimetres.

The results may be seen in Fig. 32, curves A A A and B B B.
Curve A A A, which was obtained with a solid positive carbon,
starts with a P.D. of 44 volts, while B B B, for which the
positive carbon was cored, starts with one of 25 volts.   The
low P.D. at starting *is*, then, caused by the positive carbon
being cored.   But why should there be such a great difference,
namely, 19 volts, in the P.Ds. at starting, when the difference
between the steady values of the P.Ds. is only about four volts?

To settle this question an arc was started with a cored positive carbon shaped as when it came from the makers, and a
'normal' negative, and, after burning for less than one second,
the current was suddenly turned off.   On examining the carbons it was found that the core had been torn out of the positive carbon to the depth of one eighth of an inch, while the
negative carbon was covered with the finely-powdered material.
All this extra loose and easily volatilised carbon would, of course,
much enlarge the cross-section of the arc, and thus lower its apparent resistance, and also the current probably flows with a much
smaller P.D. when it is conveyed by means of small particles

of carbon actually travelling across the arc than when it flows simply through volatilised carbon. For this reason, it is probable that the fact that particles of unvolatilised carbon fall in showers from the positive carbon when there is *hissing* partly accounts for the fall of P.D. in the arc in that case.

In curve C C C (Fig. 32) the crater of the positive carbon had already been formed mechanically, and therefore the P.D. started higher than in curve B B B, and remained higher for about 45 minutes; in fact, for half an hour it more nearly

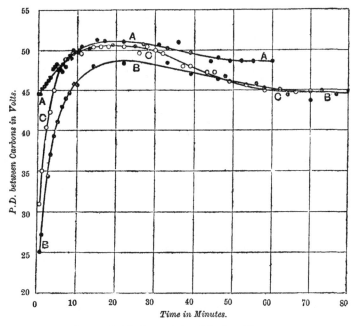

FIG. 32.—Current suddenly started and kept at 10 amperes. Length of arc kept at 3mm. Carbons: Positive, 18mm. solid or cored; negative, 15mm. solid. Negative carbon in each case shaped by being previously used for a long time to form an arc 3mm. long, with a current of 10 amperes. A A A: Positive carbon solid, end filed flat. B B B: Positive carbon cored, end filed flat. C C C: Positive carbon cored, with a crater mechanically made at the end after it was filed flat.

coincided with A A A, the curve for solid carbons, than with B B B, which is not to be wondered at, seeing that the part of the carbon from which the core had been mechanically extracted, was practically solid; and thus, while it burnt away, there was very little loose soft carbon to be easily volatilised, and so lower the P.D.

The time that the P.D. takes to reach its constant value when the positive carbon is flat to start with is very remarkable. From Fig. 32 we see that with a flat-tipped positive carbon of 18mm. diameter and a 'normal' negative carbon of 15mm. diameter, when a constant current of 10 amperes was flowing through an arc of 3mm., *the P.D. did not acquire its constant value till 50 minutes after the arc was struck with a solid positive carbon, and an hour after with a cored positive carbon.* How important a fact this is may be gathered from the consideration that the P.D. which would send a current of 10 amperes through an arc of 3mm. when the positive carbon was cored, might have been put down as being anything from 25 to 48 volts (curve B B B, Fig. 32), according to the time that had been allowed to elapse after striking the arc before the reading was taken. Of course such a wide range of P.D. for such a current would only be possible when the positive carbon was cored and had been flat before striking the arc, but curve A A A, Fig. 32, shows that even with a solid positive carbon the P.D. may range from about 44 to 52 volts. With small currents of 2 or 3 amperes I have found that with both carbons solid the P.D. may take as much as two and a half hours to acquire its constant value, even when the positive carbon has not been filed flat, but has been shaped beforehand by some such current as 5 or 6 amperes.

It is possible that the extra low P.D. at starting in Fig. 31 may have been caused by the positive carbon with a pointed tip being easier to volatilise than the one with a flat tip. The reason that the P.D. took such a short time to reach its steady value—only five minutes—must have been that when the carbon from the crater had been volatilised it only came in contact with the hot tip, and was not cooled down by the mass of comparatively cold carbon surrounding the crater, as happened when the tip was flat. This mass of cold carbon is evidently the cause of the P.D.'s not retaining its constant value when it first reaches it, but rising to a higher value and then falling again, as it does in all the curves in Figs.

31, 32, and 33. For, from the time of starting the arc, this mass
of carbon was heated sufficiently for it to gradually burn away,
therefore part of the heat of the crater was used in warming
up this surrounding carbon until it was all burnt away, and
the positive carbon had become normal. This took place
between 50 minutes and an hour after the arc was started.
The extra amount of heat needed during this time meant, of
course, that a higher P.D. was required to send the same
current through the arc; for, since the rate of production of
heat depends upon the current multiplied by the P.D., and the
current was kept constant, the P.D. was bound to be higher.
Hence the "hump," which will be found in every curve for
which a *flat* positive carbon has been used.

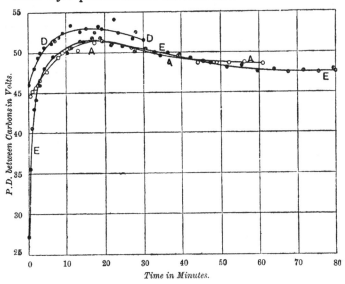

FIG. 33.—Current suddenly started and kept at 10 amperes. Length of
arc kept at 3mm. Carbons :—Positive, 18mm. solid ; negative, 15mm.
solid. Negative carbon normal in each case. A A A : End of solid
positive carbon filed flat. D D : Hole 2in. deep drilled in end of
solid positive carbon. E E E : Hole 1in. deep drilled in end of solid
positive carbon and packed tightly with soft carbon, end filed flat.

The curves in Fig. 33 are interesting as showing how
completely and certainly the core of the positive carbon is
responsible for a very low P.D. on starting the arc.

With the curve E E E, in which the P.D. started at about 27 volts, the positive carbon was solid, but drilled to the depth of one inch, and packed moderately tightly with the carbon from a soft core. For the first three minutes, therefore, it behaved exactly as if it were cored in the usual way, and after that it acted like a completely solid carbon. It is a little curious that the effect of a whole inch of soft core should apparently have exhausted itself in three minutes, and that from that time onwards the curve obtained with the drilled and packed positive carbon should be almost identical with that obtained with an ordinary solid positive carbon (*see* curve A A A, Fig. 33). One would have expected the 'hump' to be lower with the drilled and packed positive, on account of the loose soft carbon. Probably, however, the carbon was not as tightly packed as in a manufactured core, and, therefore, most of it was shot out during the first three minutes after starting the arc, only enough of it being left to keep the positive carbon from behaving as if it had merely a hole and no core at all.

Curve D D, Fig. 33, shows what that behaviour would have been. The positive carbon in this case had a hole two inches deep drilled in it, and left hollow, and this kept the P.D. slightly higher throughout the whole variable period than when an ordinary solid positive carbon was used as in curve A A A, Fig. 33.

In order to find what influence, if any, a flat negative carbon would have in retarding the period at which the P.D. became constant, cored and uncored normal positive carbons were used, with flat uncored negatives. In Fig. 34, A A is the curve obtained with the uncored, B B that obtained with the cored positive carbon.

From these curves it is evident that the shape of the negative carbon plays a very small part in the change of P.D. that takes place on starting the arc, for in each case the P.D. reached its steady value in about 8 minutes, and in neither case did it deviate more than about 5 volts from that steady value, whereas, as has been shown, with a flat positive carbon, the P.D. took about 50 minutes to reach its steady value, and it deviated by from 7 to 33 volts from that steady value before reaching it.

Fig. 35 gives a sort of bird's-eye view of the differences caused in the change of P.D. after starting the arc by using

(1) A A A, both carbons solid and flat.

(2) B B B, positive carbon cored and both flat.

(3) B′ B′, positive carbon cored, both normal, arc started with cold carbons.

(4) B″, positive carbon cored, both normal, arc started with hot carbons.

After three minutes from starting the arc the curve A A A (Fig. 35) differs very little from the curve A A A (Fig. 32), in which the negative carbon was normal instead of flat, and all the other conditions were the same. The reason there is so little appearance of 'hump' is, I find on referring to the laboratory note books, that certain somewhat high P.Ds. obtained

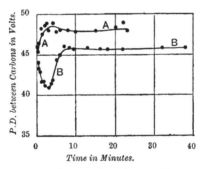

Fig. 34.—Current suddenly started and kept at 10 amperes. Length of arc kept at 3mm. A A, Carbons: Positive, 18mm. solid; negative, 15mm. solid. B B, Carbons : Positive, 18mm. cored ; negative, 15mm. solid. Positive normal, negative filed flat, in both instances.

between 10 and 35 minutes after the arc had been started have been omitted in drawing this curve, presumably because it was supposed that these observations were wrong. Had points corresponding with these somewhat higher P.Ds. been plotted, and the curve A A A (Fig. 35) drawn through the average position of all the points, it would have shown a hump such as exists in all the other curves obtained from experiments with a flat positive carbon.

B B B (Fig. 35) also differs very slightly from B B B (Fig. 32) in which the negative was normal instead of flat ; in Fig. 35 the curve starts with rather a higher P.D., 27 instead of

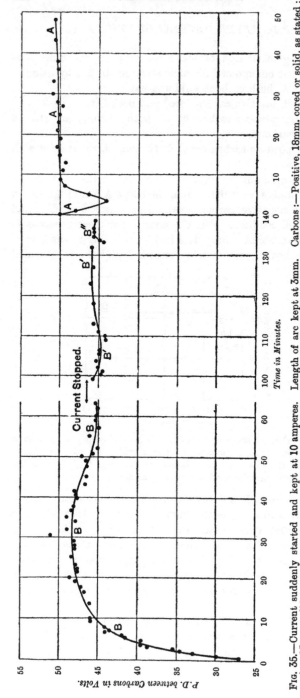

FIG. 35.—Current suddenly started and kept at 10 amperes. Length of arc kept at 3mm. Carbons :—Positive, 18mm. cored or solid, as stated ; negative, 15mm. solid.
Ends of both carbons filed flat. A A A : Positive carbon solid. B B B : Positive carbon cored.
Both carbons shaped by being previously used for over one hour to give an arc 3mm. long with 10 amperes. Positive carbon cored.
B′ B′ : Arc started when carbons were cold. B″ : Arc started when carbons were hot.

25 volts, and rises a little more slowly. Thus in both this and the preceding case it is evident that the difference made by the shape of the negative carbon is very small.

B' B' B' and B" (Fig. 35) show how small is the change that takes place in the P.Ds. when the arc is started with both carbons normal, whether they be hot or cold beforehand. Started cold the P.D. is about $1\frac{1}{2}$ volts higher than started hot, which is what one might have expected. The whole change of P.D., however, is very small, under both circumstances, not more than about $1\frac{1}{2}$ volts altogether.

Thus we may gather from Figs. 31 to 35 that the changes that take place in the P.D. of an arc just after it is started are due in order of importance :—

(1) To the core of the positive carbon, if it has one—very low P.D. at starting.

(2) To the shape of the tip of the positive carbon—the 'hump.'

(3) To the shape of the tip of the negative carbon.

(4) Very slightly to the temperature of the carbons before starting the arc.

Coming now to Question (4), to see how changing the length of the arc affected the time during which the P.D. remained variable after the arc was started, an 18mm. cored positive carbon and a 15mm. solid negative were again used, the ends of both carbons were filed flat, and the current was again kept constant at 10 amperes; but the length of the arc was kept constant at 6mm. instead of at 3mm. as before.

It was found that in this case, as with the 3mm. arc, there was a very low P.D. at starting, a rise to a maximum, and then a slight fall, and finally the steady P.D. But the whole series of changes, which extended, as we have seen, over a period of *55* minutes with the 3mm. arc, took only *20* minutes with the 6mm. arc.

Next, to see how varying the current affected the time during which the P.D. remained variable after the arc was started, an 18mm. cored positive carbon was again used with a 15mm. solid negative, the ends of both carbons were filed flat, the arc was kept at the constant length of 3mm.; but a constant current of 20 amperes was maintained, instead of one of 10 amperes, as in all the previous experiments on the variation of the P.D. with the time after starting the arc.

Again the curve obtained was found to be of the same character as curve BBB (Fig. 35), with which, also, both carbons were flat, and the positive cored; only, with the current of 20 amperes the changes were more rapid than with that of 10 amperes, and the P.D. became constant *38* minutes after starting the arc, instead of *55* minutes after.

We may gather from these last two experiments that the time during which the P.D. between the carbons remains variable after starting the arc with flat carbons is longer

(*a*) The shorter the arc,

(*b*) The smaller the current;

and thus Question (4) is answered.

The next question to determine was what change took place in the P.D. between the ends of the carbons when the current was suddenly changed from a lower to a higher and from a higher to a lower value, the length of arc being kept constant during each series of experiments.   The experiments, the results of which are noted in the curves in Figs. 36 and 37, were made in order to answer this question.

These curves, as well as all the others published in this chapter, are merely specimens of a number of sets of curves that have been obtained under similar conditions.

Since, at the very first instant that a change of current is made the arc cannot have had time to change its cross section, it would seem as if at that first moment the arc should act like a wire, and a rise of potential should accompany an increase of current, and a fall of potential a decrease.   In 1893 I made some experiments to see if I could detect this first momentary similarity of sign between the change of P.D. and change of current, and found that in some cases it could be easily detected, and in others not at all.   Being pressed for time, I did not then continue the investigation; but when the discussion about a negative resistance in the arc arose in 1896 (*see* p. 75), I repeated the experiments with Mr. Frith, for they seemed to have some bearing on the question.   We then found that in all cases where the first momentary similarity of sign could be perceived, either one or both carbons were cored.   The whole question of the *instantaneous* change of P.D. with change of current will be discussed later on, in the chapter on the resistance of the arc, &c.

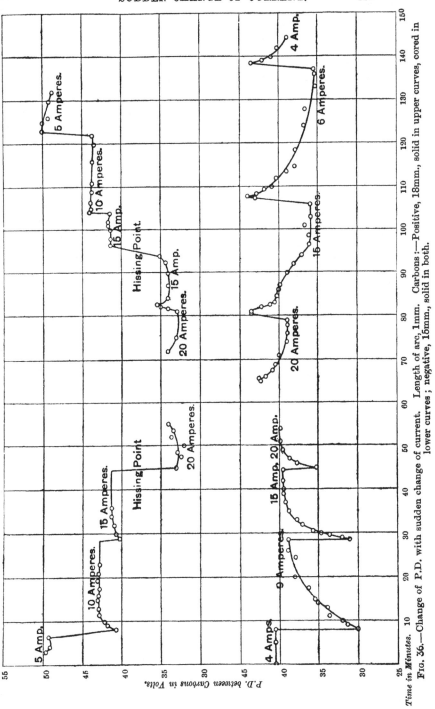

Time in Minutes.

FIG. 36.—Change of P.D. with sudden change of current. Length of arc, 1mm. Carbons:—Positive, 18mm, solid in upper curves, cored in lower curves; negative, 15mm, solid in both.

The changes of P.D. in Figs. 36 and 37 are very striking. With wires one is accustomed to associate an increase of current

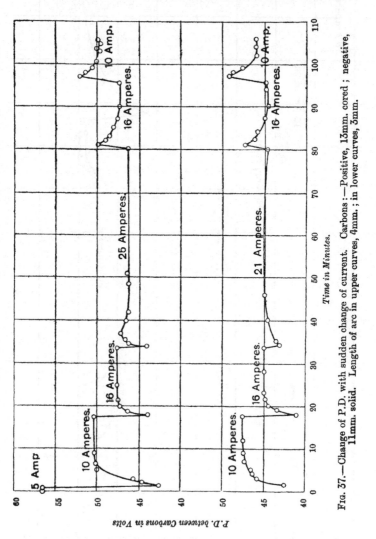

FIG. 37.—Change of P.D. with sudden change of current. Carbons:—Positive, 13mm. cored ; negative, 11mm. solid. Length of arc in upper curves, 4mm. ; in lower curves, 3mm.

with a *rise* of potential and a diminution of current with a *fall*. But with the arc, except, perhaps, in the very first instant, exactly the reverse takes place, and no experiments that I know of are better calculated to impress upon one the immense difference between the way in which a current flows through the arc and the way in which it flows through a wire than those from which these curves were taken.

With the arc (with the above exception), a sudden *rise* of current is in every case accompanied by a sudden *fall* of potential, and a sudden *fall* of current by a sudden *rise* of potential, even when, as in the case of an arc of 1mm. with a cored positive carbon (lower curves, Fig. 36), the final steady value of the P.D. is higher with the larger current than with the smaller.

Indeed, in nearly every case the P.D. overshoots the mark, as it were, and goes much lower with an increase of current, and much higher with a diminution, than its own final steady value. This exaggeration of the decrease and increase of P.D., which is very marked when the currents are small, becomes less and less marked as the currents increase in value, until finally, with currents of 30 amperes and over, it ceases to exist with the carbons we have tried, and the P.D. remains practically constant, whether the current is changed suddenly or gradually. In fact, when we get on to the flat part of the curves connecting P.D. with current for constant lengths of arc (Chap. IV., pp. 121 to 130) the P.D. is practically a constant, however quickly or slowly the current may be changed.

The curves in Fig. 36, the upper of which is for a solid and the lower for a cored positive carbon, show that the excessive sudden rise and fall of P.D. with a sudden diminution and increase of current does not depend entirely upon the core, for it takes place in both sets of curves alike, although it is more marked in the lower.

The upper curves are a little deceptive, because the hissing, which took place in this particular experiment at 15 amperes, lowered the P.D. considerably, quite apart from the sudden change of current.

Some experiments I have made since these, as well as the above curves, amply verify the following deductions. With a sudden change of current :—

(1) The sudden change of P.D. is greater with a cored than with a solid positive carbon;

I 2

(2) The subsequent slow rise, or fall, of P.D. is greater with a cored than with a solid positive carbon;

(3) The time during which this slow change of P.D. takes place is greater with a cored than with a solid positive carbon.

The curves in Fig. 37 show, although not to a very marked extent, that the time the P.D. takes to reach its steady value after a sudden change of current is less with a longer than with a shorter arc. Experiments made later prove this point quite conclusively. For instance, with an 18mm. cored positive carbon and a 15mm. solid negative, arcs of 6mm. and 1mm., respectively, were maintained. The current was, in each case, kept first at 4 amperes, and, when the P.D. had become steady for that current, it was suddenly changed to 9 amperes, and kept constant at that value till the P.D. had become steady. It was found that, after suddenly altering the current from 4 to 9 amperes, the time that elapsed before the P.D. assumed its steady value for 9 amperes was 9 minutes in the case of the 6mm. arc, and 16 minutes in that of the 1 mm. arc.

These sudden exaggerated changes of P.D. probably depend upon the difference between the shapes of the carbons and craters with small and large currents, an idea which is strengthened by the fact that with very large currents the shapes of the carbons alter very slightly with a change of current, and, as we have just seen, the P.D. also scarcely alters.

It is probable that the action is as follows : It has been shown (Chap. I., p. 13) that with a large current both carbons are burnt away much farther down than with a small current, thus making the lengths of the pointed parts of the carbons shorter the smaller the current. Hence, when a small current has been flowing through the arc for some little time, the carbons are very blunt, and the larger amount of volatile carbon sent off by the larger current when it is suddenly switched on will be squeezed out laterally, thus making the cross section of the arc abnormally great, and its resistance exceptionally small ; therefore, the P.D. necessary to keep the current flowing will be below the steady value. Then as the points of the carbons burn away and become tapered under the influence of the larger current, the volatile carbon can stretch out more lengthwise, and gradually take its normal form for the larger current, and at the same time the P.D. *rises* to its final steady value.

Similarly, when a small current was turned on after a large current had been flowing for some time, the volatile carbon would at first be too much elongated owing to the ends of the carbons being much tapered, and as these burnt away and became blunter the volatile carbon would be squeezed out more laterally, the cross section of the arc would be increased, and the P.D. would *fall* to its final steady value.

The exaggerated change of P.D. when the current is suddenly altered depends in another way also on the tips of the carbons being blunter with small currents than with large ones. For the mean distance of the tips of the carbons from one another is less when they are blunt, *i.e.*, normal for a small current, than when they are pointed, *i.e.*, normal for a large one, even although the length of the arc is maintained the same in both cases (*see* definition of length of arc, p. 99). Hence the mean length of the arc must also be less in the first case than in the second. But if the mean length of the arc is less, the P.D. necessary to maintain a given current flowing through the arc will be less, all other conditions being the same. Therefore the P.D. necessary to maintain a large current flowing through an arc of given length will be less when the carbons are normal for a small current than when they are normal for a large one, and *vice versâ*.

SUMMARY.

I. After the arc has been maintained of a constant length and with a constant current flowing for a certain time, the P.D. between the carbons finally becomes constant also.

II. The time that elapses before the P.D. becomes constant is less :

(1) The more nearly the original shapes of the carbons approximate to the shapes they finally take when the P.D. becomes constant ;

(2) The longer the given arc ;

(3) The greater the value of the constant current

III. The time that elapses before the P.D. assumes its constant value is less and the P.D. is greater with solid than with cored carbons.

IV. When the current is suddenly changed from a higher to a lower, or a lower to a higher value, the P.D. between the carbons increases or diminishes to a value greater or less respectively than its final constant value for the new current, and then gradually falls or rises to that constant value.

V. This first excessive increase or diminution is greater the greater the difference between the original and final current.

VI. For a given sudden *change* of current the first excessive increase or decrease of P.D. is greater the smaller the original current, while with large currents it is practically non-existent.

VII. When a cored positive carbon is used the P.D. is sometimes as low as 16 volts for a short time, with a perfectly silent arc.

VIII. With a cored positive carbon the change of P.D. is sometimes observed to be in the same direction as the corresponding sudden change of current, for the first instant. This first *increase* of P.D. with an *increase* of current and *decrease* of P.D. with a *diminution* of current has never been observed with solid carbons.

CHAPTER IV.

CURVES CONNECTING THE P.D. BETWEEN THE CARBONS WITH THE
CURRENT FLOWING FOR CONSTANT LENGTHS OF ARC, AND
CURVES CONNECTING THE P.D. BETWEEN THE CARBONS WITH
THE LENGTH OF THE ARC FOR CONSTANT CURRENTS.

When Prof. Ayrton first directed his attention to obtaining
a series of observations of the arc which would enable him to
form curves connecting any two of the variables, while a third
was kept constant, the three variables of which direct observations were made were the P.D. between the carbons, the
current flowing, and the length of the arc. From these the
apparent resistance and the power consumed in the arc could
also be found. Most of the experiments were made with
cored positive and solid negative carbons; but a single set of
results was obtained with both carbons cored, and another with
both carbons solid, in order to see what variations in the
curves these changes produced.

When beginning to study the arc on my own account, it
appeared to me that it would be better to avoid the complications arising from the use of cored carbons, and to study
the problems under their simplest conditions by employing
none but solid carbons. Accordingly, my experiments were
conducted with solid carbons for both positive and negative,
the positive carbon being 11mm. and the negative 9mm. in
diameter in all cases.

The make of carbon employed was the "Apostle," the same
as had been used in all the investigations carried out under
Prof. Ayrton's direction, so that the results obtained with
solid carbons might be compared with those obtained when a
cored positive carbon was used.

As the relations existing between the variables of the arc

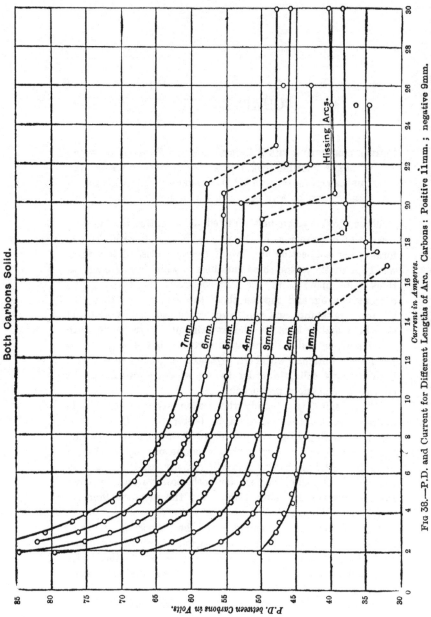

Fig 38.—P.D. and Current for Different Lengths of Arc. Carbons: Positive 11mm.; negative 9mm.

are undoubtedly simpler when both carbons are solid, it will be best to examine the curves for solid carbons (Fig. 38) first.

The values used in plotting the curves in Fig. 38 were the means of the results obtained on *different* days with *different* pairs of 11mm. and 9mm. solid carbons, and, in order to indicate to what extent these means differed from the actual observations, a sample is appended, in Table XI., of the actual results that were obtained when the arc was 5mm. in length :—

Table XI.—*Specimen of the Actual Daily Results obtained when the Arc was 5mm. long.*

Carbons both solid. Positive, 11mm. ; negative, 9mm.

Current in amps.	Potential Difference between Carbons in Volts.							
	I.	II.	III.	IV.	V.	VI.	VII.	Mean of the 7 days' results.
1·96	84·4	...	...	...	...	...	84·9	84·65
2·45	77·1	...	...	73·1	...	...	...	75·1
2·95	73·1	...	...	70·2	...	...	...	71·65
3·45	67·7	...	...	...	...	...	...	67·7
3·96	66·3	65·5	...	...	...	...	...	65·9
4·46	64·8	64·5	...	...	64·8	63·7	...	64·45
4·97	62·8	63·0	62·8	...	63·0	61·6	...	62·64
5·47	61·3	...	61·7	...	62·0	60·7	...	61·4
5·97	58·9	...	60·7	...	...	59·7	...	59·77
6·47	...	...	59·9	...	...	58·9	...	59·4
6·97	57·9	...	59·2	...	...	58·4	...	58·5
7·97	57·4	...	58·0	...	...	57·0	57·2	57·4
9·0	56·0	...	57·0	...	...	56·0	56·5	56·4
10·07	...	...	56·4	...	...	...	55·5	56·0
11·07	...	...	...	...	...	...	55·0	55·0
12·07	...	...	55·2	...	...	...	54·3	54·8
14·06	...	...	54·4	...	...	...	53·5	54·0
16·04	...	...	...	...	...	...	52·5	52·5
18·03	...	...	53·5	...	...	...	...	53·5
20·0	...	...	53·0	...	...	...	...	53·0
22·0	...	...	*43·0*	...	...	...	...	*43·0*
26·0	...	...	*43·0*	...	...	...	...	*43·0*

The numbers in italics refer to hissing arcs.

The numbers in any one column in Table XI. are the results of the experiments carried out in a single day with a 5mm. arc, and, although the carbons did not require to be changed each day, sometimes a new positive and sometimes a new negative carbon had to be inserted, so that, on the whole, about three different positive and three different negative carbons were used in obtaining the numbers given in the last column of this Table.

Table XII. gives the means of the whole series of results from which the curves in Fig. 38 were plotted.

Table XII.—*Means of the Experimental Results used in Plotting the Curves in Fig. 38. Carbons both solid. Positive, 11mm.; negative, 9mm.*

Current in amps.	l=1. P.D. in volts.	l=2. P.D. in volts.	l=3. P.D. in volts.	l=4. P.D. in volts.	l=5. P.D. in volts.	l=6. P.D. in volts.	l=7. P.D. in volts.
1·96	50·25	60·0	67·0	79·5	84·6	...	...
2·46	48·7	55·75	62·75	67·7	75·1	82·0	85·9
2·97	47·9	53·5	59·75	65·0	71·7	76·1	81·0
3·45	47·5	52·0	58·5	63·0	67·7	72·4	77·0
3·96	46·8	51·2	56·0	61·0	65·9	69·6	75·1
4·46	45·5	50·6	54·5	59·0	64·45	67·5	71·25
4·97	45·7	49·8	53·5	58·25	62·6	65·9	70·25
5·47	...	...	52·75	57·25	61·4	64·6	68·2
5·97	45·0	49·0	52·0	56·25	59·75	63·1	67·3
6·47	...	...	...	...	59·4	62·4	66·55
6·97	44·0	48·1	51·4	55·1	58·5	61·4	65·65
7·47	...	...	...	...	...	61·1	64·65
7·97	43·6	47·4	50·6	54·3	57·4	60·5	64·2
8·48	...	...	...	...	...	...	63·25
9·0	43·5	...	50·2	53·5	56·3	59·5	62·6
10·07	42·8	46·0	49·8	53·0	56·0	58·8	61·5
11·07	...	...	...	...	55·0	58·2	...
12·07	42·35	45·5	48·5	51·75	54·8	57·6	60·35
14·06	42·2	45·0	...	50·6	54·0	56·8	59·5
16·05	...	...	...	...	52·5	56·0	58·75
16·55	...	44·5	...	...	...	...	...
16·85	*32·0*	...	...	...	...	...	...
17·54	...	*33·4*	47·5	...	...	...	...
17·64	...	...	...	49·4	...	...	...
18·03	...	*35·0*	...	...	53·5	...	...
18·53	...	...	*38·5*	...	...	...	...
19·0	...	...	*38·0*	...	...	...	...
19·22	...	...	...	50·0	...	...	...
19·42	...	...	...	...	...	55·5	...
20·0	...	*34·5*	*38·0*	...	53·0	...	...
20·5	...	...	...	*39·5*	...	55·5	...
21·0	..	...	...	...	...	...	56·9
22·0	...	...	...	...	*43·0*	46·5	...
23·0	...	...	...	...	...	...	*48·0*
25·0	...	*34·5*	*36·5*	*40·0*	...	...	...
26·0	...	...	...	...	*43·0*	*47·0*	...
29·97	...	...	...	*40·5*	...	*46·0*	*48·0*

The numbers in *italics* refer to hissing arcs.

These curves connect the P.D. between the carbons with the current flowing for the various constant lengths of arc, with solid carbons. Each point on each curve represents the P.D. between the carbons *after* the current had been kept flowing at its specified value for a considerable time, and the length of the arc kept at its specified value during the *whole* of that time. The carbons had thus acquired their normal shape for the particular current and length of arc. The time required for this varies from about 10 minutes to over two hours under different circumstances. It was the want of appreciation of the very long time that it is necessary in certain cases to keep the current and length of arc constant before the carbons acquire their final shape, that led to so much labour being wasted in 1890, in obtaining the looped curves for ascending and descending currents, of which Fig. 30, Chap. III., is a specimen.

The general character of the curves indicates an inverse connection between the P.D. and current for any given length of arc. That is to say, the P.D. diminishes as the current increases. It diminishes rapidly with small currents, and more and more slowly as the current increases, never, however, becoming constant with solid carbons. Take, for example, the 5mm. arc. With an increase of current of 4 amperes—from 2 to 6 amperes—the P.D. falls 23 volts—from 83 to 60 volts ; with a further increase of 5 amperes—from 6 to 11 amperes— the P.D. falls only 5 volts, and with a still further increase of 9 amperes—from 11 to 20 amperes—it falls only 2·5 volts. Thus the P.D. never becomes constant, but it falls very slowly as the hissing point—the point at which the current is so large that any increase would cause the arc to hiss—is approached. The position of these hissing points, which evidently varies with the length of the arc, will be discussed in the chapter on hissing.

The curves for the shorter arcs are much less steep than those for the longer ones ; but from their shape it is evident that this is only because the steeper parts of the curves for short arcs would belong to smaller currents than 2 amperes, the smallest used. In other words, the ratio of change of P.D. to change of current is greatest, not only when the current is least, but also when the arc is longest.

For instance, in the 5mm. arc (Fig. 38), when the current starts at 2 amperes, an increase of 4 amperes is accompanied by a fall of 23 volts in the P.D. between the carbons, but in the 1mm. arc, when the current starts at 2 amperes, a rise of 4 amperes is accompanied by a fall of only 6 volts. It would be necessary, indeed, for the current to start at 1 ampere, or even less, with an arc of 1mm., for a rise of 4 amperes to be accompanied with so great a fall of the steady P.D. as 23 volts.

In their Paper read before the Physical Society in 1882 (*Proc.* Phys. Soc., Vol. V., p. 197) on "The Resistance of the Electric Arc," Messrs. Ayrton and Perry said, in speaking of the curve given in that Paper, which they had drawn from the results of their experiments connecting the P.D. between the carbons with the length of the arc with solid carbons for arcs up to 31mm. long: "The curve we have obtained is also strikingly like that obtained by Drs. W. De La Rue and Hugo Müller for the connection between the electromotive force and the distance across which it would send a spark. Those gentlemen also made experiments on the electric arc with their large battery. . . . The result of an experiment in air between two brass points is given; but, according to that, when the arc was half an inch in length the difference of potential between the brass points was about 700 volts. How far the very high electromotive force found by Drs. W. De La Rue and Hugo Müller to be necessary in this case arose from a combination of the material employed for the electrodes and the smallness of the diameter of the brass electrodes, or whether the law that 'the electromotive force necessary to maintain an arc depends mainly on the length of the arc, and hardly at all on the strength of the current,' fails when the current is below a certain small limit, we are unable to say; but of course both the diameter of the brass electrodes they employed and the strength of the current that was passing (0·025 ampere) in the arc was very much less than that used in any ordinary electric light, to which the experiments of Mr. Schwendler and ourselves especially refer. It is very probable that the difference in the material of the electrodes has mainly to do with the difference between their results and ours; and we think it very probable that, with very soft carbons, an arc of a given length could be maintained with a much less difference of potentials than that found by us, since it would be more easy for a shower of carbon particles to be maintained between the ends of the carbons."

Two results are here foreshadowed, which have both since been verified, the one that the P.D. for a given current and length of arc would be found to be smaller if the carbons were made softer, the other that the P.D. for a given length of arc would be far higher with very small currents than with those which are used practically with an arc lamp. The first result

Table XIII.—*Means of the Experimental Results used in Plotting the Curves in Fig. 39. Positive carbon, 18mm., cored; negative carbon, 15mm., solid.*

Current in amps.	$l=0.5$ P.D. in volts.	$l=1$ P.D. in volts.	$l=2$ P.D. in volts.	$l=3$ P.D. in volts.	$l=4$ P.D. in volts.	$l=5$ P.D. in volts.	$l=6$ P.D. in volts.
4·0	...	37·0	...	...	...	...	...
5·0	...	...	49·3	...	...	...	...
6·0	33·0	36·0	47·0	50·5	53·0	54·75	59·5
7·8	34·5	...	...	...	...	...	...
8·0	34·5	36·0	...	...	...	...	55·0
10·0	...	...	43·0	46·5	47·5	49·75	...
11·0	...	36·1	...	...	...	...	...
12·0	35·0	...	...	...	...	...	...
15·0	...	...	42·0	44·5	46·5 ⎱ 46·0 ⎰	48·0	49·5
16·0	...	...	...	44·0	...	...	...
17·0	...	38·1	...	...	...	...	...
18·0	38·0	...	...	...	...	...	...
20·0	...	38·75	42·0	...	44·0	...	48·0
22·0	...	...	...	43·5	...	...	...
23·0	...	39·75	...	...	...	...	...
24·0	39·0	...	...	...	...	...	...
25·0	...	...	...	...	44·2	46·0	48·0
27·5	...	40·0	...	...	...	...	...
28·0	...	...	...	43·0	...	...	...
30·0	...	40·0	41·5	43·0	44·2	45·5	48·0
33·0	...	40·1	...	...	...	...	...
34·0	39·5	...	...	...	...	...	...
35·0	...	...	...	...	44·0	45·4	47·7
40·0	40·0	40·9	...	...	43·9	45·75	47·5
41·0	...	...	...	43·0	...	...	...
45·0	...	...	...	...	...	46·0	47·5
47·0	...	...	...	43·0	43·9	...	...
47·25	...	...	...	43·2	...	...	...
47·5	*33·0*	...	...	...	...	...	...
48·0	...	*35·0*	...	...	...	...	...
48·75	...	...	42·5	...	...	...	...

The numbers in italics refer to hissing arcs.

was found to be true when *cored* carbons were subsequently manufactured, and the second is borne out by the strikingly rapid rise in the curves in Figs. 38, 39, 40 and 41, as the current becomes very small.

In fact, for very small currents, whether one, or both, or neither of the carbons be cored, the P.D. falls rapidly with increase of current, probably on account of a small current arc presenting a large cooling surface in proportion to its cross-section.

Tables XIII., XIV. and XV. give the results of the experiments made by students of Professor Ayrton at the Central Technical College, London, with cored positive and solid negative carbons. Each number gives the mean of several observations.

Table XIV.—*Experimental Results used in Plotting the Curves in Fig. 40. Positive carbon, 13mm., cored ; negative carbon, 11mm., solid.*

Current in amps.	$l=0.$ P.D. in volts.	$l=1.$ P.D. in volts.	$l=2.$ P.D. in volts.	$l=3.$ P.D. in volts.	$l=4.$ P.D. in volts.	$l=5.$ P.D. in volts.	$l=8.$ P.D. in volts.
2·0	...	47·1	...	...	...	71·2	...
2·5	56·0	...	63·0	...	...	...	...
2·8	42·0	...	...	...	...	...	...
3·0	40·5	...	56·5	61·5	62·0	64·5	76·2
4·0	...	39·5	...	...	58·25	...	69·5
4·5	...	...	49·0	...	...	...	...
5·0	...	...	...	...	56·5	...	...
6·0	35·0	37·5	46·0	50·2	$\left\{\begin{matrix}55·0\\54·0\\52·7\\52·4\end{matrix}\right\}$	56·0	63·3
8·0	35·5	...	...	...	...	...	...
9·9	...	39·0	43·0	$\left\{\begin{matrix}47·5\\46·5\end{matrix}\right\}$	$\left\{\begin{matrix}49·2\\49·7\\50·2\end{matrix}\right\}$	52·0	$\left\{\begin{matrix}60·7\\59·2\end{matrix}\right\}$
12·0	37·0	...	...	...	...	...	...
13·0	...	...	43·0	...	...	...	...
15·0	..	...	...	...	$\left\{\begin{matrix}47·4\\47·7\end{matrix}\right\}$	...	...
16·0	38·0	...	...	$\left\{\begin{matrix}45·2\\44·7\end{matrix}\right\}$	...	50·2	54·7
19·4	...	...	43·0	...	...	...	...
19·9	40·0	41·0	...	...	47·5	49·2	$\left\{\begin{matrix}54·0\\52·5\end{matrix}\right\}$
21·0	...	...	...	44·5	...	...	...
23·7	...	...	...	...	46·5	...	...
25·0	...	...	...	...	46·5	...	53·5
27·5	...	...	...	45·2	...	...	...
27·9	41·5	42·5	43·5	...	...	...	...
29·9	...	...	...	45·2	46·6	48·6	...

Table XV.—*Experimental Results used in Plotting the Curves in Fig. 41. Positive carbon, 9mm., cored ; negative carbon, 8mm., solid.*

$l=0.$		$l=0\cdot5$		$l=1.$		$l=2.$	
Current in amps.	P.D. in volts.	Current in amps.	P.D. in volts.	Current in amps.	P.D. in volts.	Current in amps.	P.D. in volts.
1·8	33·5	2·4	38·5	3·0	48·4	3·0	52·5
2·2	30·0	3·0	35·8	3·5	46·0	3·1	51·5
2·3	29·2	4·0	35·6	4·9	42·0	4·0	48·5
3·0	29·0	5·9	37·5	6·9	42·0	5·9	44·3
5·3	30·0	7·8	37·5	8·8	41·5	8·8	43·5
7·8	32·5	9·8	39·5	12·1	42·6	10·0	44·1
11·3	36·0	12·1	39·0	15·2	43·0	12·1	43·6
13·4	37·5	13·2	40·5	15·5	43·0	15·2	44·2
15·0	38·6	14·5	40·5	19·0	*34·0*	15·5	44·0
15·5	38·6	21·3	*31·7*	22·3	*33·8*	19·2	*36·0*
18·3	*29·8*	30·3	*32·5*	25·3	*34·4*	22·3	*35·5*
25·0	*29·8*	...	...	30·3	*34·5*	26·3	*35·5*
30·0	*29·8*	...	...	...	...	30·3	*36·0*

$l=3.$		$l=4.$		$l=5.$		$l=8.$	
Current in amps.	P.D. in volts.	Current in amps.	P.D. in volts.	Current in amps.	P.D. in volts.	Current in amps.	P.D. in volts.
3·5	55·2	3·0	59·0	3·5	61·1	3·5	69·0
4·5	51·7	4·0	57·7	4·0	60·1	4·9	64·5
5·9	48·5	5·9	54·0	5·9	55·8	7·8	60·0
8·3	47·0	8·8	51·7	8·8	52·6	12·2	56·0
12·1	45·5	12·0	49·0	12·2	50·5	16·4	55·0
17·2	46·0	15·5	48·7	16·2	50·2	17·0	55·0
17·5	46·0	17·0	48·7	17·2	50·0	20·0	55·0
21·0	*37·5*	17·2	48·4	18·3	50·5	22·0	55·5
25·1	*37·5*	19·0	48·4	19·2	50·9	25·0	55·5
26·3	*37·5*	20·5	48·3	19·5	51·0	28·3	*49·2*
30·3	*37·5*	24·0	*40·2*	23·3	*41·5*	30·3	*49·2*
...	...	25·5	*40·2*	25·3	*41·6*	34·0	*49·2*
...	...	30·3	*40·0*	30·3	*41·5*	...	...

The numbers in italics refer to hissing arcs.

On comparing the curves in Figs. 39, 40, and 41 with those for solid carbons in Fig. 38, one very curious point in common may be noticed, viz., *if the P.D. between the carbons be kept constant, and the arc be lengthened and maintained at the greater*

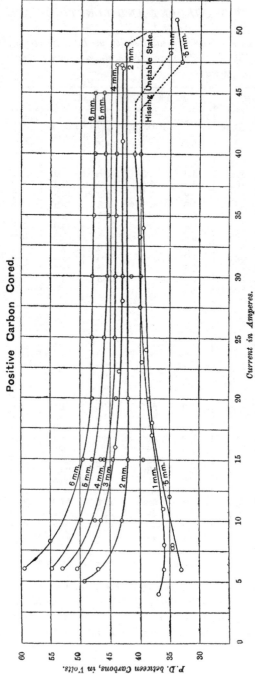

FIG. 39.—P.D. and Current for Different Lengths of Arc. Carbons :—Positive, 18mm., cored ; negative, 15mm., solid.

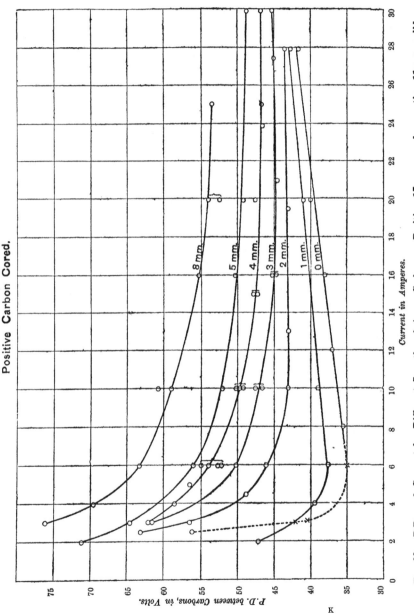

Positive Carbon Cored.

FIG. 40.—P.D. and Current for Different Lengths of Arc. Carbons :—Positive, 13mm., cored ; negative, 11mm., solid.

K

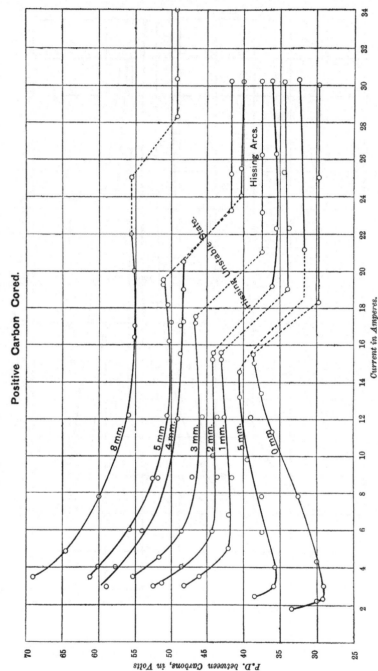

FIG. 41.—P.D. and Current for Different Lengths of Arc. Carbons :—Positive, 9mm., cored ; negative, 8mm., solid.

length, the current is larger and not smaller than it was for the shorter length of arc. For example, a given P.D., say 50 volts, sends a current of 4·1 amperes through an arc 2mm. long (Fig. 40), 6·2 amperes through an arc 3mm. long, 9·6 amperes through one of 4mm., and 16·4 amperes through one of 5mm., all the arcs being silent. Or, dividing this P.D. of 50 volts by these currents, it follows that the apparent resistances of the 2, 3, 4, and 5mm. arcs for this P.D. are 12·2, 8·1, 5·2 and 3 ohms respectively. Hence, for a *constant P.D.*, *the apparent resistance of the silent arc, when in the normal condition, diminishes rapidly as the arc is lengthened.*

Although the curves in Figs. 39, 40, and 41 bear a certain resemblance to those for solid carbons in Fig. 38, yet there are important differences between the two, the chief of which concerns the change of P.D. with change of current. With a *solid* positive carbon (Fig. 38), as already mentioned (p. 123), the P.D. falls continuously as the current increases, but with a *cored* positive carbon the P.D. generally falls to a minimum and then *rises* again, as was first shown by Nebel. Even when the P.D. does not appear to rise after reaching a minimum, as in the longer arcs with the larger sizes of carbons (Figs. 39 and 40), it is probable that if the curves had been continued till the hissing point was reached, an increase of P.D., as the curve neared the hissing point, would have been observed, for it is noticeable that all the curves in Figs. 39 and 41, which have been completed up to the hissing point, except the 4mm. curve in Fig. 41, which is probably erroneous, show the P.D. falling to a minimum and then rising.

This minimum corresponds with a larger and larger current the longer the arc, as was also pointed out by Nebel in 1886, although he had no idea that this form of curve depended upon the use of a cored positive carbon. The minimum also seems to correspond with a larger current the greater the diameters of the carbons. For instance, the minimum P.D. for a 2mm. arc seems to be reached with a current of about 15 amperes when carbons 18mm. and 15mm. are used, but with a current of about 7 amperes when carbons 9mm. and 8mm. in diameter are used. Thus it is possible that with long arcs between thick carbons the minimum P.D. might correspond with such a large current that hissing would begin before that

current was reached. In this case the P.D. would continue to
fall or remain practically constant till hissing began.

Whether the P.D. ever attains its minimum and rises again
or not, with long arcs and large carbons it alters so slowly that
it may be considered constant over a wide range of current.

In 1879 the late Mr. Schwendler published the then new
statement that the P.D. between the carbons for a fixed
length of arc was independent of the current. This result was
confirmed by Profs. Ayrton and Perry, for currents that were
fairly large for the carbons employed and for the length of
the arc (*Proc.* Phys. Soc., 1882, Vol. V., p. 197). That it is not
quite true for solid carbons we have already seen, but the
curves in Figs. 39, 40 and 41 show that Mr. Schwendler's result
is true for *long* arcs when the positive carbon is cored.

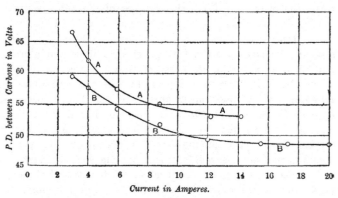

FIG. 42.—4mm. arc—A A A : Positive, 9mm., solid ; negative, 8mm., solid.
B B B, Positive, 9mm., cored ; Negative, 8mm., solid.

The curves in Fig. 42 were drawn in order to see at a glance
what was the general character of the change produced in the
curves connecting P.D. and current for a given length of arc,
when a solid positive carbon was substituted for a cored one of
the same diameter, the negative carbons in both cases being
solid.

These curves connect P.D. with current for a silent arc of
4mm., A A A when both carbons were solid, B B B when the
positive was cored, the positive carbon being 9mm. and the
negative 8mm. in diameter in both cases. We see that the

curve with a cored positive carbon is from three to six volts
lower than that with a solid positive.   This diminution of P.D.
with a cored carbon does not appear to be uniform whatever
the length of arc, but is much greater with short arcs than with
long ones; evidently also it is, on the whole, greater with
smaller currents than with larger ones.

The position of the hissing point on each curve differs
according as a cored or a solid positive carbon is used, as may
be seen by comparing Fig. 38 with Figs. 39, 40, &c.   The reason
of this will be discussed in the chapter on hissing arcs.

The two most important points of difference in the connection
between the P.D. and the current for a *silent* arc of fixed length
are, then—

(1) With a solid carbon the P.D. continually diminishes as
the current increases ; with a cored carbon the P.D. either
diminishes much less than with a solid carbon, or remains
constant for all currents above a given value, or actually
increases with the current after falling to a minimum.

(2) The P.D. is in all cases lower with the cored than with
the solid carbon.

This second variation probably depends upon the greater
ease with which the softer carbon is volatilised ; the first is an
apparently complicated effect, the explanation of which becomes
perfectly simple, however, on the hypothesis that *with a given
negative carbon the P.D. required to send a given current through
a fixed length of arc depends principally, if not entirely, on the
nature of the surface of the crater, being greater or less according
as the carbon of which that surface is composed is harder or softer.*

This is tantamount to saying that the only part of the positive
carbon that exerts much influence over the P.D. between the
carbons is that which forms the surface of the crater.

Consider, now, the nature of the surface of the crater of a
cored positive carbon when the length of the arc is fixed, but
the current may be varied.   While the current is very small
the area of the crater will be less than that of the core, and its
surface will therefore be composed entirely of soft carbon.
Hence, by the above hypothesis, the P.D. for each current will
be much the same as if the whole carbon were soft like the core.
Thus, until the current is so large that the crater exactly covers
the core, the curve connecting P.D. and current for a fixed

length of arc will be practically the same as if the whole positive
carbon were as soft as the core.

When the crater extends further than the core, however, part
of its surface will be formed of the hard outer carbon, and, by
the hypothesis, the P.D. between the carbons will then be
higher than it would be if the whole surface of the crater were
of soft carbon. Hence, after this point is reached, the curve
connecting P.D. and current for a fixed length of arc will no
longer be the same as if the whole positive carbon were soft,
but will diverge from that curve, the distance between the two
becoming greater as the current, and therefore the area of the
crater, increases. For, after it has once covered the core, any

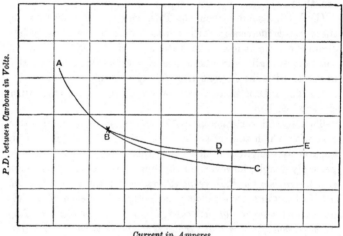

Current in Amperes.

A B C, Carbons: Positive solid, but soft as a core; negative solid and
   hard. A B D E, Carbons: Positive cored; negative solid and hard.

Fig. 43.—Ideal Curves of P.D. and Current for Constant Length of Arc.

addition to the area of the crater must be an addition to the
hard carbon part, hence the ratio of hard carbon to the total
amount of carbon in this surface must continually increase.

Fig. 43 gives a rough idea of the two curves connecting the
P.D. between the carbons with the current for a fixed length
of arc—A B C with the positive carbon *uncored* but as soft as
a core, A B D E with it *cored*, the negative carbon being solid,
and the same in both cases.

The two curves coincide up to the point B, at which the crater exactly covered the core. After this point was reached the crater began to cover some of the hard carbon; the P.D. was therefore greater with the cored than with the uncored carbon, but it still diminished as the current increased, though more and more slowly. At last the tendency for the P.D. to fall, caused by the increase of the current, was entirely counter-balanced by the tendency to rise, caused by the increasing amount of hard carbon covered by the crater. At this point (D on the curve) the P.D. became a minimum, and as the current was further increased the P.D. slowly rose till E, the hissing point, was reached.

It is obvious that the position of the point B must be determined by the area of the cross section of the core: the larger this cross section, the larger the crater can be without touching the hard carbon, and consequently the greater will be the current that can flow before the crater will exactly cover the core.

The position of D, the point of minimum P.D. with a cored positive carbon, has been shown to depend upon the length of the arc, and apparently also on the diameters of the carbons (p. 131). As a matter of fact, however, it is not so much upon the diameters of the carbons that it depends as upon the diameter of the positive carbon compared with the diameter of its core. For, evidently, if the cross section of the core were very large compared with the cross section of the whole carbon, so that the outer hard carbon was a mere shell, the current would be so large before the crater touched the hard carbon at all, that hissing would take place before the minimum P.D. had been reached. With a smaller core, the diameter of the carbon still remaining the same, the point of minimum P.D. and the hissing point might coincide, while with a still thinner core the minimum P.D. would be reached before hissing began, as in the curves in Fig. 41. Thus the minimum P.D. for any given length of arc with a cored positive carbon of given diameter might be made to correspond with any current we chose by making the cross section of the core of a suitable size. The position of D must also depend upon the degree of hardness of the outer shell. For the harder this is, the greater will be the P.D. with any given proportion of hard carbon in the surface of the crater; therefore, all other things being equal, D will correspond with a smaller current the harder the outer cylinder of the carbon.

An interesting point that follows from the particular form of the curves in Figs. 40 and 41 is that with a cored positive carbon certain P.Ds. will send two distinct currents through the arc. For example, a P.D. of 42 volts (Fig. 40) will send a current of either 3 or 21 amperes through a 1mm. arc. This points to the complete equation connecting P.D. current, and length of arc for silent arcs produced with a cored positive and a solid negative carbon, being a quadratic in terms of A, the current flowing.

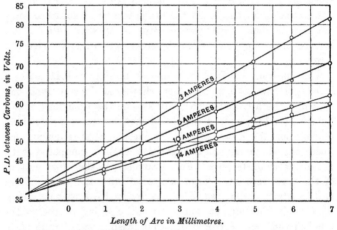

Fig. 44. — P.D. and Length of Arc for Various Constant Currents.
Carbons : Positive, 11mm., solid ; negative, 9mm., solid.

The curves connecting the steady P.D. between the carbons with the *current* flowing for certain *constant lengths of arc* having been discussed, it is now proposed to examine the connection between the steady P.D. and the *length of the arc* with certain *constant currents*.

The curves in Fig. 44, showing this connection when two solid carbons were used—the positive 11mm. and the negative 9mm. in diameter—were obtained, not from the numbers given in Table XII., but from the curves in Fig. 38, plotted with those numbers. The ordinates are the P.Ds. between the carbons and the abscissæ the lengths of the arc, the current being a constant for each curve. The result is a series of straight lines making a smaller angle with the axis of length the larger the value of the constant current flowing. These lines all meet at one point, which is *to the left* of the axis of P.D., showing that it corresponds with a *negative* length of arc, but *above* the axis of length of arc, and therefore indicating a *positive* P.D. This point shows, of course, the length of arc for which the P.D. is constant, whatever the current flowing. That this length is negative implies no absurdity, for our definition of the length of the arc is the perpendicular distance between the extreme tip of the negative carbon and the plane through the edge of the crater. If there were a deep crater, and the point of the negative carbon were thrust into it, the distance between that point and the edge of the crater would have to be measured in a direction opposite from that in which it was measured when the point of the negative carbon was outside the crater. Hence a negative sign must be given to the length of the arc when the tip of the negative carbon is inside the crater.

As a matter of fact, however, I have never succeeded in getting a permanently silent arc with the tip of the negative carbon inside the crater, however small a current was used, and since all the conditions change as soon as the arc begins to hiss, the point at which all the lines in Fig. 44 meet must be considered as simply an ideal point, at which all the lines *would* meet if the conditions remained the same when the arc was very short, or even negative, as they were when the arc was 1mm. long and more. Thus, *with both carbons solid there is no real length of arc with which the P.D. is constant for all currents,* a fact which has already been deduced from the curves in Fig. 38.

The lines in Fig. 44 being straight shows that, with a constant current, a fixed change in the length of the arc involves a fixed change in the P.D. between the carbons. For instance, with a constant current of 5 amperes, a change of length from 1mm. to 2mm. is accompanied by a change of P.D. from 45·4 to 49·6 volts, that is 4·2 volts; and a change of length from 6mm. to 7mm. is accompanied by a change of P.D. from 65·9

to 70·1 volts, or 4·2 volts. With a constant current of another
value, however, the change in the P.D. for the same amount of
change in the length of the arc would be different, being
greater for a smaller current and less for a larger one, as is
shown by the lines in Fig. 44 making smaller angles with the
axis of length the larger the current. Thus, with a current of
14 amperes, the constant change of P.D. for a change of 1 mm.
in the length of the arc is about 2·8 volts. Hence *with solid
carbons a given increase in the length of the arc involves an
increase in the P.D. between the carbons which is constant for a
constant current, but which diminishes as the value of the constant
current increases.*

The fact that with solid carbons a linear law connects the
P.D. between the carbons with the length of the arc *for constant currents* follows directly from Edlund's straight line law
connecting the apparent resistance of the arc with its length
for constant currents (*see* p. 34). For if we put Edlund's law
in the form

$$\frac{V}{A} = a + bl,$$

it is quite evident that, when A is constant,

$$V = aA + blA ;$$

or $$V = m + nl$$

follows directly from it. The discovery of this law is generally
attributed to Frölich, but, so far from discovering it, Frölich
missed the whole point of Edlund's work, and did not see that
a and *b* in the first equation and *m* and *n* in the last *varied
with each current that was employed.*

If in the last equation we make *l* equal to 0, we have

$$V = m,$$

an equation which has given rise to the idea of the existence
of a constant back E.M.F. in the arc. But in this equation *m*
is only constant for one current, and diminishes as the current
increases, as may be seen by a comparison of the points for
which $l = 0$ in Fig. 44. For instance, the P.D. is about
39·8 volts for a current of 14 amperes, 41·2 volts for one of
5 amperes, and so on. It may be contended that if the depth
of the crater were taken into account in the length of the arc,
in which case length of arc 0 would really mean that there
was *no* distance between the positive and negative carbons, the

P.D. might really be constant for all currents with length of arc 0; but Peukert, who used two solid carbons, and who took his measurements before a crater had had time to form, by filing the carbons flat before each experiment, obtained lines very like those in Fig. 44, and found also that the P.D. for length of arc 0 diminished as the current increased.

The curves connecting P.D. and length of arc for constant currents are by no means so simple when a cored positive carbon is used as when both carbons are solid, as may be seen from the curves in Figs. 45, 46 and 47. To begin with, none of them are straight lines, although as the value of the constant current increases they approximate more and more closely to straight lines. This divergence from the straight

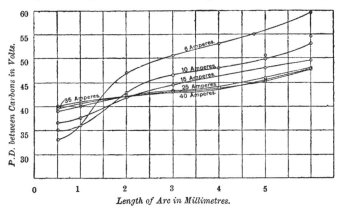

FIG. 45.—P.D. and Length of Arc for Various Constant Currents.
Carbons: Positive, 18mm., cored; negative, 15mm., solid.

line law shows that with a constant current the influence of the core is greatest with the shortest arc, and becomes less as the arc increases in length, for otherwise the lowering of the P.D. would be exactly the same for each length of arc, and the curves for solid and cored carbons would be at a constant distance from one another. That this is not so is apparent even in the curves for constant lengths of arc (Figs. 39, 40 and 41). Take, for instance, the curves for arcs of 1mm., 2mm. and 3mm. in Fig. 39. With a current of 10 amperes the distance between the first two curves is about twice as great as the distance between the second and third. But in Fig. 38, for solid carbons, the distances between the two sets of points are the same. For large currents the distances are much more nearly equal, and that is why the curves in Figs. 45, 46 and 47 are so much nearer to straight lines for large currents than for small ones.

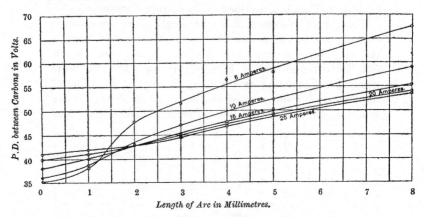

FIG. 45.—P.D. and Length of Arc for Various Constant Currents. Carbons : Positive, 13mm., cored ; negative, 11mm., solid.

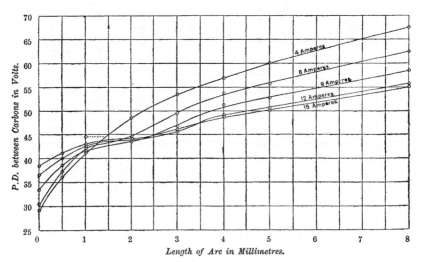

FIG. 47.—P.D. and Length of Arc for Various Constant Currents. Carbons : Positive, 9mm., cored ; negative, 8mm., solid.

We have seen that when *both* carbons are *solid* there is no positive length of silent arc, for which the P.D. is independent of the current, but with a cored positive carbon this is otherwise. Owing to the fact that the P.D. is most changed by the core with small currents and short arcs, the curve connecting P.D. and length of arc for the smallest current bends down the most towards the axis of length, and the other curves bend less and less as the current increases, so that the curves all cross one another to the *right* of the axis of P.D., instead of meeting at a point to the *left*, as they do with solid carbons. The curves do not actually cut one another at the *same* point, but at points which are very near together for all but the smallest currents, and the region in which these points are is, of course, the region in which Mr. Schwendler's observation is true that the P.D. is practically independent of the current (*see* page 132), for in this region, with a given length of arc, there is the same P.D. between the carbons whatever current is flowing.

The particular length of silent arc for which the P.D. is practically independent of the current is best seen from the points of intersection of the curves in Figs. 45, 46 and 47. This length is evidently a little under 2mm. When *both* the positive and negative carbons are *cored* the point at which the curves connecting P.D. and length of arc for different currents cut one another becomes very marked (Fig. 48), and the length of arc corresponding with this intersecting point diminishes to 0·75mm. If the curves corresponding with those in Figs. 39, 40 and 41 had been drawn for these carbons, the point for 3 amperes would have been found to be on the first steep part of the curve for each length of arc. Hence the high position of the 3-ampere curve in Fig. 48.

One result of this crossing of the curves in Figs. 45, 46 and 47 is very interesting, and it is seen particularly well in Figs. 46 and 47. Instead of the P.D. for length of arc 0 diminishing as the current increases, as it has been shown to do with solid carbons, it increases with the current. This fact was first observed by Nebel in 1886, but at that time the differences occasioned by the use of cored instead of solid positive carbons were not understood, and he did not, therefore, realise that this increase of the P.D. for length of arc 0, with increase of current, depended on the use of a cored positive carbon.

The likenesses and differences between the curves in Fig. 44 for solid carbons, and those in Figs. 45, 46 and 47, for which a cored positive carbon was used, may be summed up in the following way :—

With a Constant Current.

LIKENESSES.

(1) The P.D. increases as the length of the arc increases.

(2) The P.D. increases less for a given increase in the length of the arc the larger the value of the constant current that is flowing.

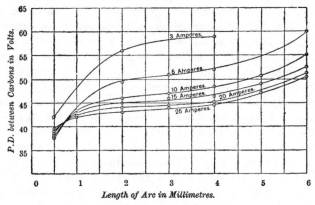

FIG. 48.—P.D. and Length of Arc for Various Constant Currents. Carbons : Positive, 18mm., cored ; negative, 15mm., cored.

DIFFERENCES.

(1) The P.D. is always higher with solid carbons than with a cored positive carbon, but the difference between the two diminishes as the arc increases in length.

(2) The rate of change of P.D. with change of length is constant with solid carbons, but diminishes as the length of the arc increases with a cored positive carbon.

(3) This rate of change with a cored positive becomes smaller and more nearly constant for all lengths of arc as the value of the constant current increases.

(4) The P.D. corresponding with length of arc 0 diminishes

as the current increases with solid carbons, but increases with the current with a cored positive carbon.

These differences have now to be explained.

When it was found above that the law of the relation between the P.D. and the current with constant length of arc varied according as the positive carbon was solid or cored, it was shown that this variation could be reasonably explained on the hypothesis that *with a given negative carbon the P.D. required to send a given current through a fixed length of arc depends principally, if not entirely, on the nature of the surface of the crater, being greater or less according as the carbon of which this surface is composed is harder or softer.*

If this hypothesis be true, it should explain *all* the changes in the laws connecting P.D., current and length of arc when a cored instead of a solid positive carbon is used, whichever one of the three variables we may choose to consider as constant. Therefore, it should explain the differences between the curves in Fig. 44 and those in Figs. 45, 46 and 47. We will see how far it does so.

Let A P Q R (Fig. 49) be the straight line representing the connection between the P.D. between the carbons and the length of the arc, with a given current flowing, when both carbons are solid, and let A' P' Q' R' be the curve representing the same connection when the positive carbon has a soft core. Let P, Q and R be the points on the upper line for arcs of 1mm., 2mm. and 3mm. respectively, and let P', Q' and R' be the corresponding points on the lower curve. Then P P', Q Q' and R R', represent the diminution in the P.D. caused by the soft core when the arcs are 1mm., 2mm. and 3mm. respectively, with the given current flowing. According to the above hypothesis, this diminution depends on how much of the carbon in the surface of the crater is soft; thus P P', Q Q' and R R' may be taken to represent the different proportions of soft carbon in the surface of the crater with the different lengths of arc. If the proportion is great the diminution of P.D. is great, if small it is small. Hence, since experiment shows that P P' is greater than Q Q', and Q Q' than R R', we must conclude that *the proportion of soft carbon in the surface of the crater diminishes as the length of the arc increases with a constant current flowing.* Now the only way in which the proportion of

soft carbon in the surface of the crater can change is by
different amounts of hard carbon being added to it, for as long
as the area of the crater is less than that of the core the
proportion of soft carbon is a constant, namely, the whole,
and as soon as the crater covers the core the *amount* of soft
carbon is a constant, but the *amount* of hard carbon can
change by the enlargement of the area of the crater; hence
the *proportion* of soft carbon in the crater becomes less as
the crater enlarges itself, for more and more hard carbon is
added, and the soft carbon remains the same. But we have
seen that by the hypothesis the proportion of soft carbon in
the surface of the crater must diminish as the arc increases

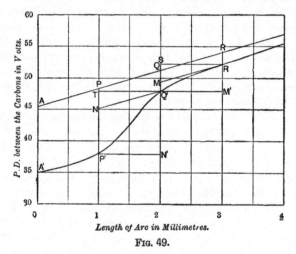

FIG. 49.

in length with a constant current, therefore, since the only
way in which this can happen is by the area of the crater
becoming enlarged, it follows that, if the hypothesis is correct,
the area of the crater must increase as the length of the arc
increases, when a cored positive carbon is used and a constant
current is flowing.

Before proceeding further it will be well to define accurately
what is meant by the expressions "area of crater," "proportion of soft carbon in the surface of the crater." It must
be clearly understood that, for the present, the depth of the
crater has been and will be left entirely out of account, for the

results of experiments show that perfectly clear and definite laws can be obtained without considering it, and the probability is therefore that it has very little effect on the action of the arc. By the term *area of the crater*, then, we mean *the area of the mouth of the crater*, or, still more accurately, the plane area of that region of the end of the positive carbon which is sharply cut off from the rest by its peculiar brilliance and whiteness. The area of the soft carbon in the surface of the crater is taken to be the projection on the mouth of the crater of that area of the crater that is composed of soft carbon, and *the proportion of soft carbon in the surface of the crater is measured by the ratio of the area of the soft carbon to the total area of the crater*. This ratio for each current and length of arc we shall call its "*soft crater ratio*," while the ratio of the amount of hard carbon to the total amount of carbon in the surface of the crater will be called the "*hard crater ratio*."

It has been shown that of the "*differences*" mentioned on page 142 the first must arise, according to hypothesis, from the area of the crater increasing as the length of the arc increases. To see the meaning of the second we must turn again to Fig. 49. In this figure Q' N', R' M' represent the increase of P.D. accompanying a change of 1mm. in the length of the arc when the positive carbon is cored, the first from 1mm. to 2mm. and the second from 2mm. to 3mm. The accompanying changes in the *soft crater ratios* are represented by N P' and M Q'. For, since R' M and Q' N are both parallel to A P Q R, P P' – Q Q' = N P' and Q Q' – R R' = M Q', and P P', Q Q' and R R' have been shown to represent the soft crater ratios for arcs of 1mm., 2mm. and 3mm. respectively. Now *difference* (2) (p. 142) says that the change of P.D. with change of length with a cored positive carbon diminishes as the length of the arc increases; that is to say, we know from experiment that R' M' is less than Q' N' (Fig. 49). We can easily see what this must mean in crater ratios according to our hypothesis. We have

$$R' M' = S Q',$$
$$Q' N' = T P' ;$$

and, since

$$R' M' < Q' N',$$
$$\therefore S Q' < T P'.$$

That is

$$S M + M Q' < T N + N P'.$$

But S R′ M and T Q′ N are equal triangles,

$$\therefore \text{S M} = \text{T N},$$
$$\therefore \text{M Q}' < \text{N P}'.$$

But M Q′, N P′ are the changes of *soft crater ratio* accompany-ing changes in the length of the arc as that length is increased by equal increments. Hence it has been shown that, since the change of P.D. with change of length diminishes as the arc increases in length, it follows, if our hypothesis be correct, that the rate of change of soft crater ratio with change of length must also diminish as the arc increases in length. In other words, we have that

(2) *When a cored positive carbon is used, and a constant current is flowing, the rate of the change that takes place with change of length in the ratio of the soft carbon to the total amount of carbon in the surface of the crater must diminish as the arc increases in length.*

Difference (3) translated into crater ratios evidently means that the rate of change of soft crater ratio with change of length is smaller and becomes more nearly equal for all lengths of arc as the value of the constant current increases ; or,

(3) *When a cored positive carbon is used the rate of the change that takes place with change of length in the ratio of the soft carbon to the total amount of carbon in the surface of the crater is smaller and becomes more nearly equal for all lengths of arc as the value of the constant current increases.*

Difference (4) can hardly be called a separate phenomenon, for not only with length of arc 0, but also with lengths of arc 1mm. and 2mm., the P.D. increases with the current with the cored positive carbons used for these experiments (*see* Figs. 45, 46 and 47), but the increase with length of arc 0 has so important a bearing on the question of a back E.M.F. in the arc that it is quite necessary to show that this increase is, so to say, accidental, and depends merely on the presence of the core in the positive carbon. This increase has already been explained above, as being due to the rise of P.D., caused by the greater amount of hard carbon in the surface of the crater with a larger current more than counterbalancing the fall due to the increase of the current. In other words, with short arcs the rise of P.D. due to the diminution of soft crater ratio with an increase of current more than counterbalances the fall of P.D. due to that increase of current.

One more point must still be mentioned with regard to the curves in Figs. 45, 46 and 47, before proceeding to justify the hypothesis upon which the explanation of the difference between the arc curves for solid and cored positive carbons rests, by an examination of the curves connecting the area of the crater with the other variables of the arc. The curves for 6 amperes in Fig. 45 and for 5 amperes in Fig. 46 change their curvature completely for arcs between 0 and 1mm. This change can only be caused by the complete absence of hard carbon in the surface of the crater, so that those portions of the curve really belong to the straight lines that connect P.D. and length of arc for a solid soft positive carbon and a hard negative carbon. Those portions of the curves, in fact, are the portions for which the crater is so small that it is entirely in the core. With the smaller sized carbons the core is smaller, and with a current of 4 amperes the crater was not small enough with the 9 mm. cored positive carbon to consist entirely of soft carbon; therefore there is no such change in the curve in Fig. 47.

SUMMARY.

The Steady P.D. between the Carbons when the Length of the Arc is Fixed and the Current Varied.

With Both Carbons Solid.

I. With the smallest current the P.D. is the highest; as the current is increased the P.D. falls, rapidly at first, and then more and more slowly, until the hissing point is reached

II. With a given range of current, the P.D. changes much more with a long arc than with a short one.

III. From I. and II. it follows that the ratio of change of P.D. to change of current is greater the smaller the current and the longer the arc.

With the Positive Carbon either Cored or Solid, and the Negative Carbon Solid.

IV. With any given P.D. the longer the arc the larger is the current, and consequently

V. With a given P.D. the longer the arc the smaller is its apparent resistance.

With the Positive Carbon Cored and the Negative Carbon Solid.

VI. The curves connecting the P.D. between the carbons with the current for a fixed length of arc vary in form according to the ratio that the diameter of the core of the positive carbon bears to the whole diameter of the carbon. When the carbons are such as are ordinarily used for electric lighting purposes, the P.D. generally falls to a minimum as the current is increased, and then rises again as the current is still further increased, till hissing takes place.

VII. With the same pair of carbons the minimum P.D. corresponds with a larger current the longer the arc.

VIII. With arcs of from 4 to 7mm. with the larger sizes of carbons, as the current is increased above a certain value, the P.D., as it falls to a minimum and then rises, changes so slightly that it may be considered practically constant for a wide range of current.

IX. The P.D. is always lower for any given current and length of arc when a cored positive carbon is used than with a hard solid positive carbon.

THE STEADY P.D. BETWEEN THE CARBONS WHEN THE CURRENT IS FIXED AND THE LENGTH OF THE ARC IS VARIED.

With both Carbons Solid.

I. A straight line law connects the P.D. between the carbons with the length of the arc, and hence the rate of change of P.D. with change of length is a constant for each current.

II. The straight lines meet at a point *to the left* of the axis of P.D. and *above* the axis of length, therefore there is no *real* length of arc for which the P.D. is constant for all currents.

III. The straight lines make a smaller angle with the axis of length the larger the current, hence the P.D. for length of arc 0 diminishes as the current increases.

With the Positive Carbon either Cored or Solid and the Negative Carbon Solid.

IV. The P.D. increases as the length of the arc increases.

V. The rate of change of P.D. with change of length is smaller the larger the current.

With the Positive Carbon Cored and the Negative Carbon Solid.

VI. The P.D. is lower than with a solid positive carbon.

VII. The rate of change of P.D. with change of length diminishes as the length of the arc increases.

VIII. The rate of change of P.D. becomes more nearly constant as the current increases.

IX. The P.D. for length of arc 0 increases with the current.

On the hypothesis that, *with a given negative carbon the P.D. required to send a given current through a fixed length of arc depends principally, if not entirely, on the nature of the surface of the crater, being greater or less according as the carbon of which this surface is composed is harder or softer*, these changes show that—

X. The area of the crater must *increase*, and consequently the soft crater ratio must *diminish* as the length of the arc increases.

XI. The rate of change of soft crater ratio with change of length must diminish as the length of the arc increases.

XII. The rate of change of soft crater ratio with change of length must be smaller and become more nearly constant the larger the current.

CHAPTER V.

IF we accept the definition of the area of the crater given in Chapter IV., it is clear that that area can only be measured while the arc is burning, for when the arc is extinguished we have no certain means of determining the boundary of the white-hot part of the carbon. Table XVI. gives the diameter of the crater measured in this way and the square of the diameter for different currents and lengths of arc. To measure the diameter of the crater, strips of paper were placed across the enlarged image of the crater, when it was properly formed with a given current and length of arc, and marks were made with a very fine pencil at the points where the paper cut the diameter of the image of the crater.

Table XVI.—*Diameter of Crater, Square of Diameter, P.D. and Current for Different Lengths of Arc.*

Carbons : Positive, 13mm., cored ; negative, 11mm., solid.

Current in amp	$l=1$			$l=2$			$l=3$			$l=4$		
	Dia. in mm.	Sq. of dia.	P.D. in volts.	Dia. in mm.	Sq. of dia.	P.D. in volts.	Dia. in mm.	Sq. of dia.	P.D. in volts	Dia. in mm.	Sq of dia.	P.D. in volts.
4	3·1	9·6	38·75	3·8	14·4	51·0	3·55	12·6	56·0	3·55	12·6	58·6
7	4·2	17·64	38·0	4·2	17·64	45·15	4·2	17·64	49·0	4·4	19·36	52·1
10	4·25	18·06	38·6	4·75	22·56	43·3	...	...	46·8	4·9	24·0	49·7
15	5·45	29·7	39·8	5·6	31·36	43·2	5·35	28·62	45·2	5·8	33·64	47·75
20	6·4	40·96	40·9	6·4	40·96	43·3	6·4	40·96	44·9	6·6	43·56	46·6

Each set of numbers in Table XVI., and, indeed, in all such tables, is liable to three sources of error, quite apart from all actual errors of observation :

(1) The carbons may not be of uniform constitution and density.

(2) They may not have been burnt long enough to acquire their characteristic shape for each current and length of arc.

(3) They may not have been perfectly in line.

Errors of the first and second kind would affect the reading of the P.D. ; the third would render it difficult, if not impossible, to measure either the length of the arc or the diameter of the crater accurately. When to all these possible errors are added those arising from the difficulty of keeping either the length of the arc or the current absolutely constant, and the possibility of measuring some sector of the crater other than its diameter, as well as the possibility of the actual boundary of the crater being obscured, it will be seen that, even if both the ammeter and the voltmeter employed could be always depended on to give perfectly accurate readings under all circumstances, the difficulties of finding the laws which connect variables thus obtained would be very great.

Each new variable introduces a fresh difficulty, and thus the law connecting the area of the mouth of the crater with the length of the arc—which really requires a correct measurement of *four* variables, the P.D., the current, the length of the arc and the diameter of the crater—is harder to determine than the law which connects the P.D., the current and the length of the arc only, for which no measurement of the crater is necessary.

As, in consequence of this difficulty, the numbers in Table XVI were too irregular to allow the shape of the curves connecting the area of the crater with the length of the arc to be determined with any certainty, curves connecting the area of the crater with the other variables of the arc were made in turn, to see if any of these were more regular. The best results were obtained with the curves connecting the P.D. between the carbons for each length of arc with the square of the diameter of the crater for the same length, when the current was constant for all the four lengths of arc in each case. These curves are given in Fig. 50, the lines having

been drawn as nearly as possible through the average position of
the points. It is evident that they are straight lines, becoming
more steep as the constant current increases. From the values
taken from these curves the curves in Fig. 54(p.159) were plotted
connecting the current with the area of the crater for constant
lengths of arc. These lines, which are very interesting on their
own account, will be discussed later, and will for the present
only be used as a means of obtaining values for plotting the

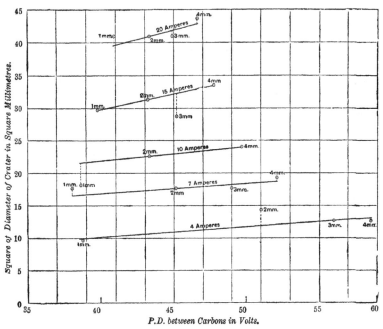

Fig. 50.—Square of Diameter of Crater and P.D. between Carbons for
various Currents and Lengths of Arc.

Carbons : Positive, 13mm., cored ; negative, 11mm., solid.

curves connecting the area of the crater and the crater ratios
with the length of the arc. Table XVII. gives the squares of
the diameters of the crater obtained from these curves, together
with the square roots of these squares, and the differences
between the diameters thus calculated and the observed
diameters given in Table XVI.

Table XVII.—*Squares of Diameters of Crater from Fig. 54, Diameters calculated from those Squares, and Differences between those Diameters and Observed Diameters.*

Carbons : Positive, 13mm., cored ; negative, 11mm., solid.

Current in Amperes.	$l=1.$			$l=2.$			$l=3.$			$l=4.$		
	d^2	d	Difference of Diam.	d^2	d	Difference of Diam.	d^2	d	Difference of Diam.	d^2	d	Difference of Diam.
4	9·9	3·14	+0·04	11·7	3·42	−0·38	12·6	3·55	0	13·2	3·63	+0·08
7	15·6	3·95	−0·25	17·4	4·17	−0·03	18·3	4·28	+0·08	18·9	4·35	−0·05
10	21·1	4·59	+0·34	22·9	4·78	+0·03	23·8	4·88	...	24·4	4·94	+0·04
15	30·3	5·5	+0·05	32·1	5·67	+0·07	33·0	5·74	+0·39	33·6	5·80	0
20	39·5	6·28	−0·12	41·3	6·43	+0·03	42·2	6·49	+0·09	42·8	6·54	−0·06

Fourteen of the nineteen diameters taken from the curves differ from the observed diameters by less than 0·1mm. Of the other five one differs by 0·12mm., one by 0·25mm. and the other three by between three and four tenths of a millimetre. Hence there are four really bad points, two belonging to the 1mm. arc, one to the 2mm. and one to the 3mm., and all four belonging to different currents. It seems pretty certain then, that the curves in Fig. 54 do really represent the relation between the current and the square of the diameter of the crater for the different lengths of arc, and that we may safely use the values obtained from them to plot the curves connecting the *length of the arc* with the square of the diameter of the crater for constant currents. These curves are given in Fig. 51, and verify the first prediction made about the area of the crater in Chap. IV., namely, that it would increase as the length of the arc increased with a constant current.

In order to calculate the crater ratios for each current and length of arc from Table XVII. it is necessary to know the diameter of the core of the positive carbon employed. This was 3mm., and therefore to obtain the ratio of the area of the core to the area of the crater—that is, the soft crater ratio—the number 9 must be divided by each of the squares of diameter given in Table XVII. The *hard* crater ratio, that is, the ratio of the area of *hard* carbon in the surface of the crater to the area of the crater, is obtained by subtracting the soft crater ratio from 1 in each case, for if s be the area of soft carbon and h the area

of hard carbon in the surface of the crater, and if a be the area
of the crater, then

$$a = s + h$$
$$\frac{h}{a} = \frac{a - s}{a} = 1 - \frac{s}{a}.$$

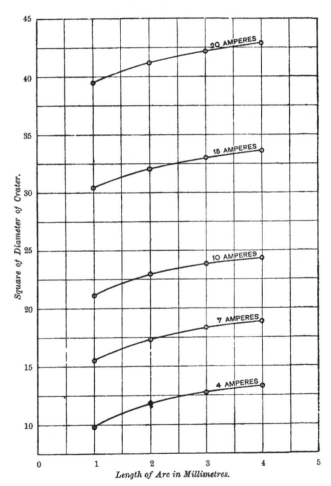

FIG. 51.—Square of Diameter of Crater and Length of Arc for various
Constant Currents.

Carbons : Positive. 13mm., cored ; negative, 11mm., solid.

Table XVIII. gives the soft and hard crater ratios with the corresponding currents, P.D.s, and lengths of arc.

Table XVIII.—*Crater Ratios calculated from Table XVII., with corresponding Currents, P.Ds., and Lengths of Arc.*

Carbons : Positive, 13mm., cored , negative, 11mm., solid.

Current in Amperes.	l=1.			l=2.			l=3.			l=4.		
	P.D. in volts.	Soft crater ratio.	Hard crater ratio.	P.D. in volts.	Soft crater ratio.	Hard crater ratio.	P.D. in Volts.	Soft crater ratio.	Hard crater ratio.	P.D. in volts.	Soft crater ratio.	Hard crater ratio.
4	38·75	0·909	0·091	51·0	0·769	0·231	56·0	0·714	0·286	58·6	0·682	0·318
7	38·0	0·577	0·423	45·15	0·517	0·483	49·0	0·492	0·508	52·1	0·476	0·524
10	38·6	0·427	0·573	43·3	0·393	0·607	46·8	0·378	0·622	49·7	0·369	0·631
15	39·8	0·297	0·703	43·2	0·280	0·720	45·2	0·273	0·727	47·75	0·268	0·732
20	40·9	0·228	0·772	43·3	0·218	0·782	44·9	0·213	0·787	46·6	0·210	0·790

From this table the curves in Fig. 52 have been constructed, connecting the soft crater ratio with the length of the arc for

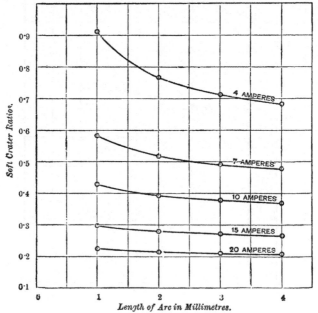

Fig. 52.—Soft Crater Ratio and Length of Arc for various Constant Currents.

Carbons : Positive, 13mm., cored; negative, 11mm., solid.

the various constant currents. These curves completely verify the second and third predictions made in Chap. IV., namely, that the rate of change of soft crater ratio with change of length would diminish as the length of the arc increased, and that the change of soft crater ratio with change of length would be smaller, and the rate of change more nearly constant, the larger the current.

For, firstly, all the curves are steeper with short arcs than with long ones, so that an increase of length from 1mm. to 2mm. is marked by a much greater change in the soft crater ratio than an increase from 3mm. to 4mm. Secondly, the 20 ampere

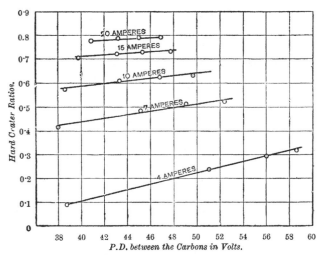

Fig. 53.—Hard Crater Ratio and P.D. between Carbons for various Constant Currents.

Carbons : Positive, 13mm., cored ; negative 11mm., solid.

curve is far less steep than the 4 ampere curve, showing that, for a given change in the length of the arc, the soft crater ratio alters less with a large than with a small current. Thirdly, the 20 ampere curve is much nearer to a straight line than the 4 ampere curve, showing that the change of soft crater ratio with change of length is more nearly constant the larger the current.

Hence, the theory of the dependence of the P.D. on the soft

crater ratio, with a cored positive carbon, is fully confirmed by these curves, obtained from observations on the area of the crater.

The relation between the hard crater ratio and the P.D. between the carbons is shown in Fig. 53. The connection evidently follows a straight line law, and is such that the P.D. increases as the hard crater ratio increases, that is to say, as the proportion of hard carbon in the surface of the crater increases, which was to be expected from what has gone before.

Many other curves might be drawn by using the numbers in Table XVIII. in various ways, but as those already drawn fully confirm the theory given in Chap. IV., concerning the interdependence of the crater ratios and the P.D. between the carbons, the remainder, with the deductions to be made from them, may be left to the ingenuity of the reader.

Before leaving the subject of the area of the crater the curves in Fig. 54, showing the connection between the area of the crater and the current flowing when the length of the arc is constant, must be examined. A straight line law connects these two variables, but the area of the crater is not, as Andrews concluded (*see* p. 39), from his measurements of the crater, proportional to the current. It is the area of the crater *minus a constant depending on the length of the arc* that varies as the current, at any rate when a cored positive carbon is used. From the figure it appears very likely that Andrews' conclusion may be true when the length of the arc is 0, for the distance between the lines for length of arc 1 and length of arc 0 would be just about right if the latter passed through the origin.

The fact that the lines connecting the area of the crater with the current for the various constant lengths of arc are all parallel to one another leads to two very interesting conclusions. The first, that the change of crater area accompanying a change of current is independent of the length of the arc, and the second, that the change of crater area accompanying a change in the length of the arc is independent of the current. For example, an increase of current from 4 to 10 amperes is accompanied by an increase of 11·2 square millimetres in the square of the diameter of the crater, whether the arc be 1mm., 2mm., 3mm. or 4mm. in length; similarly an increase in the ength from 1mm. to 4mm. is accompanied by an increase of 3·3 square millimetres in the square of the diameter of the crater, whatever the current may be, from 4 to 20 amperes. Thus, although the *area* of the crater increases, both with an increase of the current and an increase of the length of the arc, the *change of area* with change of current is *independent* of the length of the arc, and the *change of area* with change of length is *independent* of the current flowing.

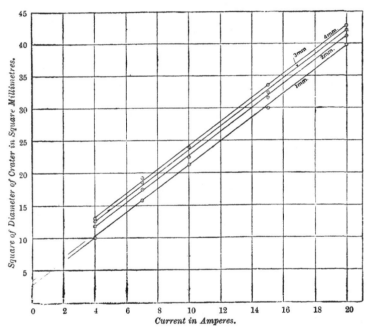

Fig. 54.—Square of Diameter of Crater and Current for various Constant Lengths of Arc.

Carbons : Positive, 13mm., cored ; negative, 11mm., solid.

It has been observed (p. 144) that the depth of the crater had very little, if any, influence on the P.D. between the carbons. In order to see whether this hollowing out of the crater (which is very great in cored carbons) really had any influence on the P.D., measurements were taken of the depths of craters with different currents and lengths of arc. The results are given in Table XIX. The carbons used were the same as those used for measuring the diameter of the crater, 13mm. cored positive, and 11mm. solid negative, in fact the measurements of diameter and depth were made on the *same* craters. The measurements were made by means of a microscope with a micrometer screw arrangement for focussing. The edge of the crater was first brought into focus and the reading taken on the micrometer, then the bottom of the crater was brought into focus, and the reading again taken. The difference of the two readings gave the depth of crater. Great care was taken that the craters should be completely formed with the given current and length of arc, before the depth was taken.

Table XIX.—*Depth of Crater with Different Currents and Lengths of Arc.*

Carbons : Positive, 13mm., cored ; negative, 11mm., solid.

Length of arc in millimetres	Current in amperes.	Depth of crater in millimetres.	Length of arc in millimetres	Current in amperes.	Depth of crater in millimetres.
1	6	0·96	3	20	1·06
1	10	1·33	3	30	0·98
1	15	1·6	6	6	0·72
1	20	1·38	6	10	0·55
1	28	1·266	6	15	0·63
3	6	0·82	6	20	0·77
3	10	0·95	6	25	0·65
3	15	0·96	6	30	0·65

It is not, of course, pretended that these measurements of the depth of the crater were absolutely accurate, a small excrescence upon the edge of the crater or a slight irregularity of form, such as those who are familiar with arcs will recognise as being of constant occurrence, would tend to obscure the real depth of crater; but the general character of the change in the depth, with variation of current and length of arc, is very apparent.

It is evident that with a given current the depth of the crater is greater the shorter the arc, and therefore the percentage increase of length of arc caused by the depth of the crater increases rapidly as the arc is shortened. For example, with a current of 6 amperes the depth of the crater increases the apparent length of the 6mm. arc by 12 per cent., the apparent length of the 3mm. arc by 27 per cent. and the apparent length of the 1mm. arc by 96 per cent.

With a given length of arc the depth of the crater appears to increase as the current increases, to reach a maximum with currents of from 15 to 20 amperes and then to diminish.

Whether the depth of the crater *per se* has any influence either in raising or lowering the P.D. it would require much more detailed experiments to determine, but at any rate the effect is so completely masked by the far greater influence of the *area* of the crater (which has been shown to be sufficient alone to account for the differences between the curves for cored and solid carbons) that it may be altogether neglected. For example, there is nothing in the curves for constant length of arc (Fig. 40) to connect the P.D. between the carbons with the depth of the crater, for there is no maximum P.D. with a current of 15 amperes for the 1mm. arc as there is a maximum depth of crater, nor is there any maximum P.D. with any particular current for the other lengths of arc. We may therefore conclude that *there is no evidence to prove that the depth of the crater has any influence on the P.D. between the carbons.*

A glance at Figs. 39, 40 and 41 (pp. 128-130) is sufficient to show the general character of the changes in the curves caused by varying the diameters of the carbons employed. The chief of these is the change in the largest silent current, which will be dealt with in the chapter on hissing arcs.

As regards those portions of the curves which refer to *silent* arcs, it will be seen that the P.D. required to send a particular current through a given length of arc depends comparatively little on the cross-section of the carbons, for although the carbons for the curves in Fig. 39 had about four times the cross-section of the carbons for curves in Fig. 41, yet the P.D. for any given current and length of arc does not in any case differ with the two sets of carbons, by as much as 14 per cent., and in most cases it differs by far less than that.

In Table XX. some examples are given illustrating this point.

Some of the P.Ds. for a given length of arc and current in this table are almost identical for the three different pairs of carbons ; for example, 43, 43 and 43·7 volts for a 2mm. arc with 10 amperes, or, again, 44·5, 45 and 45·8 for a 3mm. arc with 15 amperes. In many other cases, however, there is a marked difference in the P.Ds.; for example, 36, 37·5 and 41·5 volts for a 1mm. arc with 6 amperes. We must therefore conclude that the steady value of the P.D. between the carbons while depending principally on the particular current and length of arc, does also depend on the diameters of the carbons,

Table XX.—*Influence of Diameters of Carbons on P.D. between them.*

Length of arc in millimetres.	Current in amperes.	P.D. between carbons for		
		+18mm. cored. −15mm. solid.	+13mm.cored. −11mm. solid.	+9mm. cored. −8mm. solid.
0·5	6	33·0	...	36·7
1·0	6	36·0	37·5	41·5
2·0	6	47·0	46·0	44·2
3·0	6	50·5	50·2	48·4
4·0	6	53·0	54·0	53·8
5·0	6	54·75	56·0	55·8
0·5	10	34·9	...	39·0
1·0	10	36·0	38·5	42·0
2·0	10	43·0	43·0	43·7
3·0	10	46·5	46·7	46·2
4·0	10	47·5	49·7	50·0
5·0	10	49·75	52·0	51·8
0·5	15	36·5	...	hissing.
1·0	15	37·5	39·7	43·0
2·0	15	42·0	43·0	44·0
3·0	15	44·5	45·0	45·8
4·0	15	45·8	47·5	48·5
5·0	15	48·0	50·5	50·2

This is seen still more plainly from the curves in Fig. 55, which connect P.D. with length of arc, for constant currents of 6 and 15 amperes, with each of the three pairs of carbons.

These curves show that while, with a current of 6 amperes, the P.D. is sometimes the same for all three carbons, sometimes highest with the largest carbons, and sometimes with the smallest; with the larger current of 15 amperes the P.D. is uniformly highest with the smallest carbons and lowest with the largest.

Such variations in the P.D. for the same current and length of arc with different sized carbons have probably two causes,

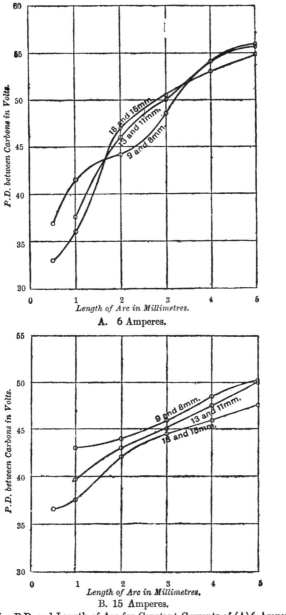

A. 6 Amperes.

B. 15 Amperes.

Fig. 55.—P.D. and Length of Arc for Constant Currents of (A) 6 Amperes, and (B) 15 Amperes.

Carbons : Positive, 18mm., cored ; negative, 15mm., solid.
     ,,   13mm.   ,,   ,, 11mm. ,,
     ,,   9mm.   ,,   ,, 8mm.   ,,

which may sometimes aid and sometimes counteract one
another :—(1) The different degrees of cooling to which the
crater is subjected by the difference of mass of the cold carbon
round it ; (2) The difference in the crater ratios caused by the
core having a different diameter in each of the three carbons ;
for the core of the 18mm. carbon was 4mm., that of the 13mm.
was 3mm., and that of the 9mm. was 2mm. in diameter. Hence,
with the same area of crater, the hard crater ratio, to which the
P.D. has been shown to have some sort of proportionality,
would be greatest in the smallest carbon, and least in the
greatest. If, therefore, the area of the crater were independent
of the size of the positive carbon, we should expect the P.D.
for a given current and length of arc to be greatest with the
smallest carbon and least with the greatest, as, indeed, it is
with a current of 15 amperes.

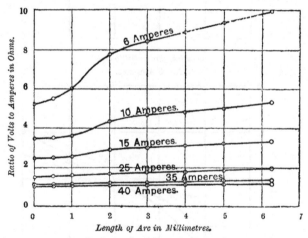

Fig. 56.—Apparent Resistance and Length of Arc for various Constant
Currents.

Carbons: Positive, 18mm., cored ; negative, 13mm., solid.

The curves connecting apparent resistance with length of
arc (Figs. 56, 57 and 58) have next to be considered.
The ordinates of these curves were obtained by dividing the
P.D. for each length of arc by each current used, after the
current had in each case been kept constant for a sufficient

length of time for the carbons to be properly shaped for
the particular current and length of arc. The abscissæ
are the corresponding lengths of arc, the current being the
same for all the points on any one curve.

These curves are naturally of the same *form* as those in
Figs. 45, 46 and 47, for the abscissæ are the same in both cases,

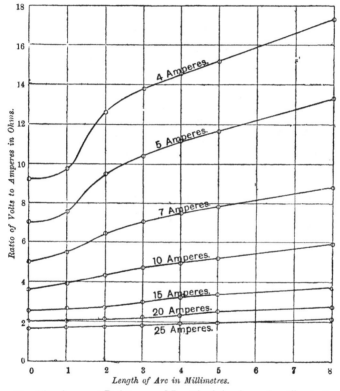

FIG. 57.—Apparent Resistance and Length of Arc for various Constant
Currents.

Carbons : Positive, 13mm., cored ; negative, 11mm., solid.

and the ordinates of the curves in Figs. 56, 57 and 58 may be
obtained from those in Figs. 45, 46 and 47 by dividing those
belonging to each curve by the corresponding constant
current.

The meaning of the peculiar shapes taken by these curves has already been fully discussed (pp. 139–147), and need not therefore, be further enlarged upon, for everything that has been said concerning the curves connecting the P.D. between the carbons with the length of the arc for constant currents with a cored positive carbon, applies equally to the apparent resistance curves in Figs. 56, 57 and 58.

In 1886 Messrs. Cross and Shepard constructed resistance curves of this kind for *solid* carbons,* and found that in every case they obtained straight lines for both silent and hissing arcs, showing that the apparent resistance of the arc

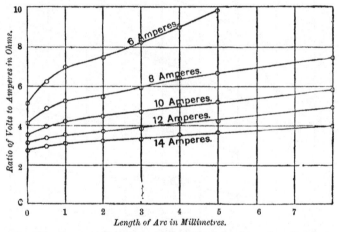

FIG. 58.—Apparent Resistance and Length of Arc for various Constant Currents.

Carbons : Positive, 9mm., cored ; negative, 8mm., solid.

minus a constant varied directly as the length of arc. Thus they showed that Edlund's formula

$$r = a + b\,l,$$

where r is the apparent resistance of the arc, l the length of arc,.and a and b are constants which depend on the currents and the carbons, is applicable to the hissing as well as to the silent arc.

---

* *Proceedings* of the American Academy of Arts and Sciences, June 16, 1886.

Th₃ results of my own experiments with solid carbons agree with this equation ; as a reference to Fig. 44 will show for *silent* arcs, and, as will be seen in Chap. X., for *hissing* arcs. It has already been mentioned that the ordinates of each curve in Fig. 44 have only to be divided by the number representing the constant current corresponding with that curve to give the apparent resistance of the arc in each case, therefore it has not been thought necessary to plot separate resistance curves for solid carbons.

The sets of curves in Figs. 45, 46 and 47 were drawn by taking a number of *vertical* sections of the curves in Figs. 39, 40 and 41 (pp. 128–130). In a similar way a set of curves can be drawn for each of the three pairs of carbons by taking a number of *horizontal* sections of the curves in those figures, and the results so obtained show the connection between current and length of arc for various constant P.Ds.

Dealing with Fig. 40 in this way, we obtain Fig. 59 for the 13mm. cored positive carbon and the 11mm. solid negative carbon ; but it is to be remembered that the points on these curves were *not* obtained by keeping the P.D. constant until the current became constant for each length of arc, but by keeping the current constant until the P.D. became constant for each length of arc. In fact, *current* and length of arc were kept constant, and not P.D. and length of arc. From the curves in this figure we can see the current that each particular P.D. will *finally* send through each length of silent arc with this pair of carbons, as well as the maximum length of silent arc that can be maintained with any given P.D. between the carbons.

To obtain this maximum length for any constant P.D. we have merely to draw a vertical tangent to the corresponding curve, and observe where it cuts the axis of length of arc. For example, the vertical tangent to the curve corresponding with a constant P.D. of 43·5 volts between the carbons, cuts the axis of. length at about 2·4mm. Hence, while a P.D. of 43·5 volts can produce a silent arc of *any* length between 1 and about 2·4mm., it can produce *no permanent silent* arc longer than 2·4mm. with these carbons.

If, starting with a small current, the P.D. between the carbons be kept constant and the arc gradually lengthened,

the current will gradually increase until a certain value is
arrived at, when either the current begins to decrease or the
arc becomes unstable; for example, with the constant P.D.
of 48·5 volts this happens with a current of about 32 amperes
if the current has been kept at a fixed value for some time
for each length of arc.

When the constant P.D. is less than about 46 volts the
curves in Fig. 59 bend back before the condition of instability
is reached, so that for some lengths of silent arc there
are two very different currents *permanently* produced with
the same P.D.   For example, with the constant P.D. of
45 volts a current of about either 15 or of 30 amperes may
pass and maintain a silent arc of 3mm.   With a constant
P.D. of 43·5 volts a silent arc of 2mm. can be produced with
a current of about 8·8 or 28 amperes, and a constant P.D.
of 41·5 volts will send a current of about 2·8 or of 23 amperes
silently through an arc of 1mm.

For a constant P.D. of 48·5 volts, on the contrary, there
is only one current that will pass silently for each length of arc
and remain constant, this current being about 4·5 amperes
when the arc is 2mm. long, about 7·5 amperes when it is 3mm.,
about 12 amperes for a 4mm. arc, and about 32 amperes when
the arc is lengthened to 5mm.

The arc now becomes unstable, and the curve shifts *sideways*,
and continues as a vertical straight line for a much longer
hissing arc.

From the curves in Fig. 59 it follows that when the
P.D. is kept constant at, say, 46·5 volts and the arc is
lengthened from 1mm. to 5mm. the current *increases* in value ;
whereas when it is kept at about 44 volts a wide range of current can be obtained with the same length of arc, and, lastly,
that when the constant value of the P.D. is, say, 41·5 volts,
lengthening the arc sometimes *diminishes* the current.

These facts have been already stated and explained in connection with the curves in Fig. 40 (p. 129) from which the curves
in Fig. 59 have been deduced.   But at the time the experiments
were made it appeared to be so astonishing, as well as novel,
that a change in the steady value of the P.D. from, say, 46
to 41 volts should entirely alter the way in which the current
varied with length of arc, *increasing* in the former case with the

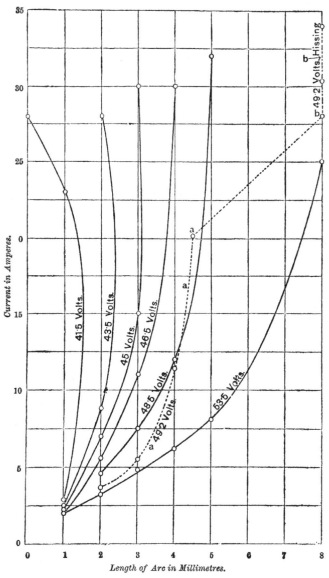

FIG. 59.—Current and Length of Arc for various Constant P.Ds.

Carbons : Positive, 13mm., cored ; negative, 11mm., solid.

lengthening of the arc, and *decreasing* in the latter, that it seemed worth while to test these facts by actually experimenting on arcs with various constant P.Ds.

The experiments were tried in two ways—(1) the P.D. was kept constant by using accumulators with no external resistance beyond that of the arc; (2) a dynamo was used, and the P.D. between the ends of the carbons was kept constant by means of a wide range of external resistance.

With the dynamo, a constant P.D. of 55 volts was maintained between the ends of the carbons in the following way : The arc was struck with a large amount of resistance in circuit, the carbons were then separated to the desired extent, the resistance being altered meantime until a P.D. of 55 volts was obtained ; then the arc was kept at the given length for some time, and the P.D. kept at 55 volts by means of the external resistance, until the current appeared to have reached its steady value, when the reading was taken. The carbons used were 13mm. cored positive and 11mm. solid negative.

It was found that with a P.D. of 55 volts a current of 2·7 amperes could be maintained steadily flowing through an arc of 2mm. ; a current of 3·3 amperes would flow steadily through an arc of 3mm., and that the current that could be maintained *permanently* flowing through the arc *increased* steadily as the length of the arc was *increased*, thus proving that the curves for constant P.D. of 46·5 volts and upwards in Fig. 59 were correct in character when a cored positive carbon was used.

The second problem, Does the current sometimes *decrease* as the length of arc is *increased* with short arcs and large steady currents ? was tried with the dynamos and external resistance in the same way as before, and an answer was obtained in the affirmative. It was found that with a constant P.D. of 42·5 volts a current of 22 amperes could be maintained steadily flowing through an arc of 1·3mm., while a current of 19·5 amperes only would flow permanently through an arc of 1·5mm. Thus it was shown experimentally that when a cored positive carbon is used with certain constant P.Ds. the current increased as the arc was *lengthened*, and that with other constant P.Ds. the current increased as the arc was *shortened*, it was evident therefore that there must be an intermediate constant P.D which would permanently maintain many different currents with the same length of arc ; therefore, all the three cases suggested by the curves in Fig. 59 were shown to be correct.

These experiments, in which the P.D. and length of arc are kept constant and the current allowed to vary, are very difficult to carry out, because each P.D. will only send a current permanently through certain lengths of arc, and if you happen to hit upon a length of arc to which the P.D. you are trying does not belong, you may have all your pains for nothing. It may take an hour to find out that, although you are sending a current through the given length of arc with the fixed P.D.

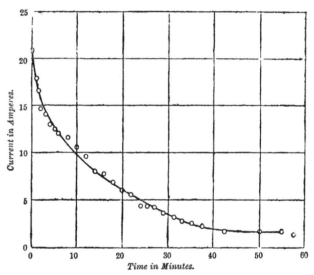

Fig. 60.—Time record of Current with Constant P.D. of 42·5 volts, and Constant Length of Arc of 1mm.

Carbons: Positive, 13mm., cored ; negative, 11mm., solid.

all the time, yet that current will not remain constant, but will change continually. If the current is increasing, it rushes up rapidly, and then if the P.D. is maintained constant by means of accumulators with no interposed resistance the cut-out goes. Or if, to prevent this occurring, you endeavour to maintain the P.D. between the carbons constant by means of a dynamo with interposed resistance, the arc suddenly hisses and the

P.D. falls in spite of all attempts to prevent it by rapidly adjusting the interposed resistance. If, on the other hand, the current is decreasing, the arc flickers and goes out.

Fig. 60 shows very well what happens when the P.D. and length of arc are kept constant and the current is allowed to vary. The current started at 20·6 amperes and gradually diminished till it reached its steady value of about 1·6 amperes nearly 50 minutes after starting. The *larger* steady current of about 28 amperes that might have been maintained with the same P.D. and length of arc according to Fig. 40 was too near the hissing point to be attempted by this method. The meaning of the difficulty experienced in maintaining the arc with a constant P.D. and a small external resistance will be explained when the mathematical relations of the variables of the arc are considered in Chap. VIII.

SUMMARY.

The Area of the Crater and Crater Ratios

Silent Arcs.

With Constant Currents.

I. A straight line law connects the area of the crater with the P.D. between the carbons.

II. The area of the crater increases as the length of the arc is increased.

III. The rate of change of soft crater ratio with change of length diminishes as the length of the arc increases.

IV. The change of soft crater ratio with change of length is smaller and the rate of change more nearly constant the larger the current.

V. The soft crater ratio diminishes and the hard crater ratio increases as the length of the arc increases.

VI. The change in the area of the crater with change of length is independent of the current flowing.

With Constant Lengths of Arc.

VII. The area of the crater increases as the current increases.

VIII. The change in the area of the crater with change of current is independent of the length of the arc.

The Depth of the Crater.

IX. There is no evidence to prove that the depth of the crater has any influence on the P.D. between the carbons.

Influence of the Diameters of the Carbons on the P.D. between them with Cored Positive Carbons.

X. With cored carbons the P.D. accompanying a given current and length of arc is in some measure influenced by the diameters of the carbons, partly because of the greater or less mass of carbon to be warmed up, partly because the diameter of the core is different with different sized carbons.

Curves Connecting Apparent Resistance and Length of Arc with Constant Currents.

XI. The curves connecting the apparent resistance of the arc with its length for constant currents are straight lines with solid carbons, but bend down towards the axis of length for short arcs and small currents when a cored positive carbon is used.

Curves Connecting Current and Length of Arc for Constant P.D.

XII. When the P.D. is constant, with a cored positive carbon, lengthening the arc may increase the current, or leave it unchanged, or diminish it. This depends on the original current and length of arc, and on the cross-section of the core.

CHAPTER VI.

CURVES CONNECTING POWER WITH LENGTH OF ARC FOR CONSTANT
CURRENTS, AND POWER WITH CURRENT FOR CONSTANT LENGTHS
OF ARC. EQUATION CONNECTING P.D., CURRENT, AND LENGTH
OF ARC WITH SOLID CARBONS. ANALYSIS OF RESULTS OBTAINED
BY EARLIER EXPERIMENTERS.

Having found it a very tedious and lengthy process to
obtain a steady P.D. with small currents, I started my experiments for finding the true connection between the P.D., the
length of the arc, and the current, when the arc was in a stable
condition, by using only currents of 5 amperes and upwards,
and I found the steady P.D. that would send each current
through lengths of arc varying from 1mm. to 7mm. The
currents used were 5, 8, 10, and 14 amperes respectively, the
largest current that did *not* cause hissing for each length of
arc, and two or three of the currents that did cause hissing; I
then proceeded to draw curves connecting P.D. and current
for the various constant lengths of arc, similar to those in Figs.
39, 40 and 41.

On attempting, with the data thus obtained, to find an
equation that would fit all the curves, to my dismay I discovered that there were *two* equations, either of which might
be correct, since all the curves could be deduced with a very
fair amount of accuracy from each of them. One of these was
the equation to a series of ellipses, the other the equation to a
series of hyperbolas. If the curves were really ellipses, no
current of less than about 4 amperes ought to be able to
maintain an arc continuously with the carbons used; whereas
if they were hyperbolas, any current, however small, ought to
maintain an arc, provided a large enough P.D. could be
supplied.

Now the difficulty, already alluded to, that had always been found in maintaining small currents, made it appear as if the ellipse equation were the correct one ; for it seemed possible that, although these currents could be maintained for a short time after striking the arc, or changing the current from a larger to a smaller one, yet that, when the excess of loose carbon supplied in the previous state was exhausted, the remaining supply, kept up by the small steady current, would be insufficient to maintain the arc, and so it would go out. As a matter of fact, it had been found in former experiments that the arc *did* go out again and again after small currents had been flowing for a short time.

It became necessary, therefore, to decide whether the conditions of the E.M.F. of the dynamo and the resistance of the circuit were such as to cause this instability, or whether it really was impossible, with the size of carbons I was using, to maintain an arc, with less than about 4 amperes, for a sufficient length of time for the P.D. to reach its steady value.

Using a dynamo producing a P.D. of about 150 volts between its terminals, and a large resistance in circuit, I found that currents of 2 amperes *could* be kept flowing long enough for the P.D. to become perfectly steady, although, with such small currents, the carbons took between one and two hours to form, and therefore the P.D. took the same length of time to become steady.

Thus it was evident that the curves were not ellipses. It remained to prove whether they were really hyperbolas, or only fairly close approximations to hyperbolas ; whether, in fact, there was not some extra term in the true equation which remained too small to be noticed with currents of 5 amperes and upwards, but which would be large enough, with smaller currents, to show some discrepancy between the curves and the equation I had obtained. It seemed advisable, also, to determine experimentally a great many more points on the curves than I had hitherto done. I therefore decided to make an entirely new set of experiments, using the same lengths of arc as before, but taking many more currents, and beginning with 2 amperes, so that the points on the curves should be so numerous as to preclude the possibility of any mistake being made as to their true shape.

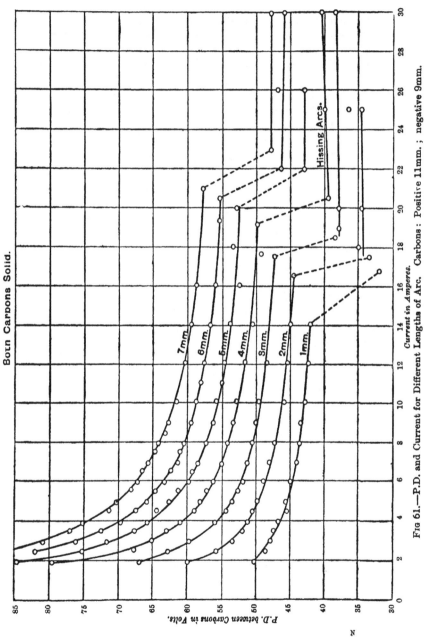

FIG 61.—P.D. and Current Curves for Different Lengths of Arc. Carbons: Positive 11mm.; negative 9mm.

N

Finding that even a slight difference in the hardness of the carbons caused a corresponding small change in the steady P.D. necessary to send a given current through an arc of given length, I generally took two or three readings of the P.D. for each current and length of arc, with different carbons, at different times, and used the means of these readings in plotting the curves. The current was kept flowing at a perfectly constant value through the given length of arc for periods varying from a quarter of an hour for large currents to nearly two hours for very small currents, before each reading was taken, so that there might be no doubt as to the P.D. having reached its final steady value for each current and length of arc, when the reading was taken. In this way the 137 points were obtained, through which the curves in Fig. 61, (which is a reproduction of Fig. 38), were plotted.

There are various ways of finding the equation to the family of curves in Fig. 61. For example, if

$$V = f(A, l)$$

be this equation, where V is the P.D. between the carbons, A the current, and l the length of the *silent* arc as measured on the projected image, we may (1) begin by finding the connection between V and l when A is constant, then the connection between V and A when l is constant, and combine the two; or (2) first find the connection between the apparent resistance, $\dfrac{V}{A}$, and l when A is constant, then the connection between $\dfrac{V}{A}$ and A when l is constant, and combine the two; or (3) start by finding the relation between the power, $V \times A$, and l when A is constant, next find the connection between $V \times A$ and A when l is constant, and combine the two, &c.

If both the connections used for any one of these methods followed straight-line laws, it was obvious that the equation would be very much easier to evolve, and the results would be more certain. It was found that this was the case with the third method only, and that was, therefore, the one employed. Plotting curves connecting *power* and length of arc for various constant currents, the series of *perfectly straight lines* seen in Fig. 62, was obtained, and plotting curves connecting *power* and current for constant lengths of arc the *perfectly straight lines* seen in Fig. 63 were obtained.

In the case of carbons the positive of which was cored, it was found by the students working under Prof. Ayrton in 1893 that the curves connecting *power* and *current* for various constant lengths of arc were straight lines when the current was not less than about 5 amperes. With the *solid* carbons that I used, on the other hand, these lines are quite straight for all currents—even the smallest that I have tried, viz., 2 amperes.

As already stated, almost every point in Fig. 61 represents the mean of several results obtained at different times with different pairs of carbons. But the curves in Fig. 61, having been drawn through the *average* position of these mean points, give still more accurately than the points themselves the ordinate corresponding with any given abscissa, *i.e.*, the number of volts corresponding with any given number of amperes. Hence, to obtain the number of watts for each point in Fig. 62, points were taken that were actually on the curves in Fig. 61, and the number of volts represented by their ordinates was multiplied by the number of amperes represented by their abscissæ. And it is because greater accuracy was obtained in this way that the points lie so well on the straight lines in Fig. 62. The ordinates of points in Fig. 63 were taken direct from the lines in Fig. 62.

From Figs. 62 and 63 we can ascertain the exact form of the general equation

$$V = f(A, l)$$

connecting the P.D. between the carbons with any current and any apparent length of arc. For, in any one of the curves in Fig. 62, let

l equal the apparent length of the arc in millimetres,

W equal the power in watts expended in sending a given current through an arc l millimetres long,

W_0 be the power in watts that would be expended in sending the same current through an arc of 0mm., as shown by the curve, and

W_7 be the power in watts expended in sending the same current through an arc 7mm. long.

N 2

Both Carbons Solid.

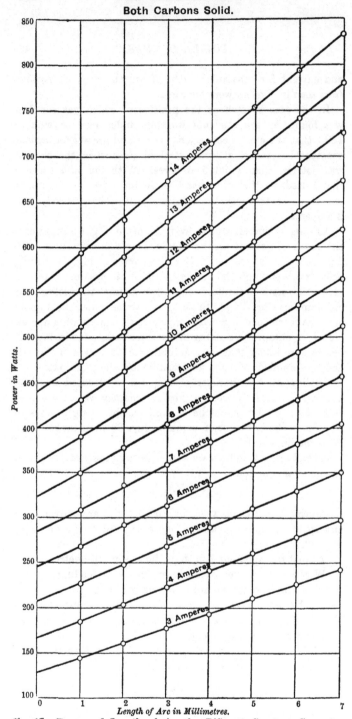

FIG. 62.—Power and Length of Arc for Different Constant Currents. Carbons: Positive 11mm.; negative 9mm.

Then, by similar triangles,

$$\frac{W - W_0}{l} = \frac{W_7 - W_0}{7}. \quad \cdot \quad \cdot \quad \cdot \quad \cdot \quad (1)$$

This, therefore, is the general equation to the lines in Fig. 62. To obtain the equation to any particular line, W_0 and W_7 must be replaced by their actual values in the particular curve.

To give an instance. Take the equation to the line for a constant current of 6 amperes. In this case

$$W_7 = 406,$$

and

$$W_0 = 245.$$

Therefore we have,

$$\frac{W - 245}{l} = \frac{406 - 245}{7},$$

$$= \frac{161}{7},$$

$$= 23.$$

Hence $W = 245 + 23l$ is the equation to the line representing the connection between power in watts and apparent length of arc in millimetres for the constant current of 6 amperes.

Thus, in order to apply equation (1) to any line in Fig. 62, we must know the two constants for that line, W_7 and W_0, and, as there are 11 such lines, it would be necessary to have 22 constants for those 11 lines. In order to get the complete equation, therefore, we must find some relation between all these constants.

Now the relation between W_7 with one current and W_7 with another current is simply the relation between power in watts and current in amperes for a constant length of arc of 7mm. Hence the next thing to be done is to find the law connecting power with current for various constant lengths of arc. This law, as already explained, is strictly linear for solid carbons, and in Fig. 63 are given the straight lines for lengths of arc 0mm. and 7mm.

The top straight line in Fig. 63 for an arc of constant length of 7mm. cuts the axis along which current is measured at a distance to the left of the zero point which represents 1·6 amperes.

Therefore, calling A the current in amperes, and using the

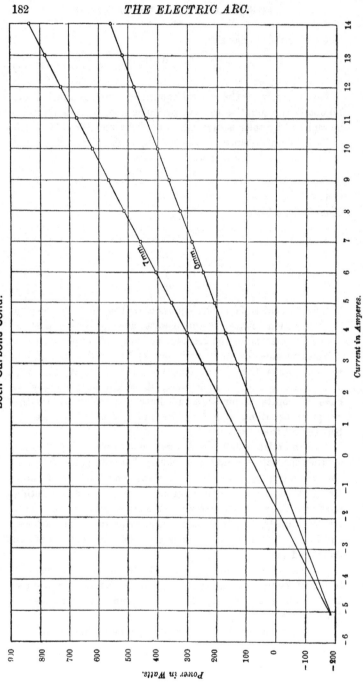

FIG. 63.—Power and Current for Arcs of 7mm. and 0mm.   Carbons : Positive 11mm., negative 9mm.

notation already adopted for W, we have, by similar triangles,

$$\frac{W_7}{A+1\cdot6} = \frac{\text{the watts corresponding with a current of 14 amperes}}{14+1\cdot6}$$

$$= \frac{833}{15\cdot6},$$

$$= 53\cdot397;$$

or $\qquad W_7 = 53\cdot397\,A + 1\cdot6 \times 53\cdot397,$

$$= 53\cdot397\,A + 85\cdot435.$$

Hence $\qquad W_7 = 53\cdot397\,A + 85\cdot435$

is the equation to the upper line in Fig. 63.

Using the same notation for the lower curve, and remarking that it cuts the axis of current at a distance from the zero point which represents a current of $0\cdot3$ ampere, we have

$$\frac{W_0}{A+0\cdot3} = \frac{\text{the watts corresponding with a current of 14 amperes}}{14+0\cdot3}$$

$$= \frac{556}{14\cdot3},$$

$$= 38\cdot881;$$

or $\qquad W_0 = 38\cdot881\,A + 0\cdot3 \times 38\cdot881,$

$$= 38\cdot881\,A + 11\cdot664.$$

Hence $\qquad W_0 = 38\cdot881\,A + 11\cdot664$

is the equation to the lower line in Fig. 63.

Substituting the above values for W_7 and W_0 in equation (1) we have

$$\frac{W - (38\cdot881\,A + 11\cdot664)}{l}$$

$$= \frac{53\cdot397\,A + 85\cdot435 - (38\cdot881\,A + 11\cdot664)}{7},$$

$$= \frac{14\cdot516\,A + 73\cdot771}{7},$$

$$= 2\cdot074\,A + 10\cdot54.$$

Therefore $\quad W = 38\cdot881\,A + 11\cdot664 + l\,(2\cdot074\,A + 10\cdot54).$ (2)

is the general equation connecting the power expended in a silent arc in watts, the current flowing in amperes, and the apparent length of the arc in millimetres for the *solid* carbons used.   But W = AV, therefore, dividing both sides of the preceding equation by A, and omitting the fifth significant figure, we obtain

$$V = 38\cdot88 + 2\cdot074\,l + \frac{11\cdot66 + 10\cdot54\,l}{A} \quad . \quad . \quad . \quad (3)$$

This is the equation which we have been seeking, to the family of curves in Fig. 61, and it enables us to calculate V, the P.D. in volts that it is necessary to maintain between a pair of *solid* carbons of the size and hardness that I used, in order that a current of A amperes may flow through a silent arc whose apparent length, as measured on the projected image, is *l* millimetres.

In order to see how nearly this equation actually represents the curves in Fig. 61, I have used it to calculate the P.D. that should correspond with currents of 3, 4, 5, 6, 7, 8, 9, 10, 12, 14 amperes, and in some cases with currents of 2 and 16 amperes for all the arcs from 1mm. to 7mm. in length ; and in Table XXI. are given the results of 81 such calculations, as well as the actual P.Ds. taken from the curves themselves.

An examination of this table shows that 58 out of the 81 calculated values of the P.D. do not differ by more than 0·05 of a volt from the values of the P.D. taken from the curves, while of the remaining 23 values

    8 differ by 0·05 to 0·15 volts from the curve values,
    5 differ by 0·15 to 0·25 volts   ,,   ,,
    4 differ by 0·25 to 0·35 volts   ,,   ,,
    4 differ by 0·35 to 0·45 volts   ,,   ,,
    1 differs by 0·8 volt   ,,   ,,
    1 differs by 1·5 volt   ,,   ,,

so that only two of the 81 calculated values differ from the curve values by as much as 0·5 volt, and, since these two values, viz., one belonging to a current of 2 amperes with an arc of 1mm., and the other to a current of 3 amperes with an arc of 7mm, are on the steepest parts of the curves, the wonder is, not that there are two values somewhat wrong, but that the agreement between the calculated and the curve values of the P.D. should be so marked for all the other points even in the steep parts of the curves.   For it must be remembered that a slight mistake either in the reading of the

Table XXI.—*Comparison of P.D. taken from the Curves in Fig. 16, with the P.D. calculated from Equation (3).* Carbons solid. Positive: 11mm.; negative 9mm.

Amperes.	$l=1$		$l=2$		$l=3$		$l=4$		$l=5$		$l=6$		$l=7$	
	Volts taken from curves.	Calculated volts.	Volts taken from curves.	Calculated volts.	Volts taken from curves.	Calculated volts.	Volts taken from curves.	Calculated volts.	Volts taken from curves.	Calculated volts.	Volts taken from curves.	Calculated volts.	Volts taken from curves.	Calculated volts.
2	50·6	52·05	59·4	59·4	66·75	66·74	74·1	74·09	81·4	81·43	...	...	...	...
3	48·3	48·35	53·75	53·94	59·5	59·53	65·1	65·12	70·7	70·7	76·1	76·29	81·1	81·88
4	46·5	46·5	51·2	51·21	55·9	55·92	60·6	60·63	65·3	65·34	69·6	70·04	74·4	74·76
5	45·4	45·39	49·6	49·58	53·6	53·76	57·9	57·94	62·15	62·12	65·9	66·3	70·1	70·49
6	44·65	44·65	48·5	48·49	52·25	52·31	56·15	56·15	60·0	59·98	63·5	63·8	67·6	67·64
7	44·1	44·12	47·8	47·71	51·3	51·28	54·9	54·87	58·4	58·44	61·7	62·02	65·6	65·61
8	43·7	43·72	47·15	47·12	50·5	50·51	54·0	53·91	57·3	57·3	60·5	60·68	64·1	64·08
9	43·4	43·42	46·7	46·67	49·9	49·91	53·3	53·16	56·4	56·4	59·6	59·64	62·9	62·89
10	43·1	43·17	46·3	46·3	49·4	49·43	52·7	52·56	55·7	55·69	58·8	58·81	61·9	61·94
12	42·6	42·8	45·65	45·76	48·7	48·71	51·7	51·67	54·6	54·61	57·6	57·56	60·5	60·52
14	42·25	42·54	45·25	45·37	48·2	48·19	51·0	51·02	53·85	53·85	56·7	56·67	59·5	59·5
16	...	...	44·75	45·08	47·8	47·8	50·5	50·54	53·25	53·27	56·0	56·0	58·75	58·74

current or of the length of the arc makes a large error in the
P.D. when the current is small.

There can be no doubt, then, that equation (3) *accurately*
gives the law connecting P.D. current, and apparent length of
arc for solid carbons of the size and hardness that I have used.

The general form of this equation is

$$V = a + bl + \frac{c + dl}{A}, \quad \cdots \cdots \quad (4)$$

which may be written

$$A\{V - (a + bl)\} = c + dl.$$

Now this is the equation to a *rectangular hyperbola* when l is
constant, and the asymptotes are—one the axis along which
P.D. is measured, the other a line parallel to the axis along
which current is measured. Hence the curves in Fig. 61 are
a series of rectangular hyperbolas, having one asymptote in
common, which is used as the axis of P.D., while their other
asymptotes are lines parallel to the axis of current and at a
distance from it depending upon the value of l. In fact, if d
be the distance of the asymptote of any one of the curves from
the axis of current, then

$$d = a + bl,$$

$$\text{or } d = 38 \cdot 88 + 2 \cdot 074 \, l,$$

where l stands for the number of millimetres in the length of
the arc, and the unit of length for d is the length that has
been arbitrarily taken to represent a volt in drawing the curves
in Fig. 61.

These curves have not the appearance of rectangular hyperbolas, but that arises from different units of length having
been taken to represent a volt and an ampere. In Fig. 64 I
have therefore redrawn the particular curve in Fig. 61 which
corresponds with a constant length of arc of 5mm., and, since
the same length has been taken to represent a volt and an
ampere in Fig. 64, the identity of the curve in this figure with
a rectangular hyperbola becomes evident. The asymptotes,
the axis, and the focus of this rectangular hyperbola are also
indicated.

The law embodied in equation (4) connecting the P.D.
between the carbons with the length of the arc and the current
flowing has been proved to be true for the *solid* carbons I used

Both Carbons Solid.

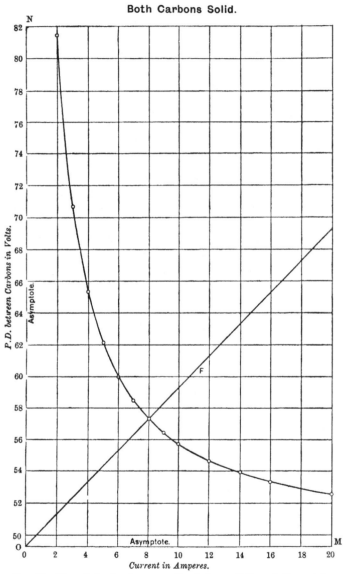

Fig. 64.--Hyperbola showing connection between P.D. and Current for constant Length of Arc of 5mm., when Volts and Amperes are drawn to the same scale. O M, O N are the two asymptotes, O F the axis, and F the focus of the hyperbola.

(it cannot be too carefully borne in mind that this law does not apply to *corcd* carbons); but before it can be accepted as a universal law, it must be shown to apply to the results obtained by *other* experimenters with *solid* carbons.

In their Paper published in the American *Electrical Engineer* for August 2, 1893, of which an abstract is given on p. 66, Messrs. Duncan, Rowland and Todd, after quoting—not quite correctly—the equations employed by various experimenters to connect the P.D. between the carbons with the current flowing and the length of the arc, said : " In fact, almost any results may be obtained by modifying the size and composition of the carbons." I shall show, however, from the results of the very experimenters whose equations they quoted that this is not the case, and that the *form* of the true equation connecting these variables is the same whatever the size of the carbons may be, and whatever their composition, as long as the carbons are solid. The values of the *constants* in the equation does, however, depend on the hardness of the carbons, and perhaps on their size.

It will be found, also, that although in certain cases my equation does not support the *conclusions* arrived at by previous experimenters, it is really more in accordance with the *results of their own experiments* than were the conclusions they themselves deduced from them.

If we take equation (2)

$$W = 38\text{·}88\,A + 11\text{·}66 + (2\text{·}074\,A + 10\text{·}54)\,l.$$

and put $\quad A = -\dfrac{10\text{·}54}{2\text{·}074}$ in it,

we have $\quad W = 38\text{·}88 \times \left(-\dfrac{10\text{·}54}{2\text{·}074}\right) + 11\text{·}66,$

that is, we have a value for A, and a corresponding one for W, both of which are independent of the value of l, showing that all the lines connecting A and W for the various constant lengths of arc must pass through one point, the co-ordinates of which are

$$A = -\frac{10\text{·}54}{2\text{·}074}$$

and $\qquad W = 38\text{·}88 \times \left(-\dfrac{10\text{·}54}{2\text{·}074}\right) + 11\text{·}66 ;$

or, $\qquad A = -5\text{·}08$

and $\qquad W = -185\text{·}92.$

Similarly, by putting $l = -\dfrac{38\cdot88}{2\cdot074}$, the coefficient of A disappears in the same equation, and we find that all the lines connecting W and l for the various constant currents must pass through a point, the co-ordinates of which are

$$l = -18\cdot7,$$
$$W = -185\cdot92.$$

Hence the laws upon which equation (4) is founded, and upon which it depends absolutely, may be put in the following form :—

1. When the current is constant, a straight-line law connects the power consumed in the arc with the length of the arc.

2. When the length of the arc is constant, a straight line law connects the power consumed in the arc with the current flowing.

3. The straight lines representing the connection between power and length of arc for constant currents all meet at a point.

4. The straight lines representing the connection between power and current for constant lengths of arc all meet at a point.

The first of these four laws follows directly from Edlund's law, that, with a constant current, the apparent resistance of the arc is equal to a constant part plus a part that varies with the length of the arc. The other three have not hitherto been enunciated, but it will be seen that the whole four laws are true for the results of all the experiments made with solid carbons of which detailed information regarding the numbers has been published.

Edlund gave only the ratios of the currents he used, so that it would not be possible to construct a complete equation, including all three variables from his results ; but the equation he discovered, $r = a + b\,l$ for constant currents, is, as stated above, really another form of the first law I used, for if we multiply this equation throughout by the constant A^2 we obtain an equation of the form

$$W = a + \beta\,l,$$

a linear equation connecting W and l for constant currents.

The resistance form of my equation (3) is

$$r = \frac{38\cdot88 + 2\cdot074\,l}{A} + \frac{11\cdot66 + 10\cdot54\,l}{A^2}, \quad . \quad . \quad (5)$$

and making A constant in this, we get an equation of the form

$$r = a + b\,l,$$

where

$$a = \frac{38\cdot88}{A} + \frac{11\cdot66}{A^2}$$

and

$$b = \frac{2\cdot074}{A} + \frac{10\cdot54}{A^2}.$$

Now Edlund concluded from the results of his experiments that a and b both diminished as the current increased, and this is borne out by the above values for a and b; but he also thought that a varied inversely as the current, which although not exactly correct, was nearly so for all but small currents, for unless A were very small in the above value for a, the term $\frac{11\cdot66}{A^2}$ would be small compared with $\frac{38\cdot88}{A}$.

Further, that the values which Edlund found for a A were not constant, as they would be if a varied inversely as A, but in every instance but one showed a decrease as the current increased, is very easily seen from the following table, which gives the results Edlund obtained from his experiments with arcs between carbon electrodes :—

Table XXII.—*Results obtained by Edlund.*

Numbers proportional to A.	Numbers proportional to a A.	Series of experiments.
1·2387	0·3239	1
1·0176	0·3416	,,
0·6661	0·3336	,,
1·1139	6·69	2
0·9435	6·877	,,
0·9618	6·34	3
0·7738	6·48	,,
1·3270	4·45	4
0·9827	5·21	,,

The experiments of one series must not be compared with those of another, for each was carried out under different circumstances. But a comparison of those in each series will show that, except for one in the first series, the product a A increases as the current A diminishes, which is exactly what would follow from my value of a A, viz. :—

$$a\,\mathrm{A} = 38 \cdot 88 + \frac{11 \cdot 66}{\mathrm{A}}.$$

Hence deviations which Edlund mistook for errors of observation were evidently caused by the presence of the term corresponding with $\dfrac{11 \cdot 66}{\mathrm{A}}$ in the value of a A, and what Edlund did not notice was that with one exception all his errors were *in the same direction*. Thus the conclusion to be drawn from my equation (4) with regard to a A, which Edlund called the back E.M.F. of the arc, is more in harmony with the results of his experiments than his own conclusion was.

It is hardly necessary to enter into Edlund's theoretical reasons for supposing the term a A to be independent of the current, since his own results really proved that it was not so.

To show the fallacy of his theory that, with a generator of constant E.M.F., the total power expended in the arc was proportional to the current, I need only point out that, if it were so, A V would be equal to k A, where k is a constant, or V would be a constant for all currents.

In the experiments used by Frölich there was no sort of order. Only in two cases were more than two currents employed with the same length of arc, and in no case was the same current used with more than two lengths of arc ; thus it is impossible to construct a complete equation from them. Fortunately, however, several currents were used with an arc of 2mm., and by calculating the power expended, from the numbers given for currents and P.Ds., it is possible to plot the curve connecting watts and amperes for length of arc 2mm., and this proves to be a straight line, as it should be according to the second law upon which my equation is founded.

The equation to this straight line obtained from the experiments used by Frölich was

$$\mathrm{W} = 40 \cdot 25\,\mathrm{A} + 68 \cdot 25,$$

from which we get the P.D. in volts,

$$\mathrm{V} = 40 \cdot 25 + \frac{68 \cdot 25}{\mathrm{A}}.$$

Table XXIII. shows the value of the P.D. for each current quoted by Frölich for the 2mm. arc, and the value calculated from the preceding equation.

Table XXIII.—*Results used by Frölich.*

Current in amperes.	P.D. from experiment.	P.D. from equation.
27·4	42·7	42·75
11·6	46·3	46·13
8·39	50·1	48·38
7·67	47·1	49·15
6·92	50·1	50·11

The general agreement of the values for V obtained by experiment and calculation is striking, considering that different pairs of carbons were used by different people in obtaining the experimental results, and this agreement renders it quite certain that V and A are connected by an equation of the form I have given.

Frölich's statement, therefore, that the P.D. between the carbons was independent of the current, and that the true equation was

$$V = m + n\,l,$$

where m and n were constant *for all currents*, is indeed astonishing, and still more so is the fact of his having given values for m and n—viz., 39 and 1·8, and having asserted that the equation

$$V = 39 + 1·8\,l$$

was true for all currents up to 100 amperes. For, to take only the numbers given in Table XXIII., which are for length of arc 2mm., his formula would give

$$V = 42·6 \text{ volts}$$

for all the currents, whereas 42·7 volts is the smallest P.D. for the largest current, and the P.D. for the smallest current is 50 volts, or $7\frac{1}{2}$ volts greater than the P.D. as calculated by Frölich. Such discrepancies as these, however, he dismissed as errors of observation, and he, like Edlund, did not notice that all or nearly all these apparent errors of observation *were in the same direction*.

Since Frölich's equation did not really express the results of the experiments upon which he founded it, it is unnecessary to dwell upon the other equations which he deduced from it, and which were equally at variance with the experimental results. Indeed, the only suggestion made by Frölich in his Paper that appears likely to prove true was that the cross-section of the arc was directly proportional to the current.

The first systematic attempt to find the P.Ds. that would send a given constant current through many lengths of arc was made by Peukert, who first, after Edlund, saw the value of eliminating one of the three variables, P.D., current and length of arc in his experiments.

Having measured the P.Ds. corresponding with various currents and lengths of arc, Peukert plotted curves connecting apparent resistance and length of arc, and found them to be straight lines. He gave the equations to these lines, and from them I have constructed the following equations connecting *power* and length of arc for each of the currents he used, by multiplying each equation through by the square of the current which corresponded with it :—

$$\text{For } A = 10 \text{ amperes, } W = 366 + 23\,l.$$
$$\text{,, } A = 15 \text{ ,, } W = 517 + 33 \cdot 75\,l.$$
$$\text{,, } A = 20 \text{ ,, } W = 720 + 32\,l.$$
$$\text{,, } A = 25 \text{ ,, } W = 812 \cdot 3 + 46 \cdot 87\,l.$$

These, when plotted, give the four straight lines seen in Fig. 65, and thus it is evident that Peukert's results follow the first of the four laws enunciated on p. 189. It is clear that the line for 20 amperes is wrong for it does not make a large enough angle with the axis of l, hence the coefficient of l, which determines the slope of the line, must be too small in the power equation for 20 amperes. That that is the case is seen from an examination of these four power equations, for while the coefficients of l with 10, 15 and 25 amperes increase as the current increases, that with 20 amperes—viz., 32—is less than 33·75, the coefficient with 15 amperes.

If from the four lines connecting power and length of arc for each current we take the number of watts corresponding with length of arc 10mm., and plot them as ordinates with their respective currents as abscissæ, we shall obtain a curve representing the connection between the power absorbed in the arc and the current flowing when the length of the arc is kept

o

constant at 10mm. Similar lines may be obtained in the same way for all the other lengths of arc down to 0mm. These lines,

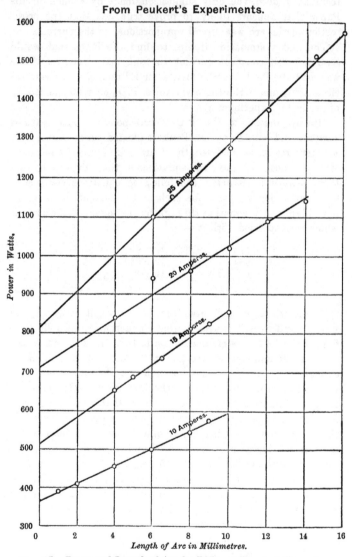

Fig. 65.—Power and Length of Arc for Different Constant Currents.

of which those for 10mm. and 0mm. are given in Fig. 66, are all *straight lines* showing that Peukert's results follow the second law given on p. 189, namely that with a constant length of arc the connection between the power absorbed in the arc and the current flowing follows a straight line law.

The equations to these lines are

$$W_{10} = 46\cdot22\ A + 129\cdot42,$$
$$W_0 = 30\ A + 66.$$

Substituting these values for W_{10} and W_0 in

$$\frac{W - W_0}{l} = \frac{W_{10} - W_0}{10},$$

which is the equation to any one of the lines in Fig. 65 connecting power with length of arc for a constant current, we find the equation

$$W = 30\ A + 66 + (1\cdot622\ A + 6\cdot342)\ l,$$

representing the connection between power, length of arc and current, when all three may vary.

Dividing throughout by A, we have—

$$V = 30 + 1\cdot622\ l + \frac{66 + 6\cdot342\ l}{A},$$

which is the general equation representing the connection between the P.D. between the carbons, the current flowing and the length of the arc with the *solid* carbons used by Peukert.

In order to test how nearly the equation given above really expresses Peukert's results, we may divide by A, and obtain the equation for the apparent resistance of the arc :—

$$r = \frac{30 + 1\cdot622\ l}{A} + \frac{66 + 6\cdot342\ l}{A^2}.$$

If we now give A the values respectively of 10, 15, 20 and 25 amperes in this resistance equation, we find the equations on the right-hand side of Table XXIV. :—

Table XXIV.—*From Peukert's Results.*

Current in amperes.	Peukert's equations.	From the above general equation.
10	$r = 3\cdot66 + 0\cdot23\,l$	$r = 3\cdot66 + 0\cdot23\,l$
15	$r = 2\cdot3 + 0\cdot15\,l$	$r = 2\cdot29 + 0\cdot14\,l$
20	$r = 1\cdot8 + 0\cdot08\,l$	$r = 1\cdot67 + 0\cdot096\,l$
25	$r = 1\cdot3 + 0\cdot075\,l$	$r = 1\cdot31 + 0\cdot075\,l$

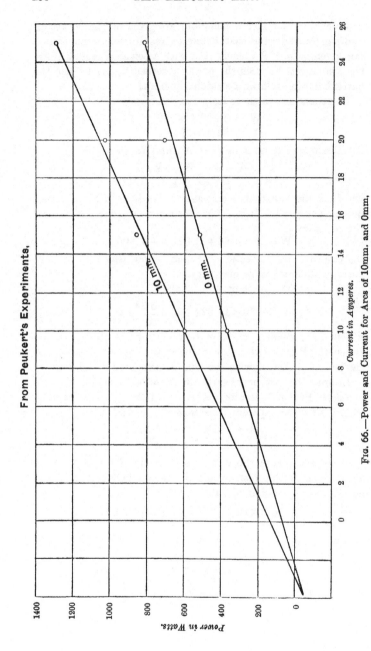

From Peukert's Experiments,

Current in Amperes.

Power in Watts.

10 mm.

0 mm.

Fig. 66.—Power and Current for Arcs of 10mm. and 0mm,

Comparing the two sets of equations, we find a striking agreement in the case of the first, second and fourth equations, while the third, which differs from that given by Peukert, expresses the result which Peukert would have obtained if he had not, as already explained, made an error when testing with 20 amperes.

Thus it is apparent that the results of Peukert's experiments lead to an equation of exactly the same form as my equation (4), differing from it only in its constants, hence his results must not only follow the two first laws upon which equation (4) was founded, but they must also follow the last two. That this is so is easily seen,

for, if we put $A = -\dfrac{6\cdot342}{1\cdot622}$ in the power equation for Peukert's results, the coefficient of l disappears, and we have

$$W = 66 - 30 \times \frac{6\cdot342}{1\cdot622}.$$

Hence $\qquad A = -3\cdot91$

and $\qquad W = -51\cdot3$

are the co-ordinates of a point where all the lines connecting power and current for the various constant lengths of arc must meet.

Similarly putting $l = -\dfrac{30}{1\cdot622}$, the coefficient of A disappears, and we get

$$l = -18\cdot5$$

and $\qquad W = -51\cdot3$

as the co-ordinates of a point at which all the lines connecting power and length of arc for the various constant currents must meet.

Thus the *results* of Peukert's experiments completely fulfil all the four laws upon which my equation (4) is founded and are, therefore, completely in harmony with my *results*. It is easy to show that where the *conclusions* he drew differ from mine, his own experiments prove him to have been mistaken.

The resistance form of the general equation deduced above from Peukert's results may be put in the form

$$r = a + b\,l,$$

where $\qquad a = \dfrac{30}{A} + \dfrac{66}{A^2}$

and $\qquad b = \dfrac{1\cdot622}{A} + \dfrac{6\cdot342}{A^2}.$

The second term in the value for a is never very big compared with the first term, with the currents Peukert used, for the numerator of the second term is only about twice as great as the numerator of the first, whereas the denominator of the second is at least ten times as great as the denominator of the first, since the smallest current Peukert used was one of 10 amperes. Hence it is not very surprising that Peukert followed Edlund in considering that the variations in the value of a caused by the presence of this term were due to errors of observation, and in asserting that a varied inversely as the current. Yet his own results show that this is not the case, for if it were, a A would be a constant, whereas multiplying the first term in each of Peukert's own equations by the corresponding current, the resulting numbers are 36·6, 34·5, 36, and 32·5, which, except in the case of the equation for 20 amperes, which I have already shown to be wrong, go in descending order as the current increases, as my equation indicates that they should. He saw, however, that b diminished more rapidly than the current, and this he was able to do because, on examining the expression I have given above for b, we see that the coefficient of the second term is four times as great as that of the first term. Hence even with fairly large currents the second term is not negligible compared with the first, and consequently Peukert was able to find an approximate law for the value of b, for which the last equation gives the full and exact law.

From the experimental observations published by Messrs. Cross and Shepard I have plotted curves connecting the apparent resistance and length of the silent arc for each current they used, and have found an equation to each of these lines which fits their results rather more closely than those which they themselves gave. From these slightly more accurate equations I constructed the following equations connecting *power* and length of arc, by multiplying each equation by the square of the corresponding current,

$$\text{For A} = 5 \cdot 04 \text{ amperes, } W = 201 \cdot 31 + 3 \cdot 34 \, l$$
$$\text{,, } A = 7 \cdot 0 \quad \text{,, } \quad W = 273 \cdot 01 + 3 \cdot 6 \, l$$
$$\text{,, } A = 7 \cdot 92 \quad \text{,, } \quad W = 310 \cdot 01 + 6 \cdot 24 \, l$$
$$\text{,, } A = 10 \cdot 04 \quad \text{,, } \quad W = 380 \cdot 01 + 8 \cdot 94 \, l$$

These, when plotted, give the straight lines in Fig. 67, showing that Cross and Shepard's results follow the first law on p. 189. As with Peukert's curves, one line—the line for 7 amperes—has the wrong slope, as may also be seen from the power equations. For evidently the coefficient of l for 7 amperes is too near in value to the coefficient for 5·04 amperes, and not near enough to that for 7·92 amperes.

From Cross & Shepard's Experiments.

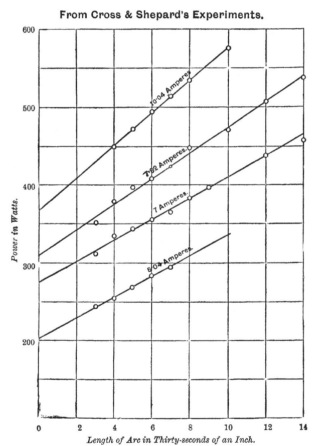

FIG. 67.— Power and Length of Arc for Different Constant Currents.

Taking the number of watts corresponding with length of arc 10 from each of the four lines in Fig. 67, and plotting them with their respective currents, we obtain the upper line in Fig. 68, and plotting the watts for length of arc 0 with their corresponding currents, we get the lower line.

That these are straight lines is very evident, although one point of each is off the line, and it is thus evident that Cross and Shepard's results, as well as Peukert's, follow the second of the four laws on p. 189.

From Cross & Shepard's Experiments.

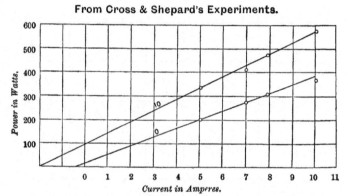

FIG. 68.—Power and Current for Arcs of 10 thirty-seconds of an inch and 0 thirty-seconds of an inch.

The equations to the two lines in Fig. 68 are

$$W_{10} = 48 \, A + 93 \cdot 6,$$
$$W_0 = 37 \, A + 14 \cdot 8.$$

Substituting these values for W_{10} and W_0 in the equation

$$\frac{W - W_0}{l} = \frac{W_{10} - W_0}{10},$$

which is the equation to any one of the lines in Fig. 67, we have

$$W = 37 \, A + 14 \cdot 8 + (1 \cdot 1 \, A + 7 \cdot 88) \, l$$

as the general equation connecting the power, current and length of arc, when any of the three may vary, with the carbons used by Cross and Shepard.

Dividing by A we have

$$V = 37 + 1 \cdot 1 \, l + \frac{14 \cdot 8 + 7 \cdot 88 \, l}{A},$$

as the general equation connecting P.D., length of arc and current for the Cross and Shepard results.

If now we divide this last equation throughout by A we get

$$r = \frac{37}{A} + \frac{14 \cdot 8}{A^2} + \left(\frac{1 \cdot 1}{A} + \frac{7 \cdot 88}{A^2} \right) l$$

for the resistance equation, and giving A the values of the currents used in the experiments, we get the equations on the right-hand side of the following table :—

Table XXV.—*From Cross and Shepard's Results.*

Current in amperes.	From experiments.	From general equation.
5·04	$r = 7 \cdot 925 + 0 \cdot 525\, l$	$r = 7 \cdot 923 + 0 \cdot 528\, l$
7·0	$r = 5 \cdot 57 + 0 \cdot 277\, l$	$r = 5 \cdot 59 + 0 \cdot 317\, l$
7·92	$r = 4 \cdot 94 + 0 \cdot 259\, l$	$r = 4 \cdot 91 + 0 \cdot 264\, l$
10·04	$r = 3 \cdot 77 + 0 \cdot 188\, l$	$r = 3 \cdot 83 + 0 \cdot 187\, l$

The two sets of equations agree very nearly as well as those obtained from Peukert's results, although the range of current was so much smaller, and the exact course of the lines was, therefore, much more difficult to obtain. This closeness of agreement shows that my general equation gives the connection between the three variables in the case of Cross and Shepard's experiments, as it does with Peukert's.

The co-ordinates of the point at which the lines in Fig. 67 meet may be found in the same way as similar co-ordinates have been found previously. They are—

$$l = -33 \cdot 64$$

and
$$W = -250 \cdot 12.$$

Similarly, the co-ordinates of the point at which the lines in Fig. 68 meet are

$$A = -7 \cdot 16$$

and
$$W = -250 \cdot 12.$$

Thus, Messrs. Cross and Shepard's results follow the four laws upon which my equation is founded.

Besides calculating the values of the constants a and b in the resistance equation $r = a + b\, l$, Messrs. Cross and Shepard gave the value of the product a A for each of the currents used. As already mentioned, Edlund, Frölich and Peukert had considered that this product was on the whole a constant, and this constancy Edlund regarded as proving the existence of a constant back E.M.F. in the arc.   Messrs. Cross and Shepard, however, showed that the product diminished as the current increased ; they did not give the actual law of variation, which, in my remarks on Peukert's equations, I have pointed out is

$$a \, \text{A} = \text{constant} + \frac{\text{constant}}{\text{A}} \, ;$$

but to them is due the credit of first noticing that as A increased a A diminished, and also of pointing out that Peukert's equations, when correctly interpreted, led to the same conclusion.

The formula given by Prof. Silvanus Thompson in 1892,

$$\text{V} = m + n \frac{l}{\text{A}},$$

where m might vary from 35 to 39 volts, and n might have values from 8 to 18, was the first real attempt to find one equation connecting the P.D., current and length of the arc when all three varied.

Multiplying his equation throughout by A, we have the equation for the power

$$\text{W} = m \, \text{A} + n \, l,$$

which gives straight lines for power and length of arc with constant currents, and straight lines for power and current with constant lengths of arc, following the two first laws on p. 189 ; but as he did not know the law of variation of the coefficients m and n, nor what their variation depended upon, his equation was necessarily incomplete.

At the meeting of the British Association at Ipswich in 1895, Prof. Thompson stated that his equation was founded on experiments made by different students at different times, with different carbons, and with different currents produced by different generators, and that the variations of m and n in it appeared to depend upon *some* of these differences, but he could not quite tell which.

The Paper by Messrs. Duncan, Rowland and Todd, from which I have already quoted, begins with a list of the equations used by other experimenters, among them

$$V = m + b \, A \, l,$$

attributed to Edlund, and

$$V = a \, A + b \, A \, l,$$

attributed to Cross and Shepard.

What Edlund, and also Cross and Shepard, really proved, however, was that for any particular value of the current the equation connecting the apparent resistance of the arc with the length was

$$r = a + b \, l,$$

where a and b were constants for the particular current. But when the current was increased they showed that a and b both diminished ; therefore, while it is perfectly correct to add together $a_1 \, A_1$ and $b_1 \, A_1 \, l_1$, in order to find V_1, the P.D. between the carbons for a length l_1 and for a current A_1, for which the values of a and b are a_1 and b_1, it is entirely wrong to attribute to Edlund, or to Cross and Shepard, as Messrs. Duncan, Rowland and Todd have done, the general equation

$$V = a \, A + b \, A \, l,$$

where a and b appear to be constant for all values of A and of l.

It seems to be desirable to draw attention to this correction, because the versions of Edlund's and of Cross and Shepard's equations given by Messrs. Duncan, Rowland and Todd have been subsequently quoted as correct by others, who have, I presume, not referred to the original publications.

Messrs. Duncan, Rowland and Todd gave the values of the P.D. and current obtained experimentally for one length of arc $\frac{1}{8}$in. long obtained with cored carbons. On calculating from these the values of the power, I find that the curve connecting power and current is very nearly a straight line, and therefore very nearly fulfils my second law. I have before mentioned that for *cored* carbons the curve connecting power and current was only *almost* a straight line.

The equation to the straight line obtained from Duncan
Rowland and Todd's experiments is

$$W = 40.6 \, A + 84.17 \, ;$$

therefore for the P.D. we have

$$V = 40.6 + \frac{84.17}{A}.$$

In Table XXVI. is given a comparison of the values of the
P.D. obtained experimentally by Messrs. Duncan, Rowland and
Todd, and the values calculated from the preceding equation.

Table XXVI.—*From Duncan, Rowland and Todd's Results.*

Current in amperes.	P.D. in volts.	
	From experiment.	From the equation.
3·1	65·0	67·8
4·6	58·5	58·9
6·15	54·8	54·3
7·7	52·5	51·5
8·0	52·0	51·1
9·82	49·2	49·2
11·26	47·5	48·1
12·75	46·5	47·2

It is, of course, impossible to tell from the table given
whether the first law connecting power and length of arc for
constant currents is borne out by Duncan, Rowland and Todd's
results, for they only gave the results for one length of arc. In
any case, however, it was amply proved by the experiments
of Prof. Ayrton's students with cored carbons, that the curve
representing the connection between power and length of
arc for constant current is *not* a straight line for cored
carbons.

Messrs. Duncan, Rowland and Todd gave as the equation
connecting V, l and A,

$$V = m + f(l) \, f(A) + n f'(l) \, f'(A),$$

but, as the form of the functions was not stated, it is, of course,
impossible to make a comparison between that equation and
the one at which I have arrived.

To sum up, then, it appears that wherever experimenters
using *solid* carbons have given the actual results of their

experiments in numbers, the law of those numbers can be expressed with remarkable accuracy by an equation of the same form as mine

$$V = a + b\,l + \frac{c + d\,l}{A}$$

and differing from it only in the values of the constants a, b, c and d. Hence, so far from the different experimenters on the arc having all obtained different laws, as has hitherto been supposed, their results present a striking unanimity when viewed by the light of the wider generalisations in which they have now been presented.

SUMMARY.

I. With solid carbons, when the current is constant, a straight line law connects the power consumed in the arc with its length, and the several lines showing this connection for given carbons, all meet at a point.

II. With solid carbons, when the length of the arc is constant, a straight line law connects the power consumed in the arc with the current, and the lines showing this connection for given carbons also meet at a point.

III. The equation representing the connection between the P.D. between the carbons, the current, and the length of the arc, with solid carbons, is

$$V = a + b\,l + \frac{c + d\,l}{A}$$

where a, b, c, and d are constants depending only upon the carbons employed.

IV. This is the equation to a series of rectangular hyperbolas having the axis of P.D. for one asymptote, and a line parallel to the axis of current at a distance from it depending on the length of the arc only, for the other.

V. The results of experiments made by Edlund, Frölich, Peukert, and Cross and Shepard all give equations of the same form as the above.

CHAPTER VII.

THE P.D. BETWEEN THE POSITIVE CARBON AND THE ARC. THE FALL OF POTENTIAL THROUGH THE ARC VAPOUR. THE P.D. BETWEEN THE ARC AND THE NEGATIVE CARBON. THE DISTURBANCE CAUSED IN THE ARC BY THE INSERTION IN IT OF A THIRD IDLE CARBON.

It has long been known that the principal fall of potential in the arc takes place between the positive carbon and the arc. Twelve years ago, Lecher found that if he placed a third carbon of 1·5mm. in an arc 2·5mm. in length, midway between the carbons, the P.D. between the positive carbon and this rod was about 35 volts, while the P.D. between the rod and the negative carbon was about 10 volts. At about the same time Uppenborn found only about 5 volts P.D. between the arc and the negative carbon. He tried rods of copper or platinum wire, embedded in clay, in steatite, and in glass tubes, for exploring, but had finally to abandon them all, in favour of bare carbon rods. With these he found that the fall of potential at the positive carbon varied between 32·5 and 38 volts in arcs of from 6mm. to 16mm.

For some experiments made for Prof. Ayrton's Chicago Paper in 1893, by Mr. Mather, Mr. Brousson and myself, a bare carbon rod was placed close to each main carbon in succession, and thus the P.Ds., between the positive carbon and the arc and between the arc and the negative carbon, were found, for several currents and lengths of arc. The results may be summed up as follows :—

(1) The curve connecting the fall of potential at the positive carbon with the current for constant lengths of arc was very much of the same shape as the similar curve for the total P.D. across the carbons.

(2) The fall of potential at the negative carbon diminished as the current increased, with a constant length of arc.

(3) It could not be determined whether the fall of potential at either of the carbons depended on the length of the arc or not.

These results were the first ever obtained showing any definite connection between the current and the falls of potential at each of the carbons, with a constant length of arc and a varying current.

With the object of obtaining still more definite information as to the laws connecting the value of the current and the length of the arc with the falls of potential at the two carbons, I have made a series of experiments on arcs varying between 1mm. and 7mm. in length, and using currents varying principally between 4 and 14 amperes, though a few experiments were made with larger currents with silent arcs, and some with hissing arcs.

The carbons used for the whole series were, as usual, Apostle carbons, 11mm. in diameter for the positive and 9mm. for the negative. With the greater part of the experiments both carbons were solid, but in order to see what was the effect of the core on the P.Ds. in question, two series of experiments were made with the positive carbon cored and negative solid, and two with both cored. For the first series in each case the current was kept constant at 10 amperes, and the length of the arc varied from 1mm to 7mm., and for the second, the length of the arc was kept constant at 5mm., and the current was varied between the limits of 4 and 24 amperes.

The exploring carbons used were mostly about 1mm. in diameter, a few were thinner, and a few were as thick as 2mm. It was found that with the 1mm. arc the most constant results were obtained with the thinnest exploring carbons, while with the longer arcs it was better to use carbons of from 1mm. to 1·5mm. in diameter; otherwise, on account of the larger amount of oxygen with which their hot parts came into contact, they burnt away so fast as to make it difficult to obtain any results.

The arrangement was as follows : The exploring carbon was fixed in a metal holder, in which it could be tilted either up or down but could not be moved sideways. The holder itself was moved by means of two racks and pinions, one of which raised and lowered it, and the other moved it horizontally in the direction of the length of the carbon. The base of the holder was firmly screwed to the table to which the lamp was fixed, in such a position that the vertical plane which bisected the exploring carbon longitudinally also bisected the arc carbons longitudinally, and was parallel to the magnifying lens. Thus the image of this carbon was thrown on to the same screen as that of the arc, in such a way that its dimensions and its distances from the arc carbons and from the arc itself were all magnified, like the arc, ten times. Hence the distance between the point of the exploring carbon and any point in the arc or on either of the other carbons could be determined by simply measuring the distance between the two corresponding points on the image, and dividing by ten. The object of tilting the carbon was to enable it to be pushed right up into the crater or to touch the extreme tip of the negative carbon.

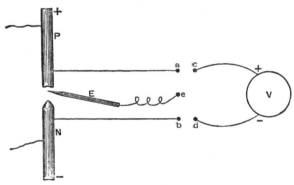

Fig. 69.—Diagrammatic Representation of Arrangement of Main Carbons Exploring Carbon and Voltmeter.

Fig. 69 shows diagrammatically the arrangement of the three carbons and the voltmeter. V was the voltmeter, a high resistance d'Arsonval galvanometer with 1,500,000 ohms in circuit. *a, b, c, d* and *e* were mercury cups, E was the exploring carbon and P and N the positive and negative carbons of the arc respectively. *e* was permanently connected with E, *a* and *b* with the positive and negative carbons, and *c* and *d* with the positive and negative poles of the voltmeter, respectively.

Now when *a* and *c* were connected by means of one wire bridge and *e* and *d* by means of another, the voltmeter measured the P.D. between P and E, that is, when E was very close to P, the fall of potential at the positive carbon. When, on the other hand, *b* and *d* were connected, and also *c* and *e*, the voltmeter then measured the P.D. between N and E, or, when E was very close to N, it measured the fall of potential at the negative carbon.

The following was the method of experimenting :

When the arc was normal with a given current and length (*see* definition of a normal arc, p. 104), the idle carbon, properly tilted, was brought into it with its point in the centre, midway between the two carbons. In a few seconds this third carbon grew beautifully pointed and its tip became white hot. It was then raised or lowered till the tip touched the crater of the positive, or the white hot spot on the negative— the "*white spot,*" as I shall call it—according to the P.D. it was desired to measure. The slow motion of the voltmeter needle, as the idle carbon was moved through the arc, was very different from the rush it made when the two carbons touched ; hence, the last reading before they touched was very easy to observe, after a little experience, and this was obviously the reading that gave most nearly the true fall of potential at the main carbon. The employment of a bare carbon for such measurements has its disadvantages, but these are much less when used as described above than when the rod is merely held in a stationary position in the arc, as has always hitherto been done. Indeed, the results of two complete series of experiments, conducted on the old stationary method, had to be abandoned, because they were so vague that no conclusions could possibly be drawn from them. But by making the rod actually *touch* the hottest part of the main carbon, the fall of potential at a perfectly definite point was always measured, and this led to very accurate and constant readings.

One great difficulty in this mode of measurement is caused by the third carbon *repelling* the arc when it approaches it. This repulsion is greater the longer the arc and the smaller the current, so that with long arcs and small currents it is difficult to get the idle carbon into the arc at all. The arc slips away from the carbon exactly as if the one were an air ball suspended by its two ends and the other were trying to penetrate it. Under certain circumstances, not yet very well defined, the carbon appears to *attract* the arc when once it has been made to dip well into it. This subject of the repulsion and attraction of the arc by the third carbon would well repay further investigation.

The third carbon disturbs the arc in other ways as well as by repelling it. If left in long enough near either carbon, it *grows* by receiving carbon from it, especially when left close to the positive pole. It also cools the arc at first, and it always disturbs the distribution of potential in it. The first difficulty is overcome by allowing the third carbon to remain for only a short time close to either carbon; the second by keeping it for a little while in the centre of the arc before bringing it up to the main carbon to take a reading. The third has to be allowed for in our interpretation of the results of the experiments.

It has been mentioned that the third carbon always becomes beautifully pointed when placed in the arc. This is the case when its diameter is small compared with those of the main carbons, but I found that when a flat rod was inserted so as to screen the hot parts of the two carbons from one another, *two arcs* formed—one between the positive carbon and the rod, and one between the rod and the negative carbon; and in that case the rod became cratered on the side opposite to the negative carbon and a small rough excrescence formed on the side opposite to the positive carbon. It was not possible to detect the presence of the two arcs from the image, which showed, apparently, only one big arc, but the voltmeter indicated it at once, for, as soon as the second arc was formed, the voltmeter deflection became nearly doubled.

In order to eliminate any error that deficiences in the hardness and texture of the carbons might introduce, even though they were all of the same make, none were employed except those of which the total P.D. across the carbons with the normal arc used was within a half a volt of that given by equation (3) (p. 184), for the same length of arc and current.

Lecher, using a bare carbon rod as I have done, but keeping it stationary in the arc instead of making it touch the main carbon, found the positive carbon P.D. to be about 35 volts. Uppenborn, by the same means, found that it varied between 32·5 and 38 volts for arcs of from 6mm. to 16mm., while Luggin obtained $33 \cdot 7 \pm 0 \cdot 46$ volts for the value with a current of $15 \cdot 5$ amperes, and could observe no change when he changed the length of the arc. All these results agree very well with one another, as far as they go; but, in order to find out what effect changes in the current and the length of the arc had on the P.Ds., it was necessary to make a complete series of experiments, using many different currents and lengths of arc, and finding the value of the P.D. between each carbon and the arc separately with each current and length of arc. Such a series of experiments I have made, and, as a check upon the results, I also observed the P.D. between each carbon and the arc *plus* the P.D., through the arc vapour itself, so that, subtracting these P.Ds. from the total P.D. across the main carbons, I might get indirect measurements of the P.D. between each carbon and the arc, as well as the direct measurements.

Thus four sets of cases were observed :

(1) The P.D. between the positive carbon and the third carbon, just before the point of the latter touched the crater.

(2) The P.D. between the third carbon and the negative carbon, just before the point of the former touched the white spot.

(3) The P.D. between the positive carbon and the third carbon, just before the point of the latter touched the white spot.

(4) The P.D. between the third carbon and the negative carbon, just before the point of the former touched the crater.

The first and second P.Ds. we may call, for convenience sake, the *positive and negative carbon P.Ds.*, respectively; the third and fourth are those same P.Ds. with the P.D. through the arc vapour itself added. This last P.D. we may call the *vapour P.D.*

The whole four cases are represented diagrammatically in Fig. 70—(1) shows how the positive carbon P.D. was measured, (2) the negative carbon P.D., (3) the positive carbon P.D. *plus* the vapour P.D., (4) the vapour P.D. *plus* the negative carbon P.D. In the first two cases the P.D. rushed down to zero when the carbons touched one another, in the last two it rushed up till it reached the value of the total P.D. across the main carbons. In each case the motion of the voltmeter needle, when the exploring carbon touched the main carbon, was very much quicker than it had been before they touched, so that the last deflection, before they touched could be read with a fair amount of ease. Each experiment was repeated at least six times, and if the readings of any particular P.D. differed much from one another, twelve or even more measurements of it were made. Tables XXVII., XXVIII., XXIX. and XXX. contain the means of the results of the four series of experiments with solid carbons and silent arcs.

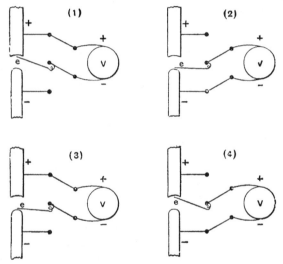

Fig. 70.—Diagrammatic Representation of Arrangement of Main Carbons and Exploring Carbon. (1) Positive Carbon P.D., (2) Negative Carbon P.D., (3) Positive Carbon P.D. + Vapour P.D., (4) Vapour P.D. + Negative Carbon P.D.

Table XXVII. gives the positive carbon P.D., as nearly as it can be measured, by means of a bare carbon placed in the arc. All such measurements are subject to the possibility that the idle carbon may not take up the potential of the part of the arc in which it is placed, that is, there may be a contact P.D. between this solid carbon and the vaporous stuff of which the arc consists. But, even if it exists, this P.D. is probably very small compared with those under consideration, and need not, therefore, be taken into account.

Table XXVII.—*P.D. between Positive Carbon and Third Carbon with Point of Latter close to Crater of Former.*

All carbons solid. Positive, 11mm.; negative, 9mm.; third carbon, 0·5mm. to 2mm.

Current in Amperes.	P.D. in Volts.						
	$l=1$.	$l=2$.	$l=3$.	$l=4$.	$l=5$.	$l=6$.	$l=7$.
4	34·3	37 0	38·6	38·5	36·7	39·7	39·0
5	33·6	33 0	34·2	34·5	35 5	36·9	37·7
6	34·95	35·2	31·0	33·9	35·55	36·6	36·1
7	33·25	33·8	33·3	33·1	35·0	36·1	36·9
8	32·7	32·75	32·6	33·3	34·9	33·8	35·2
9	33·3	33·9	34·85	35·1	35·4	36·1	34·5
10	32·6	32·4	31·5	33·75	33·5	32·7	34·7
12	31·9	32·8	32·2	32·7	33·8	33·4	34·4
14	...	31·4	31·8	32·5	31·8	34·8	36·2
16	...	...	...	32·5	32·7	33·1	33·0
20	..	...	...	...	...	34·2	33·6

Table XXVII. shows that the positive carbon P.D. is not a constant, but that, like the total P.D. between the main carbons of the arc, it diminishes as the current increases, and increases as the length of the arc increases. But the *degree* of diminution and of increase is very different in the two cases, for while the total P.D. between the carbons ranges from 42·3 volts for a 1mm. 12 ampere arc to 74·4 volts for a 7mm. 4 ampere arc (Fig. 38, p. 120), the positive carbon P.D. given in Table XXVII. varies only from 31·9 volts to 39 volts for the same range of current and length of arc. That is, the total P.D. between the carbons has a range of about 32 volts, while the positive carbon P.D. has a range of only about 7 volts for the same variation of length and current. Hence, at any rate, when measured with a bare carbon rod (and no better means of measurement has yet been devised), *the fall of potential at the positive carbon is not a constant, but varies both with the length of the arc and the current, in the same way, but to a far less extent than the total P.D. between the carbons.*

This will be seen even more clearly from the curves in Figs. 71 and 72, which were plotted from the numbers in Table XXVII. In Fig. 71 the curves show the connection between the positive carbon P.D. and the current, for constant lengths of arc, and in Fig. 72 the connection between the same P.D. and the length of the arc, for various constant currents. Had these lines all been drawn from the same zero of P.D., they would have been so close together that it would have been difficult to distinguish one line from another. The zero of P.D. has, therefore, been raised 5 volts for each line, as the numbers on either side of the figures show. Thus the lines are all drawn to the *same scale*, but each has a *different* point for its *zero* of P.D.

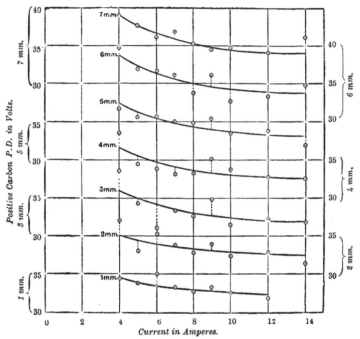

Fig. 71.—Positive Carbon P.D. and Current for Various Constant Lengths of Arc.

Solid Carbons: Positive, 11mm.; negative, 9mm.; third carbon, 0·5mm. to 2mm.

The curves in Fig. 71 are evidently not unlike those in Fig. 38 (p. 120), for the total P.D. across the main carbons with constant lengths of arc, but they are flatter. That is, as was seen from the Table, for a given range of current the range of the P.D. across the main carbons is much greater than the range of the P.D. between the positive carbon and the arc. Similarly, comparing the continuous lines in Fig. 72 with those in Fig. 44 (p. 136), we find that both are straight lines converging towards one another on the left hand side, but

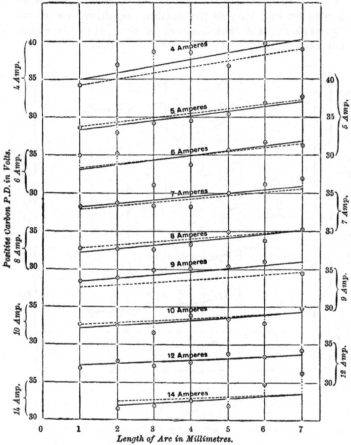

FIG. 72.—Positive Carbon P.D. and Length of Arc for Various Constant Currents.

Solid carbons : Positive, 11mm. ; negative, 9mm. ; third carbon, 0 5mm. to 2mm.

that the lines in Fig. 44 are much the steeper, showing that the range of P.D. for any given range of length of arc is considerably greater, with the same constant current, for the total P.D. across the carbons than for the positive carbon P.D. Had the lines in Fig. 72 all been drawn to the same zero, it would be found that they, as well as those in Fig. 44, all met at a point to the *left* of the axis of P.D., showing that there is no real length of arc for which the positive carbon P.D. is the same for all currents. The meaning of the dotted lines in Fig. 72 is explained later, p. 223.

Turning now to the negative carbon P.D., this has been stated by some observers to change its direction when the arc hissed, *i.e.*, instead of the arc near the negative carbon being at a higher potential than the negative carbon itself, these observers thought that with hissing arcs the negative carbon was at a higher potential than the parts of the arc nearest to it. I have never found this to be the case, either with silent or hissing arcs ; whether cored or solid carbons were used. The fall of potential at the positive pole is *always* from carbon to arc, and at the negative pole from arc to carbon, so that the potential falls continuously from the positive carbon through the arc to the negative carbon. Other observers have denied the existence of a fall of potential between the arc and the negative carbon, but these experiments conclusively prove that they are wrong, and shows that with the carbons I used this fall of potential *cannot be less than 7·6 volts.*

Table XXVIII.—*P.D. between Third Carbon and Negative Carbon with Point of Former close to White Hot Spot on Latter. All carbons solid. Positive, 11mm. ; negative, 9mm. ; third carbon, 0·5mm. to 2mm.*

Current in Amperes.	P.D. in Volts.						
	$l=1.$	$l=2.$	$l=3.$	$l=4.$	$l=5.$	$l=6.$	$l=7.$
4	12·9	11·5	12·9	12·2	10·4	10·6	10·3
5	10·1	10·1	11·2	10·4	10·8	10·7	9·9
6	8·6	10·3	9·6	9·4	9·7	9·2	9·3
7	8·8	9·8	9·6	9·4	9·4	9·5	9·6
8	8·2	8·8	10·9	9·3	10·0	9·3	8·9
9	7·8	8·8	8·9	9·1	9·6	9·0	9·0
10	8·5	8·5	9·0	8·9	9·2	9·2	9·2
12	8·5	8·4	9·2	9·3	9·6	8·8	8·7
14	...	9·7	9·2	9·3	9·2	9·0	8·9
16	...	...	...	9·3	9·2	9·1	9·1
20	...	...	...	...	...	9·6	9·2

Table XXVIII. gives the values of the negative carbon P.D.
for the same currents and lengths of arc as those for which
the P.Ds. between the positive carbon and the arc were
given in Table XXVII.

It is quite plain, from these numbers, that the negative
carbon P.D. diminishes as the current increases, following, in
this respect, the same law as the total P.D. across the carbons,

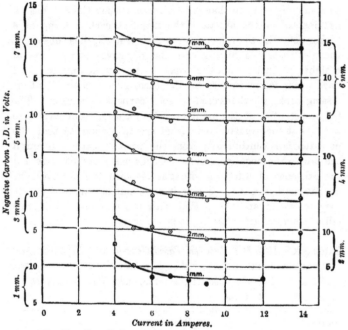

FIG. 73.—Negative Carbon P.D. and Current for Various Constant
Lengths of Arc.

Solid Carbons : Positive, 11mm. ; negative, 9mm, ; third carbon,
0·5mm. to 2mm.

and the positive carbon P.D.   But how it is affected by a
change in the length of the arc it is impossible to see without
reference to the curves plotted from the numbers.   These are
given in Figs. 73 and 74, in which, like those in Figs. 71
and 72, each curve is drawn from a separate zero of P.D. in
order to prevent overcrowding.

In Fig. 73 the curves connect negative carbon P.D. with current for constant lengths of arc. They are like the similar curves in Figs. 71 and 38, but are much flatter than either, showing that, for a given length of arc, the negative carbon P.D. varies considerably less than the total P.D. across the carbons, or even than the positive carbon P.D., for the same variation of current.

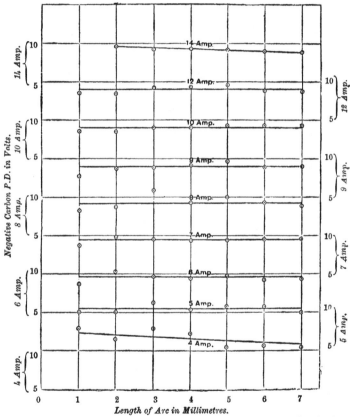

FIG. 74.—Negative Carbon P.D. and Length of Arc for Various Constant Currents.

Solid Carbons: Positive, 11mm.; negative, 9mm.; third carbon, 0·5mm. to 2mm.

The lines in Fig. 74, which connect the negative carbon P.D. with the length of the arc for constant currents, show that that P.D. is not affected at all by a change in the length of the arc, for, except the line for 4 amperes, they are all practically horizontal straight lines. Thus, when measured by means of a bare carbon rod, *the P.D. between the arc and the negative carbon varies with the current, but not with the length of the arc.*

With the help of Figs. 75, 76, and 77, it will be easy to construct equations similar to equation (3) (p. 184), for connecting each of the carbon P.Ds. with the current and the length of the arc. The lines in these figures are all power lines, and the power is measured in each case by multiplying

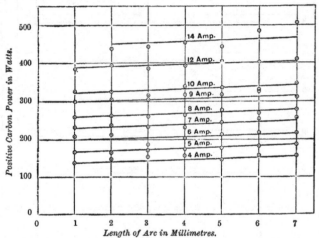

FIG. 75.—Positive Carbon Power and Length of Arc for Various Constant Currents.

Solid Carbons : Positive, 11mm. ; negative, 9mm. ; third carbon, 0·5mm. to 2mm.

the carbon P.D. by the current. For convenience sake we shall call the positive carbon P.D., multiplied by the current, the *positive carbon power*, and the negative carbon P.D., multiplied by the current, the *negative carbon power*.

Fig. 75 shows the connection between the positive carbon power and the length of the arc for constant currents. The lines are all straight, as are the similar ones for the *total* power consumed in the arc (Fig. 62, p. 180), but whereas the *total* power lines converge towards one another, the positive carbon power lines are all parallel.

Similarly, the lines in Fig. 76, which show the connection between the positive carbon power and the current for constant lengths of arc, are also parallel straight lines, while

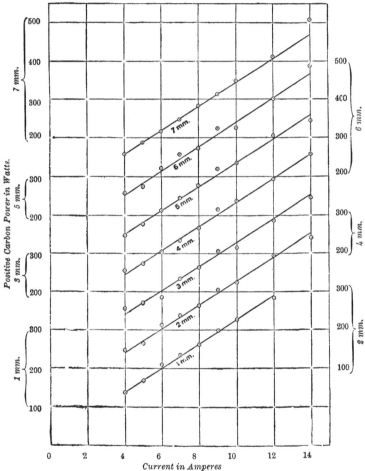

FIG. 76.—Positive Carbon Power and Current for Various Constant Lengths of Arc.

Solid Carbons: Positive, 11mm.; negative, 9mm.; third carbon, 0·5mm. to 2mm.

the similar lines for the *total* power consumed in the arc, although straight, are not parallel, but all meet at a point (Fig. 63, p. 182).

To find the equation connecting the positive carbon P.D. with the current and the length of the arc, we have the equation to any one of the lines in Fig. 75

$$\frac{W - W_1}{l - 1} = \frac{W_7 - W_1}{7 - 1},$$

where W is the positive carbon power in watts for length of arc l mm.,

W_1 is the positive carbon power in watts for length of arc 1mm.,

W_7 is the positive carbon power in watts for length of arc 7mm.

From the lines in Fig. 76, which show the connection between the positive carbon power and the current for constant lengths of arc, we can find W_1 and W_7,

$$\frac{W_1 - 137 \cdot 2}{A - 4} = \frac{387 \cdot 4 - 137 \cdot 2}{12 - 4},$$

or $$W_1 = 31 \cdot 28A + 12 \cdot 1.$$

In the same way

$$W_7 = 31 \cdot 28A + 30 \cdot 7.$$

Hence, in the equation connecting W and l, we have

$$\frac{W - 31 \cdot 28A - 12 \cdot 1}{l - 1} = \frac{31 \cdot 28A + 30 \cdot 7 - (31 \cdot 28A + 12 \cdot 1)}{6}$$
$$= \frac{18 \cdot 6}{6}$$
$$= 3 \cdot 1$$
$$\therefore \ W = 31 \cdot 28A + 12 \cdot 1 + 3 \cdot 1l - 3 \cdot 1,$$

or $$W = 31 \cdot 28A + 9 + 3 \cdot 1l.$$

Dividing all through by A, we have

$$V = 31 \cdot 28 + \frac{9 + 3 \cdot 1l}{A}. \quad . \quad . \quad . \quad (6)$$

for the equation connecting the positive carbon P.D. with the current and the length of the arc.

This equation, like the similar one for the total P.D. across the carbons (p. 184), is the equation to a series of rectangular hyperbolas, but, whereas with the total P.D. the hyperbolas have only *one* asymptote in common, with the positive carbon P.D. they have *both* asymptotes in common ; one, the axis of P.D., and the other, a line parallel to the axis of current, at a distance from it equal to 31·28 times the distance taken to represent one volt. Table XXIX. gives the positive carbon P.Ds. calculated from equation (6) for all the currents and lengths of arc for which values were obtained by experiment.

Table XXIX.—*Positive Carbon P.Ds. calculated from Equation* (6).

Solid carbons.   Positive, 11mm. ;   negative, 9mm. ;   third carbon, 0·5mm. to 2mm.

Current in Amperes.	P.D. in Volts.						
	$l=1.$	$l=2.$	$l=3.$	$l=4.$	$l=5.$	$l=6.$	$l=7.$
4	34·3	35·1	35·9	36·6	37·4	38·2	39·0
5	33·7	34·3	34·9	35·6	36·2	36·8	37·4
6	33·3	33·8	34·3	34·85	35·4	35·9	36·4
7	33·0	33·45	33·9	34·3	34·8	35·2	35·7
8	32·8	33·2	33·6	34·0	34·3	34·7	35·1
9	32·6	33·0	33·3	33·7	34·0	34·35	34·7
10	32·5	32·8	33·1	33·4	33·7	34·0	34·35
12	32·3	32·55	32·8	33·1	33·3	33·6	33·8
14	...	32·4	32·6	32·8	33·0	33·25	33·5
15	...	...	...	32·6	32·8	33·0	33·2
20	...	...	...	...	...	32·7	32·8

A comparison of the two Tables XXVII. and XXIX. shows that of the 68 calculated P.Ds., 47 differ from the observed P.Ds. by one volt or less, 18 more by less than two volts, and only three differ by more than two volts. Considering the many possibilities of error in the observation of these P.Ds., the equation must be allowed to fit the experimental results with great accuracy. To show exactly what is the degree of accuracy, the dotted lines in Fig. 72 have been drawn. These lines are the ones obtained from equation (6), while the complete lines are drawn as nearly as possible through the average positions of the observed points.

To obtain the equation connecting the negative carbon P.D. with the current and the length of the arc, the equation to the

line in Fig. 77 only is required. The points for this line were
found by taking the *average* negative carbon P.D. for each
current from Table XXVIII., and multiplying it by the current.

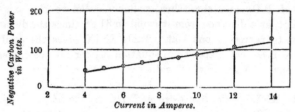

FIG. 77.—Negative Carbon Power and Current for All Lengths of Arc.
Solid Carbons: Positive 11mm.; negative 9mm.; third carbon,
0·5mm. to 2mm.

This was done because it was found from the curves (p. 219) that
with a given current the negative carbon P.D. was the same
for all lengths of arc.

The equation to the line in Fig. 77 is

$$\frac{W - 44}{A - 4} = \frac{120 - 44}{10}$$
$$= 7 \cdot 6.$$

Hence $W = 7 \cdot 6A + 13 \cdot 6.$

And, dividing all through by A, we have

$$V = 7 \cdot 6 + \frac{13 \cdot 6}{A} \quad . \quad . \quad . \quad . \quad (7)$$

for the equation connecting the negative carbon P.D. with the
current and the length of the arc. This is the equation to a
single rectangular hyperbola of which the asymptotes are the
axis of P.D., and a line parallel to the axis of current, at a
distance from it 7·6 times the distance which represents one volt.

Table XXX. gives the negative carbon P.Ds. calculated from
equation (7) for all the currents for which such P.Ds. were
obtained by experiment.

Comparing these values with those in Table XXVIII., it will
be found that, of the 68 observed values, 42 differ from the
calculated values by half a volt and under, 16 more by less
than one volt, and 9 by between one and two volts. Thus
equation (7) may fairly be taken to represent the connection
between the negative carbon P.D., the current, and the length
of the arc.

Table XXX.—*Negative Carbon P.Ds. calculated from Equation* (7).

Solid carbons. Positive, 11mm.; negative, 9mm.; third carbon, 0·5mm. to 2mm.

Current in Amperes.	P.D. in Volts.	Current in Amperes.	P.D. in Volts.
4	11·0	10	9·0
5	10·3	12	8·7
6	9·9	14	8·6
7	9·5	16	8·45
8	9·3	20	8·3
9	9·1		

From equations (6) and (7) we can find the equation for the positive carbon P.D. plus the negative carbon P.D.; *i.e.*, the whole fall of potential from carbon to carbon minus the fall of potential through the arc itself. It is

$$V = 38\cdot88 + \frac{22\cdot6 + 3\cdot1l}{A}. \quad \ldots \ldots \quad (8)$$

Now equation (3) for the *total* P.D. between the main carbons, which includes, of course, the drop of P.D. in the arc itself, is

$$V = 38\cdot88 + 2\cdot07l + \frac{11\cdot66 + 10\cdot54l}{A}. \quad \ldots \quad (3)$$

The coincidence between the first terms of equations (8) and (3) shows that this constant quantity belongs, *not* to the positive carbon alone, as has hitherto been supposed, but to *both* the positive and negative carbons in the proportion of about four-fifths to the former and one-fifth to the latter.

Hence, if, as many investigators imagine, this constant term involves the existence of a constant back E.M.F. in the arc, this back E.M.F. must be considered to reside, not at the positive carbon alone, as has hitherto been taken for granted, but at *both* carbons; and this fact will necessitate a considerable modification in any theory of the arc yet enunciated that involves the existence of a constant back E.M.F.

Table XXXI. gives the sum of the *observed* values of the two carbon P.Ds., taken from Tables XXVII. and XXVIII. Table XXXII. shows the corresponding values *calculated* from

Q

equation (8). From a comparison of these two tables it will be seen that, of the 62 calculated values, 39 differ from the observed values by one volt and under, 18 by two volts and under, and 5 by more than two volts. Although the calculated values for the sum of the two carbon P.Ds. appear to agree less perfectly with the observed values than those for each carbon separately, yet it will be found that the curves in Fig. 78, which have been taken from equation (8), go almost exactly

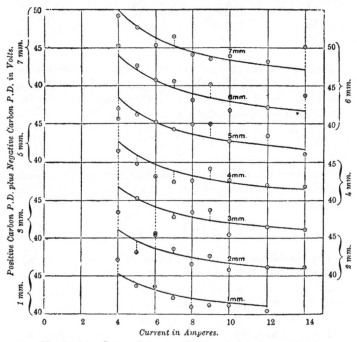

FIG. 78.—Positive Carbon P.D. plus Negative Carbon P.D., and Current.
for Various Constant Lengths of Arc.

Solid Carbons: Positive, 11mm. ; negative, 9mm. ; third carbon,
0·5mm. to 2mm.

through the average position of the *observed* points, which are the ones given in the figure ; showing, of course, that equation (8) does really represent very accurately the connection between the sum of the two carbon P.Ds., the current, and the length of the arc.

Table XXXI.—*Sum of the observed Values of the Two Carbon P.Ds. taken from Tables XXVII. and XXVIII.*

Solid carbons. Positive, 11mm.; negative, 9mm.; third carbon, 0·5mm. to 2mm.

Current in Amperes.	P.D. in Volts.						
	$l=1.$	$l=2.$	$l=3.$	$l=4.$	$l=5.$	$l=6.$	$l=7.$
4	47·2	48·5	51·5	50·7	47·1	50·3	49·3
5	43·7	43·1	45·4	44·9	46·3	47·6	47·6
6	43·55	45·5	40·6	43·3	45·25	45·8	45·4
7	42·05	43·6	42·9	42·5	44·4	45·6	46·5
8	40·9	41·55	43·5	42·6	44·9	43·1	44·1
9	41·1	42·7	43·75	44·2	45·0	45·1	43·5
10	41·1	40·9	40·5	42·65	42·7	41·9	43·9
12	40·4	41·2	41·4	42·0	43·4	42·2	43·1
14	...	41·1	41·0	41·8	41·0	43·8	45·1
16	...	...	...	41·8	41·9	42·2	42·1
20	...	...	...	...	..	43·8	42·8

Table XXXII.—*Values of the Sum of the Two Carbon P.Ds. calculated from Equation (8).*

Solid carbons. Positive, 11mm.; negative, 9mm.; third carbon, 0·5mm. to 2mm.

Current in Amperes.	P.D. in Volts.						
	$l=1.$	$l=2.$	$l=3.$	$l=4.$	$l=5.$	$l=6.$	$l=7.$
4	45·3	46·1	46·85	47·6	48·4	49·2	49·95
5	44·0	44·6	45·3	45·9	46·5	47·1	47·7
6	43·2	43·7	44·2	44·7	45·2	45·75	46·3
7	42·55	43·0	43·4	43·9	44·3	44·8	45·2
8	42·1	42·5	42·9	43·3	43·6	44·0	44·4
9	41·7	42·1	42·4	42·8	43·1	43·5	43·8
10	41·45	41·8	42·1	42·4	42·7	43·0	43·3
12	41·0	41·3	41·5	41·8	42·05	42·3	42·6
14	...	40·9	41·15	41·4	41·6	41·8	42·0
16	...	...	...	41·1	41·3	41·5	41·7
20	...	...	...	...	...	40·9	41·1

So far it has been taken for granted that the P.D. measured by the voltmeter, between the main carbon and the bare idle carbon dipping into the arc, was the actual fall of potential between the main carbon and the arc. That this is not absolutely the case is quite certain, for the bare carbon must bring all the parts of the arc that it touches to practically the same potential. And this potential will be greater

Q 2

than the least and less than the greatest of the potentials
that existed in the arc before the insertion of the exploring
carbon.

Thus it is quite possible that the *variable* part of the fall of
potential at each of the carbons, given by equations (7) and (8),
may be mainly produced by the exploring carbon being in
communication with the arc, not only at its tip, but also along
a part of its length. For, whether the fall of potential at the
carbon itself be a constant or not, it is quite certain that the
potentials of the other points at which the arc touches the
exploring carbon vary both with the current that was flowing
and with the length of the arc, before that carbon was inserted.
Hence, at least a portion of the variation in the values given by
equations (6) and (7) must have been created by the use of a
bare carbon in the arc.

It is possible to show, however, at least indirectly, that the
part of the variation of P.D. that depends upon the *current
alone*, while it is *increased* by the use of a bare carbon rod dipping into the arc, is still a genuine variation of the carbon P.Ds.
In other words, the term $\dfrac{9}{A}$ in equation (7) and $\dfrac{13\cdot6}{A}$ in equation
(7) show genuine changes with change of current in the posi-
tive carbon P.D. and the negative carbon P.D. respectively.

In order to prove this, it will be necessary to find the
equation connecting the current and the length of the arc with
the total P.D. across the carbons *when the third carbon is in
the arc.*

It has been mentioned that the insertion in the arc of a
third carbon caused the P.D. between the main carbons to be
increased by from 0·5 to 3 volts. This increase is usually
rather greater with the third carbon near the positive pole than
near the negative. Tables XXXIII. and XXXIV. give the
actual values of this P.D., the first with the exploring carbon
near the positive, and the second with it near the negative
pole.

Equation (8), which gives the *sum* of the *two* carbon P.Ds.,
was formed from equations (6) and (7), for one of which the
third carbon was near the positive pole, and for the other near
the negative. In order, therefore, to be able to compare
equation (8) with the new equation for the total P.D. across

Table XXXIII.—*P.D. between Main Carbons with point of Third Carbon in Arc close to Crater of Positive Carbon.*
Solid carbons. Positive, 11mm.; negative, 9mm.; third carbon, 0·5mm. to 2mm.

Current in Amperes.	P.D. in Volts.						
	$l=1.$	$l=2.$	$l=3.$	$l=4.$	$l=5.$	$l=6.$	$l=7.$
4	48·3	51·5	57·1	62·6	66·4	70·3	...
5	45·8	51·3	55·5	60·7	63·8	67·3	70·7
6	45·5	50·6	55·9	58·5	61·5	64·5	69·2
7	45·0	50·5	53·5	57·4	60·7	63·9	67·4
8	44·0	49·2	52·4	55·7	59·6	61·6	64·9
9	43·4	47·9	52·0	54·4	57·4	60·6	64·2
10	43·1	47·5	51·1	54·4	58·8	60·7	63·2
12	42·6	46·7	50·5	53·0	57·5	58·9	63·5
14	...	46·0	50·6	51·8	55·3	58·7	61·9
16	...	...	...	51·3	55·1	57·2	60·6
20	...	...	...	...	...	57·2	59·2

Table XXXIV.—*P.D. between Main Carbons with point of Third Carbon in Arc close to White Hot Spot on Negative Carbon.*
Solid carbons. Positive, 11mm.; negative, 9mm.; third carbon, 0·5mm. to 2mm.

Current in Amperes.	P.D. in Volts.						
	$l=1.$	$l=2.$	$l=3.$	$l=4.$	$l=5.$	$l=6.$	$l=7.$
4	49·5	52·2	57·2	63·9	68·3	70·5	...
5	46·0	51·5	55·3	60·5	64·5	68·3	70·5
6	45·5	50·7	55·0	58·8	61·8	65·5	70·1
7	45·0	50·5	53·9	57·3	61·2	65·5	69·0
8	44·0	48·7	52·2	55·7	60·0	61·6	64·9
9	43·4	49·2	51·3	54·5	57·6	61·2	65·2
10	43·5	47·3	51·2	54·7	58·8	60·7	63·8
12	42·6	46·7	50·7	53·5	58·1	59·3	63·1
14	...	46·0	51·5	53·5	56·6	59·1	60·6
16	...	...	...	51·1	55·4	57·3	60·0
20	...	...	...	...	...	58·0	59·3

the carbons with the third carbon in the arc, it will be necessary that this new equation shall connect the current and the length of the arc with the *mean* of the total P.Ds. observed when the third carbon is placed near the positive and negative carbons respectively. That is, the *mean* of each P.D. in Table XXXIII.

and the corresponding P.D. in Table XXXIV. must be taken to obtain the new equation. These mean P.Ds. with their respective currents and lengths of arc are given in Table XXXV.

Table XXXV.—*Mean Total P.D. between Main Carbons with Third Carbon in Arc near Positive and Negative Carbons respectively.*

Solid carbons. Positive, 11mm. ; negative, 9mm. ; third carbon, 0·5mm. to 2mm.

Current in Amperes.	P.D. in Volts.						
	$l=1.$	$l=2.$	$l=3.$	$l=4.$	$l=5.$	$l=6.$	$l=7.$
4	48·9	51·85	57·15	63·25	67·35	70·4	...
5	45·9	51·4	55·4	60·6	64·15	67·8	70·6
6	45·5	50·65	55·45	58·65	61·65	65·0	69·65
7	45·0	50·5	53·7	57·35	60·95	64·7	68·2
8	44·0	48·95	52·3	55·7	59·8	61·6	64·9
9	43·4	48·55	51·65	54·45	57·5	60·9	64·7
10	43·3	47·4	51·15	54·55	58·8	60·7	63·5
12	42·6	46·7	50·6	53·25	57·8	59·1	63·3
14	...	46·0	51·05	52·65	55·95	58·9	61·25
16	...	...	...	51·2	55·25	57·25	60·3
20	...	...	...	...	...	57·6	59·25

By multiplying each of the P.Ds. in Table XXXV. by the corresponding current, we obtain the total power in watts expended between the carbons when the third carbon is in the arc. The laws connecting this power with the current for constant length of arc and with the length of the arc for constant current are both straight line laws, just as they are when there is no third carbon in the arc (p. 189). By combining these two laws in the same way as when there was no third carbon we get

$$V = 38\cdot56 + 2\cdot5l + \frac{23+8l}{A}. \quad . \quad . \quad . \quad (9)$$

as the equation representing the connection between P.D., current, and length of arc, with a third carbon in the arc. Of the 67 P.Ds. calculated from this equation, 59 differ from the observed values by one volt and under, and the remaining 8 by less than 1·7 volts. These calculated P.Ds. are given in Table XXXVI.

Table XXXVI.—*Mean P.Ds. between the Main Carbons with Third Carbon in the Arc, calculated from Equation* (9). *Solid carbons. Positive, 11mm.; negative, 9mm.; third carbon, 0·5mm. to 2mm.*

Current in Amperes.	P.D. in Volts.						
	$l=1$.	$l=2$.	$l=3$.	$l=4$.	$l=5$.	$l=6$.	$l=7$.
4	48·8	53·3	57·8	62·3	66·8	71·3	...
5	47·3	51·4	55·5	59·6	63·7	67·8	71·9
6	46·2	50·1	53·9	57·7	61·5	65·4	69·2
7	45·5	49·1	52·8	56·4	60·1	63·7	67·4
8	44·9	48·4	51·9	55·4	58·9	62·4	65·9
9	44·5	47·9	51·3	54·7	58·0	61·4	64·7
10	44·2	47·5	50·8	54·1	57·3	60·7	64·0
12	43·6	46·8	50·0	53·1	56·3	59·5	62·6
14	...	46 35	49·4	52·5	55·6	58·6	61·7
16	...	...	...	52·0	55·0	58·0	61·0
20	...	...	...	...	...	57·1	60·0

To see how the disturbance of the P.D. caused by the presence of the third carbon is distributed, we must compare equation (9) with equation (3) for which there was no third carbon in the arc. We have

$$V = 38\cdot56 + 2\cdot5l + \frac{23 + 8l}{A} \quad \cdots \quad (9)$$

$$V = 38\cdot88 + 2\cdot07l + \frac{11\cdot66 + 10\cdot54l}{A}. \quad \cdots \quad (3)$$

The constant terms in the two equations may be considered to be identical, since there is only 0·32 volt difference between them. Thus the third carbon does not interfere with this constant term. It appears to increase the part of the P.D. that varies with the length alone, and to diminish that which varies with both length and current, while it practically doubles the part that varies with the current alone.

Let us now compare equation (9) with equation (8) for the sum of the carbon P.Ds :

$$V = 38\cdot88 + \frac{22\cdot6 + 3\cdot1l}{A}, \quad \cdots \quad (8)$$

and we find that, not only the constant terms, but also those which vary with the current alone are practically identical. Hence, leaving out the small variations in those two terms

(which probably arise from experimental errors), we find that
if we subtract the sum of the two carbon P.Ds. from the total
P.D. between the carbons, we have, for the fall of potential
through the carbon vapour itself, two terms, of which one
varies with the length of the arc alone, and the other with
both the current and the length of the arc, but no term varying
with the current alone. It seems, therefore, probable that that
part of the variation in the positive and negative carbon P.Ds.,
which depends on the current alone, is not caused solely by the
use of a bare idle carbon in the arc, but is a true variation
which takes place when there is no such cause of disturbance,
and is only *increased* by the insertion of the rod. In other
words, the term $\dfrac{11\cdot66}{A}$ in equation (3) seems really to belong
to the carbon P.Ds., which are *not*, therefore, constant for all
currents.

We can now, with tolerable certainty, locate three out of
the four terms in the equation which connects the P.D. across
the main carbons of the arc with the current and the length
of the arc (equation 3). Apparently the first and third terms
$38\cdot88 + \dfrac{11\cdot66}{A}$ belong *entirely* to the carbon P.Ds., and the
second term belongs *entirely* to the vapour P.D. The fourth
term $\dfrac{10\cdot54l}{A}$ is more difficult to locate; part of it certainly belongs
to the vapour P.D., but whether any of it really belongs to
the positive carbon P.D., as equation (6) appears to show,
or whether this term depends entirely on the use of the bare
third carbon, cannot be determined without fresh experiments.

So far, only cases (1) and (2) of Fig. 70 (p. 213) have been
discussed. Tables XXXVII. and XXXVIII. give the P.Ds.
found when the experiments represented by cases (3) and (4)
were made. Table XXXVII. shows the P.D. between the positive carbon and the third carbon just before it touched the
negative carbon, for each length of arc and current. That is,
the positive carbon P.D. plus the vapour P.D. Similarly,
Table XXXVIII. gives the vapour P.D. plus the negative
carbon P.D. for each length of arc and current. These
Tables may be used in various ways to confirm the deductions
made from the other Tables in this Chapter. For instance,
if the corresponding P.Ds. in each of Tables XXXVII.

and XXVIII. be added together, we shall get the total P.D. across the main carbons, for we shall have added the negative carbon P.D. of Table XXVIII. to the positive carbon P.D. plus the vapour P.D. of Table XXXVII. As both these sets of P.Ds. were obtained with the third carbon close to the *negative* pole (cases (2) and (3), Fig. 70, p. 213), their sum ought to correspond with the total P.Ds. across the main carbons, with the third carbon in the same position, given in Table XXXIV., except that the increase of the P.D. caused by having a third carbon in the arc is counted twice over in Tables XXVIII. and XXXVII. and only once in Table XXXIV. It will be found that agreement between the two sets of P.Ds. is so close that 43 of the pairs of values differ from one another by less than one volt, and the remaining 18 by less than two volts. Again, Table XXXVIII. may be used in the same way, in conjunction with Tables XXVII. and XXXIII., to find indirectly the total P.D. across the carbons with the third carbon close to the crater, and the positive carbon P.D. A comparison of these with the corresponding P.Ds. obtained by direct observation will also show that all the values agree extremely well with one another, though not quite so well as the experiments made with the third carbon near the negative pole, because the arc is always more disturbed when the third carbon is near the positive pole.

Table XXXVII.—*P.D. between Positive Carbon and Third Carbon with Point of latter close to White Spot.*

Solid carbons. Positive, 11mm.; negative, 9mm. in diameter; third carbon, 0·5mm. to 2mm. in diameter.

Current in Amperes.	P.D. in Volts.						
	$l=1.$	$l=2.$	$l=3.$	$l=4.$	$l=5.$	$l=6.$	$l=7.$
4	35·6	39·8	44·5	52·3	56·0	60·8	63·8
5	35·7	41·4	45·1	50·1	53·3	57·8	59·4
6	37·1	40·1	45·8	49·6	51·1	55·0	59·1
7	37·3	40·9	44·5	48·0	51·0	54·6	57·6
8	36·2	38·6	41·5	45·8	49·1	50·8	54·8
9	35·4	40·3	42·8	43·9	48·3	50·9	54·8
10	34·8	38·7	41·8	45·4	49·3	51·4	54·8
12	34·0	38·0	41·5	44·6	48·1	49·5	52·9
14	...	36·1	40·7	44·7	47·5	49·7	51·7
16	...	...	...	43·1	47·0	48·6	51·3
20	...	...	...	...	...	48·8	50·4

Table XXXVIII.—*P.D. between Third Carbon and Negative Carbon, with Point of latter close to Crater.*

Solid carbons. Positive, 11mm. ; negative, 9mm. in diameter ; third carbon, 0·5mm. to 2mm. in diameter.

Current in Amperes.	P.D. in Volts.						
	l=1.	*l*=2.	*l*=3.	*l*=4.	*l*=5.	*l*=6.	*l*=7.
4	14·8	14·1	19·4	24·9	30·7	30·5	34·2
5	11·2	19 3	19·1	24·4	24·5	28·9	34·4
6	11·9	15·9	22·8	21·9	26·0	29·8	31·0
7	9·9	17·7	20·9	25·8	27·1	27·5	31·7
8	10·5	15·5	19·2	20·7	23·0	28·4	28·2
9	9·4	15·6	16·6	19·7	22·8	24·1	28·7
10	11·0	15·3	20·1	20·4	24·7	27·5	27·8
12	10·4	14·2	17·7	20·8	23·0	25·2	27·1
14	...	14·1	17·8	20·0	21·9	23·1	26·8
16	...	...	...	18·6	20·6	24·0	28·0
20	...	...	..	...	...	22·4	26·3

It has been mentioned that four series of experiments were made with cored carbons, two with cored positive and solid negative carbons, and two with both carbons cored. The results of these experiments, together with the similar results obtained when both carbons were solid, are given in Tables XXXIX. and XLI.

The limited number of the experiments with cored carbons renders it difficult to judge of the effect on the various P.Ds. of changing the current and the length of the arc, as was done in the case of solid carbons. But the general result of substituting cored for solid carbons is easily seen, especially if, instead of comparing each set of P.Ds. separately with one another, we examine the *average* P. Ds. given at the end of each series. Thus the upper part of Table XXXIX. shows that with both carbons solid the average total P.D. across the carbons is 59·9 volts, while with the positive carbon cored it is 56·8, or 3·1 volts less, and with both cored it is 54·1 volts, or 5·8 volts less than with both carbons solid. Hence, coring either carbon diminishes the total P.D., but the diminution of P.D. is almost twice as great with both carbons cored as it is with the positive carbon alone cored, with an arc of 5mm. With a current of 10 amperes and various lengths of arc the diminution in the average total P.D. is exactly twice as great when both carbons are cored as when the positive alone is cored, as is seen from the lower part of Table XXXIX.

Table **XXXIX**.—*Comparison, with Cored and Solid Carbons, of P.D. between the Carbons with a Third Carbon in the Arc. Positive carbon, 11mm.; negative, 9mm.; third carbon, 0·5mm. to 2mm.*

Length of Arc, 5mm.

Current in Amperes.	P.D. between Carbons in Volts.		
	Both solid.	+ Cored, − Solid.	Both Cored.
4	67·35	65·6	62·5
5	64·15	59·9	57·8
6	61·65	57·3	54·8
7	60·95	57·6	54·0
8	59·8	55·9	53·7
9	57·5	55·6	53·6
10	58·8	54·2	53·2
12	57·8	53·9	50·9
14	55·95	54·2	50·3
16	55 25	53·7	50·0
Average.	59·9	56·8	54·1

Current, 10 Amperes.

Length of Arc in Millimetres.	P.D. between Carbons in Volts.		
	Both Solid.	+ Cored, − Solid.	Both Cored.
1	43·3	43·4	39·9
2	47·4	45·3	43·8
3	51·15	50·0	48 6
4	54·55	53·4	51·0
5	58·8	54·2	53·2
6	60·7	57·5	55·3
7	63·5	59·7	55·7
Average	54·2	51·9	49·6

The important thing to find out about this diminution of P.D. is whether it takes place in either of the carbon P.Ds. or both, or in the vapour P.D. Table XL. shows that a certain amount of it takes place in the positive carbon P.D., but not all, even when the positive carbon alone is cored, for although the average positive carbon P.D. is reduced by 3·3 volts. by the use of a cored positive carbon, which is about the same as the reduction in the total P.D., with an arc of 5mm., yet with a current of 10 amperes the average positive carbon P.D. is reduced by 1·3 volts by coring the positive carbon, whereas the average *total* P.D. is reduced by 2·3 volts. Hence it is probable that part of the reduction in the total P.D. is caused by the carbon vapour itself being rendered more conducting by the presence of the vapour from the core.

Table XL.—*Comparison, with Cored and Solid Carbons, of Positive Carbon P.Ds.*

Positive carbon, 11mm.; negative, 9mm.; third carbon, 0·5mm. to 2mm.

Length of Arc, 5mm.

Current in Amperes.	P.D. in Volts.		
	Both Solid.	+ Cored, − Solid.	Both Cored.
4	36·7	31·2	36·9
5	35·5	31·0	32·5
6	35 55	31·1	31·1
7	35·0	29·7	31·3
8	34·9	34·4	31·6
9	35·4	29·2	31·8
10	33·5	31·8	32·5
12	33·8	30·8	30·7
14	31·8	31·0	33 9
16	32·7	32·0	31·2
Average	34·5	31·2	32·3

Current, 10 Amperes.

Length of Arc in Millimetres.	P.D. in Volts.		
	Both Solid.	+ Cored, − Solid.	Both Cored.
1	32·6	34·0	31·2
2	32·4	31·2	33·4
3	31·5	31·6	33·7
4	33·75	32·4	34·3
5	33·5	31·8	32·5
6	32·7	31·0	31·9
7	34·7	30·2	32·1
Average	33·0	31·7	32·7

With both carbons cored the loss in the positive carbon P.D. appears to be less than when the positive alone is cored, but this is probably only an apparent difference caused by the core being harder in the one case than in the other. We may, I

Table **XLI.**—*Comparison, with Cored and Solid Carbons, of Negative Carbon P.Ds.*
Positive carbon, 11mm.; negative, 9mm.; third carbon, 0·5mm. to 2mm.

Length of Arc, 5mm.

Current in Amperes.	P.D. in Volts.		
	Both Solid.	+ Cored, − Solid.	Both Cored.
4	10·4	10·1	10·6
5	10·8	8·9	9·2
6	9·7	8·5	9·3
7	9·4	8·6	8·6
8	10·0	8·5	7·6
9	9·6	8·6	9·0
10	9·2	8·5	9·0
12	9·6	8·5	9·1
14	9·2	8·7	7·9
16	9·2	8·8	8·1
Average	9·6	8·8	8·8

Current, 10 Amperes.

Length of Arc in Millimetres.	P.D. in Volts.		
	Both Solid.	+ Cored, − Solid.	Both Cored.
1	8·5	8·0	7·8
2	8·5	8·8	8·9
3	9·0	8·8	8·5
4	8·9	8·6	8·7
5	9·2	8·5	9·0
6	9·2	8·7	9·0
7	9·2	8·7	8·6
Average	8·9	8·6	8·6

think, take it for granted that, other things being the same, the reduction of the positive carbon P.D. caused by coring is about the same, whether the negative carbon is cored or solid. Hence the increased reduction of the total P.D. caused by coring the negative carbon must be the result of a reduction either of the vapour P.D. or of the negative carbon P.D., or of both. That it is not in the negative carbon P.D. that the reduction takes place may be seen from Table XLI., for, with the *positive* carbon cored, the average negative carbon P.D. appears to be the same, whether the *negative* carbon be cored or not. Hence it appears that while coring the positive carbon reduces both the positive carbon P.D. and the vapour P.D., coring the negative carbon only further reduces the vapour P.D.

The reduction of the positive carbon P.D. is probably due to the ease with which the core vapourises compared with the solid carbon. The reduction of the vapour P.D. must be caused by the vapour of the core having greater conductivity than that of the solid carbon. The conductivity must indeed be very much greater, for, as was mentioned in Chapter I. (p. 16), the visible cross section of the arc is always less for the same current and length of arc, with a cored than with a solid positive carbon, and hence, all other things being equal, the P.D. required to send a given current would be *greater* and not less with the cored carbon.

It may be mentioned that a cored negative carbon always develops a deep crater just as if it were a positive carbon.

SUMMARY.

Solid Carbons.

I. The positive carbon P.D. increases as the length of the arc increases, and diminishes as the current increases.

II. The negative carbon P.D. does not vary with the length of the arc, but diminishes as the current increases. The fall of potential at the junction of this carbon with the arc is always from arc to carbon, and never in the opposite direction.

III. In the equations connecting each of the carbon P.Ds. with the current and the length of the arc, there is a constant term. The sum of these two constant terms has the same value as the constant term (commonly called the back E.M.F. of the arc) in the equation connecting the total P.D. across the carbons with the current and the length of the arc.

IV. Hence, a part of this so-called back E.M.F. belongs to the negative carbon P.D., and only about four fifths of it to the positive carbon P.D. instead of the whole, as has hitherto been supposed.

V. The term involving the current alone in the total P.D. equation belongs to the two carbon P.Ds., and the term involving the length of the arc alone belongs to the vapour P.D.

CORED CARBONS.

VI. The reduction of the total P.D., caused by using cored instead of solid carbons, is made partly in the positive carbon P.D., and partly in the vapour P.D. Very little of it is made in the negative carbon P.D., even when the negative carbon is cored.

CHAPTER VIII.

The Relations Existing between the E.M.F. of the Generator, the Resistance in the Circuit outside the Arc, the Length of the Arc, the Current and the P.D. between the Carbons with Solid Carbons.

Hitherto the arc alone has been considered, without any reference to the E.M.F. of the generator, or the resistance placed in circuit outside the arc. The influence on the arc of any change in either of these may be studied in two ways, graphically, by means of Fig. 79, the curves in which are a reproduction of those in Fig. 38, with certain additions; and analytically, by examining equation (4) (p. 186), in conjunction with an equation expressing the relation between the E.M.F. of the generator, the P.D. between the ends of the carbons, the outside resistance in circuit, and the current flowing.

The graphical method is, perhaps, the easier to follow, and may, therefore, be taken first. Let P (Fig. 79) be a point on the axis of P.D. such that its distance from the axis of current represents the E.M.F. in volts of the generator supplying the current, and let P Q be a line parallel to the axis of current, so that the distance of any point on P Q from the axis of current will represent this same E.M.F. Then, if R_1 be a point on one of the curves, the distance between R_1 and the axis of current will represent the P.D. in volts used in sending the current through the arc, and $R_1 S$, the distance between R_1 and P Q, will represent the P.D. employed in sending the current through the whole of the resistance outside the arc. Also, since P S represents the current flowing when that outside resistance is in circuit, the ratio of $S R_1$ to S P, which is tan. R_1 P Q, represents this outside resistance.

If, then, with the given generator supplying the current, any line be drawn from P making an angle Q P R with P Q, and cutting the curves at various points, the positions of those points

R

Both Carbons Solid.

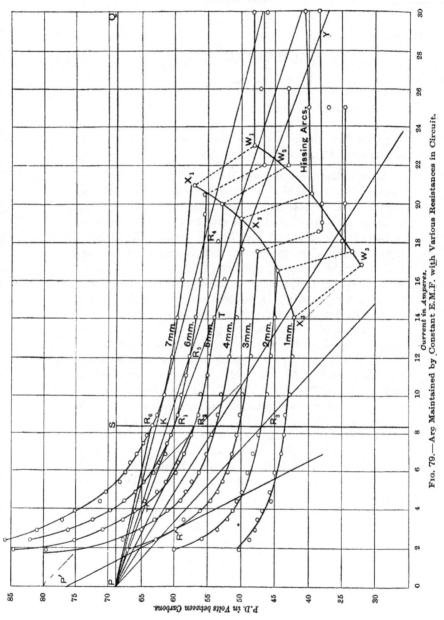

FIG. 79.—Arc Maintained by Constant E.M.F. with Various Resistances in Circuit.

will represent the relation between P.D. and current in arcs all supplied by a generator with the same E.M.F., and all having the same outside resistance in circuit—namely, that represented by the tangent of the angle R P Q. We may then, for convenience sake, call such lines as P R_1 Y resistance lines.

There are three possible ways of altering the conditions of the arc when the E.M.F. of the generator is kept constant, viz. :—

(1) The external resistance may be kept constant, and the length of the arc varied.

(2) The length of the arc may be kept constant, and the external resistance varied.

(3) Both the external resistance and the length of the arc may be varied together.

In the first case, since everything is constant except the length of the arc, lengthening the arc *must* mean increasing the resistance in the whole circuit, and consequently a diminution of current must follow on it. Similarly, shortening the arc must mean a diminution of the total resistance in circuit, and therefore an increase of current must follow on it.

In the second case, since everything is constant except the external resistance, increasing this must increase the total resistance in circuit, and therefore must diminish the current; while diminishing the external resistance must cause a diminution in the total resistance in circuit, and therefore must cause an increase in the current.

In the third case, the current may be made to vary in any manner that is desired, or to remain constant, by suitable variations of the external resistance and the length of the arc. But the sum of the effects of the variation of each of these must be the same as if each had been varied separately while the other was kept constant, so long as one of the variations does not cause the arc to be extinguished, thus making it impossible for the other to take effect after it.

While the external resistance can really be kept absolutely constant, the length of the arc cannot, for it is quite impossible, either automatically or by hand, to keep the carbons moving towards each other at exactly the same rate at which they burn away. Consequently what is *called* keeping the length of arc constant is really allowing it to become slightly longer than the desired length, and then bringing the carbons together till it is slightly shorter, so that the curve connecting time and length of arc when the length of arc was supposed to be constant, would be a zig-zag, not a straight line.

Similarly, what is called a constant current is not really constant, for it is impossible to lengthen or shorten the arc, and at the same time alter the resistance by exactly the right amount to keep the current constant. Hence there is a continual increase and decrease of current above and below the supposed constant value. Nevertheless it is possible to keep both length of arc and current constant within certain limits, which may be made very narrow by careful experimenting.

If we follow one of the resistance lines, $P R_1 Y$, for instance, we shall see what happens when the external resistance is kept constant, and the length of the arc varied.

The first thing that strikes one about this line is that the curves for silent arcs above a certain length—viz., 6mm.—are not cut by it at all; that the curve for this length is touched by the line, while those for shorter lengths of arc are cut by it in two points, or would be cut in two points, if hissing did not intervene and terminate the curve. The two points at which the line $P R_1 Y$ cuts each of the curves for lengths of arc less than 6mm. are closer together the nearer the length of the arc is to 6mm., and we may consider that it really cuts the 6mm. curve at two points which are coincident. The fact of its not cutting the curves for longer arcs at all shows that no arc of greater length than 6mm. can be maintained by a generator with the given constant E.M.F. when a resistance represented by tan $R_1 P Q$ is in the circuit outside the arc; in other words, 6mm. is the maximum length of arc that can be maintained under the given conditions of generator and external resistance.

From the resistance lines cutting each of the curves that they meet at two points it would appear as if, with a generator of constant E.M.F., and a constant resistance in circuit external to the arc, two widely different currents might be sent through two arcs of the same length. That is, apparently, under two precisely similar sets of conditions, two entirely different things may happen. This is, of course, absurd. Either the conditions cannot be precisely similar, or, if they are, the two different currents cannot permanently flow. By tracing the course of events from the moment the arc is struck it is possible to find out which of these two hypotheses is justified by facts.

We know from experience that, immediately after striking the arc, in the ordinary way, the current flowing is comparatively large, since the length of the arc is very small, and that if we lengthen the arc without altering the resistance in circuit, we diminish the current. Hence it is evident that if we want to follow the course of events after striking the arc, we must start from the right hand point of intersection with the shortest length of arc, and follow the line $P R_1 Y$ from *right* to *left*, because we then arrive successively at points of intersection which show that the current diminishes as the length of the arc increases. For instance, following $P R_1 Y$ (which corresponds with an E.M.F. of 68·88 volts and an external resistance of 1·05 ohms) from right to left, we see that when the arc is 4mm. in length the current is about 18 amperes, when it is 5mm. the current is about 14 amperes, and when it is 6mm. the current is a little over 8 amperes. But if we start from the left hand point of intersection with the shortest length of arc, and follow the line $P R_1 Y$ from *left* to *right*, we arrive successively at points of intersection which show that the current *increases* as the length of the arc *increases*. The right-hand points of intersection of the resistance lines with the curves are, then, those we are accustomed to find after striking an arc, and we know, therefore, that they are possible. Can the left-hand points also be obtained when the E.M.F. of the generator and the outside resistance are the same as for the right-hand points? is the question now to be answered.

In going from right to left along the line $P R_1 Y$, everything is perfectly easy till the arc is 6mm. in length, that is, till the resistance line is a tangent to the curve connecting P.D. and current. When this point is reached, however, trouble arises, for, since 6mm. has been shown to be the maximum length of arc that it is possible to maintain with the given resistance in circuit, and since the tendency of the arc is to lengthen, unless the carbons are brought nearer together at the very moment they have become 6mm. apart, the arc will be extinguished.

It would, of course, be very difficult, if not impossible, to bring the carbons nearer together at the precise moment when their distance apart was 6mm., but supposing that done, the question arises, which course would the point of intersection of the resistance line and the curve for the shorter arc take; would it move to the right or to the left, should we get a right-hand or a left-hand point of intersection between the resistance line and the curve; or, in other words, would the current increase or diminish? There is not much difficulty in answering that question. It has already been pointed out that, if everything else remains unchanged, shortening the arc can have only one result: it must lessen the total resistance in circuit, and therefore lead to an increase of current.

Hence the point R_1 will move to the *right* when the arc is shortened, and it is thus evident that it is not possible to obtain a left-hand point of intersection between the resistance line and the curve by keeping the external resistance constant and varying the length of the arc.

It still might be possible, however, to get to a left-hand point of intersection by varying the external resistance and keeping the arc at a constant length. Let us see. We may suppose that the arc has been struck, that its length has been increased from zero to 6mm., and that a silent arc is being maintained with a current of about 20 amperes flowing. If the length of the arc be now kept constant the external resistance will have to be increased, in order that we may pass along the curve for a silent arc of 6mm. As the resistance is increased the current will diminish, and the points of intersection between the resistance lines and the curve will be right-hand points, such as R_4 and R_5, till the resistance is represented by tan $R_1 P Q$, and the resistance line is a tangent to the curve.

When that moment is reached the resistance has of course gained its maximum value for a 6mm. arc, for if the angle $R_1 P Q$ were still further increased, the line $R_1 P$ would not meet the 6mm. curve at all. Hence, at this point, the resistance must be diminished the moment the arc is 6mm. long, and the question again arises, Which way would the point R_1 go—to the right or the left? Would the current increase or diminish when the external resistance was diminished? This question is readily answered. We know that diminishing the resistance, and leaving all else unchanged, can only cause a diminution in the total resistance in circuit, and hence an *increase* of current, which will give us over again points on the curve to the *right* of R_1. It is, therefore, impossible to obtain the left-hand points of intersection between the resistance lines and the curves by varying the external resistance and keeping the length of the arc constant.

It remains only now to see if these points can be obtained by varying both the external resistance and the length of the arc at the same moment, and this seems the most plausible method of all. It seems so possible, when the resistance line is just about to become a tangent to the curve, to suddenly diminish the resistance and lengthen the arc at the same moment, so as to pass quickly through the critical point, and arrive safely at a left-hand point of intersection on the other side of it.

That this idea is fallacious is, however, easily proved by referring to three conclusions that have already been shown to be true. With a generator of constant E.M.F. to supply the current—

(1) Changing both the length of the arc and the resistance together can only have the same effect as changing each separately in quick succession, if the order of change be such that one does not extinguish the arc before the other can take effect.

(2) Changing the length of the arc alone can give only right-hand points of intersection between the resistance lines and the curves.

(3) Changing the resistance alone can give only right-hand points of intersection between the resistance lines and the curves.

From these three facts it follows that it is impossible to obtain left-hand points of intersection between the resistance lines and the curves by altering the external resistance and the length of the arc simultaneously, and it follows that there is no possible way of varying the conditions of the arc in such a manner as to obtain left-hand points of intersection between the resistance lines and the curves connecting P.D. and current.

Hence no points can be obtained on the curves connecting P.D. and current farther to the left than those which form the points of contact between the curves and the tangents drawn to the curves from that point on the axis of P.D. which indicates the E.M.F. of the generator.

It follows, therefore, that if P be a point on the axis of P.D. such that its distance from the axis of current represents the E.M.F. of the generator, and if PR be the tangent through P to the P.D. and current curve for an arc of given length at the point indicating the given current, then the given E.M.F. is the smallest that can be used to supply that length of arc and current.   This was first pointed out by M. Blondel in 1891 (*see* p. 62).

The question arises, How can the left-hand points be obtained at all ; how *were* they obtained ?   They were obtained as *right-hand points*, by using a generator with a much larger E.M.F. than that indicated by the position of P.   It is, of course, obvious that all the points on the curves that are higher up than the line P Q must have been obtained with a generator of larger E.M.F. than that indicated by P ; but what this investigation shows is that some of the points that are much lower down in the figure than the line P Q must also have had a generator of higher E.M.F. to produce them.   The point for 2 amperes, for instance, on the 2mm. curve, although corresponding with a P.D. of only 58·5 volts, must have been found with a larger E.M.F. than 69 volts, which is indicated by the position of P, for the tangent from P to the 2mm. curve touches the curve at a point indicating a current of between 2 and 3 amperes.

On referring to Fig. 79, it is seen that a left-hand point of intersection of a resistance line and a curve—for example, the left-hand point T', where the resistance line R_1 P meets the 5mm. curve—has the following property : If the E.M.F. of the generator be kept constant, and also the length of the arc, an *increase* of the resistance of the circuit corresponds with an *increase* of the current.   Now, this is exactly the condition of the unstable solution that is obtained when a series dynamo, running at fixed speed, is in series with a set of accumulators of fixed E.M.F., the total resistance in circuit being also fixed.

For in such a case, as the late Dr. J. Hopkinson showed some years ago, there are three distinct values of the current possible, two of these corresponding with the dynamo charging the cells, while the third, which is very large and is negative, is produced when the magnetisation of the dynamo has been reversed, and the dynamo is helping the cells to discharge.   Now, the smaller of the two positive charging currents is *unstable*, for it *increases* in value with an *increase* of the resistance in circuit, so that on the slightest change in the speed of the dynamo, or in the resistance in circuit, this unstable current is suddenly changed into the larger charging current, or into the very much larger discharging current.

Further, just as I have shown that a left-hand point in Fig. 79 may, for the same current and length of arc, be changed into a right-hand one by increasing the E.M.F. of the generator and the resistance in circuit, so it may be shown that Dr. Hopkinson's unstable point may, for the same current and the same set of accumulators, be changed into a stable solution by raising the speed of the series dynamo and increasing the resistance in circuit.

The impossibility of obtaining small currents without having a comparatively large E.M.F. in the generator and a large external resistance, explains a very puzzling circumstance that occurred in carrying out the experiments of which the curves in Fig. 79 are the result—one which many other experimenters must also have noticed.

There were two dynamos at the Central Technical College, either of which it was convenient for me to use; one of these produced about 120 volts, and the other about 150 volts on open circuit. The first ran much more steadily than the second, which was driven by a single-cylinder engine; the former was, therefore, much better to employ in a general way; when, however, it was used for small currents and long arcs, the arc behaved in the most tricky manner, going out again and again for no apparent cause.

The reason of this is now obvious. The curves for long arcs are so nearly vertical in the parts for small currents, that the tangents to those curves at any of the points indicating small currents are very nearly parallel to the axis of P.D., and therefore intersect it very high up. It is this point of intersection, however, that determines the smallest E.M.F. that it is possible to have in the generator, and the smallest resistance that it is possible to have in the circuit, in order that a given current shall flow through a given length of arc. Thus, although, for the smallest current and longest arc I used, a P.D. of only about 86 volts was required for the arc itself, yet an E.M.F. of considerably over 120 volts was needed in the generator to enable the arc to be maintained at all with the given small current flowing. The remainder of this large E.M.F. had to be wasted in sending the current through the large resistance that it was absolutely necessary to have in circuit.

Thus the resistance added to an arc lamp on a constant-pressure circuit fulfils three distinct functions. It prevents an enormous current flowing when the arc is first struck; it renders it possible for a solenoid placed in series with the arc to regulate its length; and we now see that with solid carbons *this resistance fulfils a third and entirely different function, for without some resistance being placed in the circuit external to the arc it is impossible to maintain a silent arc at all.*

Hence the resistance placed in series with an arc lamp "to steady the arc," as is commonly said, fulfils the all-important function of making a silent arc possible.

In fact, an arc possesses this very curious property, viz.: that, although a certain current of, say, A amperes flowing steadily through an arc of, say, l mm. corresponds with a P.D. of, say, V volts between the carbons, the mere maintenance of this P.D. of V volts between the carbons with the arc of l mm., is not a sufficient condition to ensure that the current of A amperes shall continue to flow. There must also be a resistance in the circuit outside the arc, which can not have less than a certain minimum value.

As an example of this peculiarity, I may refer to the difficulty, alluded to on page 170, which I met with when experimenting on arcs with constant P.Ds. An examination of Fig. 79 shows that when, for example, a current of 11 amperes is steadily flowing through an arc 5mm. long, formed with solid carbons, 11mm. and 9mm. in diameter, the P.D. between them is 55 volts. It might, therefore, be imagined that this current could be sent through such an arc with a dynamo compounded and run at such a speed as to maintain exactly 55 volts between the carbons with no resistance inserted between the dynamo and the carbons. Or it might be thought that this steady current of 11 amperes could be sent through this arc by means of, say, 29 accumulators, when such a small resistance was inserted in the circuit that, with a current of 11 amperes, the accumulators maintained a P.D. of 55 volts between the carbons.

But when I tried this, and similar experiments of keeping the P.D. between the carbons constant, the arc always either went out, or the magnetic cut-out, set to open at 30 amperes, broke the circuit. The explanation of this is now clear, for what I was really endeavouring to obtain was a silent arc with a fairly *small* current, and a *small* resistance in circuit external to the arc ; in other words, I was trying to obtain the *left*-hand point of intersection between the 5mm. curve and the resistance line, and this, as was proved above, is impossible. Indeed, with a set of accumulators having an E.M.F. of about 58 volts and a small resistance in circuit, the only current that could flow steadily through a 5mm. arc between the carbons in question was a very large one far greater than 30 amperes. The value of this current would be known if we could find the intersection of the resistance line with the hissing part of the 5mm. curve, at a point far off the figure, to the right.

What we find, then, is that, *in order that a given current may be maintained flowing through a given length of arc, there must be a minimum external resistance in the circuit which determines also the least E.M.F. in the generator that would maintain such a current flowing through such a length of arc.*

For example, whatever be the nature of the generator, the *least* resistance that can be placed in the circuit to send a steady current of 11 amperes through a 5mm. arc is that given by the tangent to the 5mm. curve at the point corresponding with 11 amperes. And on drawing this tangent to the 5mm. curve in Fig. 79, we find that it corresponds with a resistance of 0·64 ohm.

This minimum resistance determines the minimum E.M.F. in the generator for each current and length of arc; for example, on continuing the resistance line just referred to for the 11-ampere 5mm. arc we find that it cuts the axis of P.D. at 62 volts. Hence a resistance exceeding 0·64 ohm must necessarily be placed in the circuit, and the E.M.F. employed must exceed 62 volts, although the arc requires a P.D. of only 55 volts. Similarly, although we see from Fig. 79 that a 12-ampere 3mm. arc requires a P.D. of only a little less than 49 volts, the preceding reasoning tells us that, with the solid carbons used, two such arcs could not be run in series off the ordinary constant-pressure 100-volt electric lighting mains, even if the supply pressure were kept absolutely constant at 100 volts. Or again, with the solid carbons I used, two 10-ampere 4mm. arcs could not be run in series off 110 volts constant-pressure mains, although each arc requires less than 53 volts. Further, the minimum resistance was so great for an arc of 7mm. and current of 2·5 amperes, with the solid carbons I used, that it made the minimum E.M.F. required more than half as large again as the P.D. used in sending the current through the arc itself, for I have calculated that the minimum E.M.F. was about 139 volts, while the P.D. needed by the arc was only 86 volts, as has been mentioned.

It may be seen from Fig. 79 that the compulsory minimum resistance outside the arc is greater—

(1) the greater the length of the arc,

(2) the smaller the current.

But the apparent resistance of the arc also increases with the length of the arc, and is greater the smaller the current, so that we arrive at this very curious fact : *When a silent arc is being maintained the smallest resistance that it is possible to have in the circuit external to the arc is greater the greater the apparent resistance of the arc, or, in other words, the more resistance you have in the arc, the more you need outside it.*

Many more facts concerning the relations between the E.M.F. of the generator, the external resistance, the length of the arc, and the current can be ascertained by treating Fig. 79 analytically, which we will now proceed to do.

Let E be the E.M.F. of a generator in volts, let r be the total resistance in ohms in the circuit outside of the arc itself, let there be a P.D. of V volts between the carbons, and let a current of A amperes be flowing through an arc of l millimetres. Then, referring to Fig. 79 (p. 242), E is represented by the distance between P and the axis of current, r by the ratio of the distance from P Q of any point on one of the resistance lines to its distance from the axis of P.D. This is the same thing, of course, as r being represented by the tangent of the angle between a resistance line and P Q. A is represented by the distance between any point at which a resistance line cuts a curve and the axis of P.D. and V by the distance between that point and the axis of current.

Then $$E = V + A\,r.$$

But from equation (4) (p. 186),

$$V = a + b\,l + \frac{c + d\,l}{A},$$

therefore $$E = a + b\,l + \frac{c + d\,l}{A} + A\,r, \quad \ldots \quad (10)$$

or $$A^2\,r - (E - a - b\,l)\,A + c + d\,l = 0;$$

hence $$A = \frac{E - a - b\,l \pm \sqrt{(E - a - b\,l)^2 - 4\,r\,(c + d\,l)}}{2\,r}. \quad (11)$$

Thus we find, as before, that with a generator of given E.M.F. two different currents, or one, or none may be sent through the arc. The conditions for the three cases in the equation are that $(E - a - b\,l)^2$ shall be greater than, equal to, or less than $4\,r\,(c + d\,l)$, and these correspond with the conditions in the figure that the resistance line shall meet the curve at two points, at one, or not at all.

Since the value for A obtained by using the positive root in equation (11) is greater for any given length of arc than that obtained by using the corresponding negative root, it follows that the positive root will give values which correspond with right-hand points of intersection between the resistance line and the curve, while the negative root will give values that correspond with the left-hand points of intersection. As it has already been shown that these left-hand points can never be obtained in practice, it will not be necessary here to discuss the negative root of the equation, and therefore henceforward when equation (11) is mentioned it will be understood that the positive root only is alluded to.

As before, we shall take a given E.M.F. in the generator, and we shall consider what happens, first, when r is constant and l is varied ; next, when l is constant and r is varied ; and finally, when r and l are both varied in such a way that A is constant. In equation (11), if A is to have a real value at all, $(E - a - b\,l)^2$ must be not less than $4\,r\,(c + d\,l)$ whatever A, E, r, and l may be. Taking the first case—r constant, *i.e.*, following one of the resistance lines in Fig. 79—with the positive sign before the root, A is evidently greatest when l is least, diminishes as increases, and is least when l is greatest—that is, A is a maximum when l is a minimum, and a minimum when l is a maximum if the resistance external to the arc is kept constant.

But, since the expression under the root cannot be negative if A is to be real, l is a maximum when

$$(E - a - b\,l)^2 - 4\,r\,(c + d\,l) = 0, \quad . \quad . \quad . \quad (12)$$

for, since E is constant, $E - a - b\,l$ is smallest when l is greatest, and since r is constant, $r\,(c + d\,l)$ is greatest when l is greatest. Hence when r is constant the condition that A shall have only one value, which is the condition that the resistance line shall be a tangent to the curve, is also the condition that the length of the arc shall be a maximum, as was shown when the subject was dealt with graphically.

Since A has been shown to be a minimum when l is a maximum, it follows that the smallest current that can flow with a constant resistance in circuit when the E.M.F. of the generator is also constant is given by the equation

$$A = \frac{E - a - b\,l}{2\,r}. \quad . \quad . \quad . \quad (13)$$

These maximum and minimum values for l and A may be found in terms of E, r, a, b, c, and d, all of which are known.

For from equation (12) we have

$$b^2\,l^2 - 2\,l\,\{b\,(E - a) + 2\,d\,r\} + (E - a)^2 - 4\,c\,r = 0,$$

hence

$$l = \frac{b(E - a) + 2\,d\,r \pm \sqrt{\{b(E - a) + 2\,dr\}^2 - b^2\,\{(E - a)^2 - 4\,c\,r\}}}{b^2}$$

or

$$l = \frac{(E - a)}{b} + \frac{2dr \pm \sqrt{\{b(E - a) + 2\,dr\}^2 - b^2\{(E - a)^2 - 4\,c\,r\}}}{b^2}.$$

But from equation (11) it may be seen that l cannot be greater than $\dfrac{E - a}{b}$ when the quantity under the root is zero, otherwise A would be negative. Therefore, in the last equation the *negative* sign must be used before the root.

The quantity under the root is

$$b^2\,(E - a)^2 + 4\,b\,d\,r\,(E - a) + 4\,d^2\,r^2 - b^2\,(E - a)^2 + 4\,b^2\,c\,r,$$

which equals

$$4\,r\{b\,d\,(E - a) + r\,d^2 + b^2\,c\}.$$

Therefore

$$l = \frac{E - a}{b} + \frac{2\,d\,r - 2\,\sqrt{r\,\{b\,d(E - a) + d^2\,r + b^2\,c\}}}{b^2}.$$

And, substituting this value for l in equation (13) we get

$$A = \frac{\sqrt{r\,\{b\,d\,(E - a) + d^2\,r + b^2\,c\}}}{b\,r} - \frac{d}{b}.$$

Hence, if the E.M.F. of the generator be known, and also the external resistance that it is desired to have in the circuit, the longest silent arc that can be maintained and the smallest current that will flow can be found in terms of the known E.M.F., the known resistance, and the known constants depending on the carbons used.

For instance, let us take

$$E = 48\!\cdot\!88 \text{ volts,}$$
$$r = 0\!\cdot\!5 \text{ ohm.}$$

Then, from the above equations we get

$$l = 1\!\cdot\!35 \text{mm,}$$

and $A = 7\!\cdot\!2$ amperes if the values of the constants are those given in equation (3) (p. 184).

Thus, with an E.M.F. of $48\!\cdot\!88$ volts in the generator, and a resistance of $0\!\cdot\!5$ ohm in the circuit external to the arc, the longest silent arc that can be maintained with the carbons in question is one of $1\!\cdot\!35$mm., and the smallest current that will flow is one of $7\!\cdot\!2$ amperes.

Next let l be constant and r variable, that is, let us follow the course of one of the curves for constant length of arc in Fig. 79. Then, of course, A will also vary, and will again be a minimum when equation (12) holds. Also r will be a maximum when equation (12) holds, since any value of r greater than $\dfrac{(E - a - bl)^2}{4\,(c + d\,l)}$ would make the quantity under the root in equation (11) negative. Hence we have for A a minimum

$$A = \frac{E - a - b\,l}{2\,r},$$

and for r a maximum

$$r = \frac{(E - a - b\,l)^2}{4\,(c + d\,l)}.$$

By combining these two equations we can obtain the following one for A in terms of known quantities only, r being already given in known terms,

$$A = \frac{2\,(c + d\,l)}{E - a - b\,l}.$$

Thus, when the E.M.F. of the generator is known, and it is required to maintain a silent arc of fixed length, the maximum external resistance that can be used, and the minimum current that can be maintained, can be found in terms of the known E.M.F., the fixed length of arc, and the known constants of the carbons.

For example, let

$$E = 58{\cdot}88 \text{ volts,}$$
$$l = 3\text{mm.}$$

Then, from the above equations we have

$$r = 1{\cdot}1 \text{ ohms,}$$
and $$A = 6{\cdot}28 \text{ amperes.}$$

Hence, with an E.M.F. of 58·88 volts in the generator, and a silent arc of 3mm. to be maintained, a resistance of 1·1 ohms is the greatest that can be placed in the circuit outside the arc, and 6·28 amperes is the smallest current that will flow.

Lastly, let r and l be both varied in such a way that A remains constant, that is, let us follow some such line as S R_1 R_2 R_3 in Fig. 79. Then, from the equation

$$r = \frac{E - a - b\,l + \sqrt{(E - a - b\,l)^2 - 4\,r\,(c + d\,l)}}{2\,A},$$

it is plain that r diminishes as l increases, and becomes a minimum when

$$r = \frac{E - a - b\,l}{2\,A},$$

—that is, when the quantity under the root vanishes—or

$$r = \frac{(E - a - b\,l)^2}{4\,(c + d\,l)},$$

which has also been found above to be the condition that l shall be a maximum for a fixed value of r.

By combining the last two equations we can determine the minimum value for r and the maximum value for l in terms of the known quantities, when E and A are constant,

$$r = \frac{d\,(E-a) + b\,c}{A\,(A\,b + 2\,d)}.$$

$$l = \frac{A\,(E-a) - 2\,c}{A\,b + 2\,d}.$$

Thus, when the E.M.F. of the generator is known and the current is fixed, the minimum external resistance that must be placed in circuit, and the maximum length of silent arc that can be maintained, can be found in terms of the known E.M.F., the fixed current, and the known constants of the carbons.

For example, let us take

$$E = 58\cdot88 \text{ volts,}$$
$$A = 10 \text{ amperes.}$$

Then from the above equations we get

$$r = 0\cdot56 \text{ ohm,}$$
and
$$l = 4\cdot2\text{mm.,}$$

which shows that if with an E.M.F. of 58·88 volts in the generator we wish a current of 10 amperes to flow, the smallest resistance that can be placed in the circuit outside of the arc is one of 0·56 ohm, and the longest arc that can be maintained is one of 4·2mm.

SUMMARY.

I. With a given E.M.F. in the generator, and a given resistance in the circuit outside the arc, no arc longer than a certain maximum length can be maintained, and with this length the P.D. between the carbons is the greatest, and the current flowing the least that can be maintained, by the given E.M.F.

II. For the maintenance of each current and length of arc a certain minimum resistance is needed in the circuit outside the arc, and this resistance determines the value of the smallest E.M.F. with which the generator could maintain the arc of the given length and with the given current flowing.

III. This minimum resistance is greater the greater the length of the arc and the smaller the current.

IV. If the E.M.F. of the generator and the external resistance be fixed, the maximum length of arc, the minimum current that can be maintained, and the maximum P.D. between the carbons, can be found in terms of that E.M.F. and resistance, and of the four constants of the carbons.

V. If the E.M.F. of the generator and the length of the arc be fixed, the maximum external resistance that can be used, and the minimum current that will flow can be found in terms of the given E.M.F. and length of arc and the four constants of the carbons.

VI. When the E.M.F. of the generator, and the current are fixed, the maximum external resistance that can be used, and the maximum length of arc that can be maintained, can be found in terms of the given E.M.F. and current, and the four constants of the carbons.

CHAPTER IX.

SOLID CARBONS.

The ratio of the power expended in a silent arc to the power developed by the generator is an important factor in determining the most efficient arrangement to use in maintaining an arc. We are now in a position to determine when this ratio of A V to A E will be greatest with solid carbons.

From equations (4) and (10) (pp. 186 and 253), we have

$$\frac{V}{E} = \frac{a + b\,l + \dfrac{c + d\,l}{A}}{a + b\,l + \dfrac{c + d\,l}{A} + A\,r}. \qquad . \quad . \quad (14)$$

Hence, for any given values of l and A, $\dfrac{V}{E}$, and therefore, $\dfrac{AV}{AE}$, is greatest or nearest to unity when Ar is least; but from equation (11) it is evident that $A\,r$ is least when

$$(E - a - b\,l)^2 = 4\,r\,(c + d\,l),$$

that is, when equations (12) and (13) hold. From those two equations we find then that when $\dfrac{V}{E}$ is greatest

$$A\,r = \frac{c + d\,l}{A}. \qquad . \quad . \quad . \quad . \quad (15)$$

or

$$r = \frac{c + d\,l}{A^2}.$$

This equation gives the value of r, when $\dfrac{V}{E}$ is as large as possible, for any given value of l and A. It is still necessary then, in order to find absolutely the largest value of $\dfrac{V}{E}$, to find what values of l and A make $\dfrac{V}{E}$ greatest. If we substitute $\dfrac{c+dl}{A}$ for A r in equation (14), and multiply numerator and denominator by A, we get

$$\frac{V}{E} = \frac{(a+bl)\,A + c + d\,l}{(a+b\,l)\,A + (c+d\,l) + (c+d\,l)},$$

from which it is plain that, whatever A may be, $\dfrac{V}{E}$ is greatest when $c+d\,l$ is least—*i.e.*, when l is least, for if the second $c+d\,l$ in the denominator were to vanish altogether, $\dfrac{V}{E}$ would be unity.

Thus *the ratio of the power expended in a silent arc to the power developed by the generator is greatest when the arc is shortest*—that is when the length of the arc is zero if we preclude negative lengths, of the existence of which we have no proof.

To find what value of A makes $\dfrac{V}{E}$ greatest for any given value of l, we may put the last equation in the form—

$$\frac{V}{E} = 1 - \frac{c+d\,l}{(a+b\,l)\,A + 2\,(c+d\,l)},$$

which shows that, for any given value of l, $\dfrac{V}{E}$ is greatest when A is as large as possible. But A is as large as possible for a silent arc of given length *when the arc is about to hiss*, so that the value of A when $\dfrac{V}{E}$ is greatest for any given length of arc must be found from the equation connecting l and A for all the points on the hissing curve X_1 X_2 X_3 Fig. 79 (p. 242). *Hence the ratio of the power expended in a silent arc to the power developed by the generator is greatest when the arc is on the point of hissing.*

Since A must be as large as possible and l as small as possible for $\dfrac{V}{E}$ to have its greatest value, it follows from equation (15) that r must be as small as possible. *Hence the ratio of the power expended in a silent arc to the power developed by the generator is greatest when the resistance in the circuit outside the arc is as small as possible.*

It is still necessary to find what value of E makes $\dfrac{V}{E}$ greatest when A and l are fixed, but this is easy, for since V and E are independent of one another evidently E must be as small as possible for $\dfrac{V}{E}$ to be as large as possible, or *the ratio of the power expended in a silent arc to the power developed by the generator is greatest when the E.M.F. of the generator is the least possible.*

Since the experiments upon which the curves in Fig. 79 are founded, and upon which the above reasoning is based, were made with arcs of not less than 1mm. in length, and with solid carbons, it is impossible to be sure that the laws which have been shown to apply to these arcs would also be true for arcs of less than 1mm. and with cored carbons. We can, however, be quite sure of the following : *In silent arcs of from 1mm. to 7mm., with solid carbons, the ratio of the power expended in the arc to the power developed by the generator is greatest when the arc is shortest, when the current is the largest that does not cause hissing, and when the resistance in the circuit outside the arc, and consequently the E.M.F. of the generator is the smallest possible.*

Although the conditions just obtained are absolutely the best where power only is concerned, such as, for instance, in the case of an electric furnace, it must not be forgotten that with arcs used for lighting purposes the final consideration must always be the greatest ratio of the *light* emitted to the power evolved by the generator. This ratio as it stands would be very difficult to find, but by combining the conditions necessary for the ratio of the power consumed in the arc to the power evolved by the generator to be the greatest possible, with the conditions necessary to make the ratio of the light emitted by the arc to the power consumed in it as large as possible, the very best arrangement of the circuit for an arc used for lighting purposes may be obtained. All the developments of the first set of conditions will be considered in the present chapter, while the second set, and the combination of the two, will be discussed in Chapter XI.

The equation to the curve $X_1 X_2 X_3$—is, as will be shown later on,

$$A = \frac{11 \cdot 66 + 10 \cdot 54\, l}{1 \cdot 17 + 0 \cdot 416\, l},$$

which, put in the general form, is

$$A = \frac{c + d\, l}{e + f\, l}, \quad \ldots \ldots \quad (16)$$

where c and d have the values already used and

$$e = 1 \cdot 17,$$
$$f = 0 \cdot 416.$$

Thus we have the three equations (12), (13), and (16) connecting the four variables l, A, E, and r, when $\dfrac{V}{E}$ is a maximum, and hence, if circumstances determine any one of the four, the other three can be found.

It would appear, from what has gone before, as if the length of the arc were the most important variable to choose first in making any arrangement to maintain an arc, for we know definitely that, if we ignore for the moment the question of the light given out, this should be as small as possible. The difference between the greatest value of $\dfrac{V}{E}$ with one length of arc and its greatest value with another is not, however, very large. For instance, for the carbons I used, with an arc of 0mm. the greatest value of $\dfrac{V}{E}$ would be 0·97, obtained with a current of 10 amperes, an E.M.F. of 41 volts, and a P.D. of 40 volts between the carbons. With an arc of 7mm. the greatest value of $\dfrac{V}{E}$ would be 0·94, obtained with a current of 21 amperes, an E.M.F. of 61·4 volts and a P.D. of 57·4 volts between the carbons. Hence, although the actual power consumed in the shorter arc (400 watts) would be less than one-third of that consumed in the longer (1205 watts), yet the ratio of the power expended in the arc to the power developed by the generator would be nearly the same—viz., 0·97 and 0·94—for the two lengths of arc when the largest non-hissing current was used in each case.

Thus it is plain that *in arranging a circuit so as to obtain the greatest value of* $\frac{V}{E}$, *the length of the arc matters very little.*

This is a fact of great importance in the consideration of the amount of light given out by the arc, for, as Prof. Ayrton showed at Chicago in 1893, the ratio of the light given out by the arc to the power consumed in it depends largely upon the length of the arc, and is a maximum when the length of the arc is increased to a certain value. Hence, since the length of the arc is practically immaterial as far as the ratio of the power consumed to the power developed by the dynamo is concerned, in order to decide what length of arc, current and E.M.F. should be used to insure the largest ratio of the light emitted to the power developed in the generator, it would simply be necessary to find what length of arc gave the greatest amount of light for the energy consumed when the current flowing was the largest that did not cause hissing, and we might then find out what resistance to place in the circuit, and what E.M.F. to use, from equations (12) and (13).

For example, let us suppose that the preceding considerations have led to an arc of 3mm. being chosen, and that the arc lamp is fitted with a series regulating coil, and supplied with current from a generator of constant E.M.F. With such an arc a current of about 18 amperes is the largest that can be used without hissing, and from equations (12) and (13) we find that the generator should have an E.M.F. of about 50 volts, and that the total resistance in the circuit outside the arc should be about 0·13 ohm.

The current actually used must not, however, be so large that the arc can have any tendency to hiss. Suppose, therefore, a normal current of 15 amperes is selected, and that when it falls to 13 amperes, and the arc has increased in length to 3·2mm., the regulator begins to feed the carbons.

Then, remembering that the current that is to flow is no longer the greatest non-hissing current, but something less, and that E and r are to be the same whatever the current and length of arc may be, we must put

$$A = 15$$
$$l = 3$$

and also
$$A = 13$$
$$l = 3\cdot2$$

in the general equation

$$A = \frac{E - a - b\,l + \sqrt{(E - a - b\,l)^2 - 4\,r\,(c + d\,l)}}{2\,r}$$

in order to find E and r. In this way we shall find that E must be 55·5 volts and r must be 0·5 ohm.

Hence we have V or $E - A\,r = 55\cdot5 - 7\cdot5$

and
$$\frac{V}{E} = \frac{48}{55\cdot5}$$
$$= 0\cdot87.$$

Thus with these carbons, if a silent arc of 3mm. be maintained with a current of 15 amperes flowing, the arrangement can be made so efficient that 87 per cent. of the power developed by the generator will be used in the arc, and only 13 per cent. will be wasted in the resistance outside the arc. This total resistance external to the arc includes, of course, the resistances of the generator, the leads, and the coil of the regulator, so that it might be practically impossible to make it as small as 0·5 ohm, but the greatest value of $\dfrac{V}{E}$ would in that case be obtained by making this outside resistance as near 0·5 ohm as possible.

Without, however, for the moment considering the new circumstance that this impossibility of using a small enough resistance introduces into the problem, we may sum up the ideal conditions to be striven for in an arc, in order that the ratio of the power expended in the arc to the power developed by the generator may be as large as possible. They are :

(1) That the arc should be the shortest that will emit the requisite amount of light.

(2) That the current should be the largest that will certainly give a silent arc.

(3) That the E.M.F of the generator and the external resistance should be the smallest with which this diminished current could be sent through the increased length of arc.

(4) That the regulator should start feeding the carbons for as small a diminution of current and as small an increase in the length of the arc as possible.

It has been shown that with silent arcs, when the length is fixed, the conditions that $\dfrac{V}{E}$ shall have the greatest value are given by equations (12), (13) and (16).

If we put $l = 0$ in these equations and give the constants the values they have for my carbons we get

$$E = a + 2\,e = 41\cdot22 \text{ volts,}$$

$$A = \frac{c}{e} = 9\cdot96 \text{ amperes,}$$

$$r = \frac{e^2}{c} = 0\cdot1175 \text{ ohm,}$$

$$\frac{V}{E} = \frac{a+e}{a+2\,e} = 0\cdot9716.$$

Having thus considered what happens when l is fixed, we will take each of the other four quantities, E, A, r and V, as being fixed in turn, and consider under what conditions $\dfrac{V}{E}$ will be the greatest in each case.

Since it will be found that the resistance external to the arc must in each case be the smallest with which the given conditions can be carried out, I will point out beforehand that in equation (11), which may be put in the form

$$r = \frac{E - a - b\,l + \sqrt{(E - a - b\,l)^2 - 4\,r\,(c + d\,l)}}{2\,A},$$

r will be least for *any* simultaneous values of A and l when

$$(E - a - b\,l)^2 = 4\,r\,(c + d\,l)$$

—that is, when equations (12) and (13) hold.

Now, first let E be fixed, then since

$$\frac{V}{E} = \frac{E - A\,r}{E},$$

$$= 1 - \frac{A\,r}{E},$$

$\dfrac{V}{E}$ will be greatest when $A\,r$ is least. But from equation (11) we have $\quad 2\,A\,r = E - a - b\,l + \sqrt{(E - a - b\,l)^2 - 4\,r\,(c + d\,l)}$,

therefore, for a fixed value of E, $A\,r$ is least when $b\,l$ has its largest value, and when also

$$\sqrt{(E - a - b\,l)^2 - 4\,r\,(c + d\,l)}$$

is nought. That is, when E is fixed, A r is least, and, therefore, $\dfrac{V}{E}$ is greatest, when we use the longest silent arc possible with the given E.M.F. and select the resistance external to the arc, so that

$$(E - a - b\,l)^2 = 4\,r\,(c + d\,l).$$

The last equation, which is (12), carries with it equation (13), and from the two we find that

$$A = \frac{2\,(c + d\,l)}{E - a - b\,l}.$$

From this last equation it follows that when E is fixed, and l is as great as possible, A is also as great as possible, or $\dfrac{V}{E}$ is greatest with a fixed E.M.F. when A is the largest silent current that can be maintained flowing through the arc of greatest length that a generator with the fixed E.M.F. can produce.

Now it was shown on p. 254 that equations (12) and (13) gave the conditions that the line representing the resistance r ohms should be a tangent to the curve connecting P.D. and current for the length of arc $l$mm. Thus it has been shown that when the E.M.F. of the generator is fixed, $\dfrac{V}{E}$ is the greatest possible when—

(1) The arc is the longest that can be maintained with the fixed E.M.F. ;

(2) The line representing the resistance is a tangent to the curve connecting P.D. and current for this greatest length of arc, and represents, therefore, the smallest resistance that can be used with the fixed E.M.F. ;

(3) The current is the largest silent current for this greatest length of arc.

But, as shown before, equation (16) gives the connection between the largest silent current for any length of arc and the length of that arc. Therefore we may say that when E is fixed, equations (12), (13) and (16) give the relations between E, A, l and r that make $\dfrac{V}{E}$ the greatest.

Next let A be fixed, and first let us suppose that l is fixed also. Then from equation (14) (p. 259) we see that for $\dfrac{V}{E}$

to be as large as possible A r must be as small as possible. But from equation (11) we have that

$$2\,\text{A}\,r = \text{E} - a - b\,l + \sqrt{(\text{E} - a - b\,l)^2 - 4\,r\,(c + d l)}.$$

Hence, whatever the length of the arc may be, if A is fixed A r is least when E is least, and when also

$$(\text{E} - a - b\,l)^2 = 4\,r\,(c + d\,l).$$

That is, A r is least—and therefore $\dfrac{\text{V}}{\text{E}}$ is greatest—when we use the smallest E.M.F. that will send the fixed current through the arc, and when we select r, so that equations (12) and (13) hold. But when these equations hold, the line representing the resistance is a tangent to the curve connecting P.D. and current for the length of arc used. Hence, when A is fixed, $\dfrac{\text{V}}{\text{E}}$ is greatest, whatever length the arc may have, when we use the smallest E.M.F. that will send the fixed current through that length of arc, and when the line representing the resistance is a tangent to the curve connecting P.D. and current for that length of arc.

Next, while A still remains fixed, let l vary. Then, since it has been shown that $\dfrac{\text{V}}{\text{E}}$ is greatest, whenever $\dfrac{\text{A}\,r}{\text{E}}$ is least and that when A is fixed, $\dfrac{\text{V}}{\text{E}}$ is greatest when E is least, we must now find what value of l makes $\dfrac{\text{A}\,r}{\text{E}}$ least, when E is least, when A is fixed, and when equations (12) and (13) hold.

Since E is to be as small as possible, A r also must evidently be as small as possible. But from equations (12) and (13) we have

$$\text{A}\,r = \frac{c + d\,l}{\text{A}},$$

which, since A is fixed, shows that $c + d\,l$ must be as small as possible for A r to be as small as possible. That is, A r is least, and therefore $\dfrac{\text{V}}{\text{E}}$ is greatest, when l is the shortest arc that can be used, with due consideration for the light to be emitted.

Thus, when A is fixed, we find that the conditions for $\dfrac{\text{V}}{\text{E}}$ to be as great as possible are—

(1) That the arc shall be the shortest that will emit the requisite amount of light;

(2) That the E.M.F. shall be the smallest that will send the fixed current through this arc;

(3) That the external resistance shall be such that the line representing it will be a tangent to the curve for this arc at the point where the current is the fixed current, and consequently the resistance must be the smallest with which the fixed current will flow through the shortest length of arc.

Next let r be fixed, and first suppose l to be fixed also. Then, since $\dfrac{A r}{E}$ must be as small as possible, we have from equation (11) that

$$1 - \frac{a + b l}{E} + \frac{1}{E} \sqrt{(E - a - b l)^2 - 4 r (c + d l)}$$

must be as small as possible. Hence, since l is fixed, E must be as small as possible, and

$$\sqrt{(E - a - b l)^2 - 4 r (c + d l)}$$

must be nought for $\dfrac{A r}{E}$ to be least. Thus, whatever the length of the arc may be, if r is fixed, $\dfrac{A r}{E}$ is least when E is least, and when

$$(E - a - b l)^2 = 4 r (c + d l)$$

—that is, when E is least, and when equations (12) and (13) hold.

Therefore, whatever the length of the arc may be with which $\dfrac{V}{E}$ is greatest when r is fixed, the E.M.F. must be the smallest that will maintain that length of arc with the given resistance in circuit, and the line representing the resistance must be a tangent to the curve connecting P.D. and current for that length of arc.

Now let the length of the arc vary, the resistance still remaining fixed. Then, since equation (12) holds, whatever the length of the arc, we get from it

$$E = a + b l + 2 \sqrt{r (c + d l)},$$

the positive sign only being taken, because from equation (10) (p. 253) it may be seen that $E - a - b l$ could not be negative. Hence, since E is to be as small as possible for the fixed resistance, l must be as small as it can be consistently with its emitting the desired amount of light. Also, since $\dfrac{\mathrm{A}\,r}{\mathrm{E}}$ has to be as small as possible, and since r is fixed and E as small as possible, A must also be as small as it can be with the given resistance in circuit.

Thus, when r is fixed, $\dfrac{\mathrm{V}}{\mathrm{E}}$ is the greatest possible when,

(1) The arc is the shortest that will give the desired amount of light ;

(2) The current is the smallest that will flow through this length of arc with the given resistance in circuit ;

(3) The E.M.F. of the generator is the smallest that will maintain that current flowing through that length of arc with the fixed resistance in circuit.

Lastly, let V be fixed, that is, let a fixed P.D. be maintained between the ends of the carbons. Then $\dfrac{\mathrm{V}}{\mathrm{E}}$ must be greatest when E is least. But, since $\mathrm{E} = \mathrm{V} + \mathrm{A}\,r$, and V is fixed, E must be least when $\mathrm{A}\,r$ is least. Now, from equation (11)

$$2\,\mathrm{A}\,r = \mathrm{E} - a - b\,l \pm \sqrt{(\mathrm{E} - a - b\,l)^2 - 4\,r(c + d\,l)},$$

it is plain that $\mathrm{A}\,r$ is least when l is greatest, and when

$$(\mathrm{E} - a - b\,l)^2 = 4\,r(c + d\,l),$$

that is, when l is greatest and when equations (12) and (13) hold. From these two equations, equation (15) is obtained

$$\mathrm{A}\,r = \frac{c + d\,l}{\mathrm{A}},$$

showing that since l is to be as large as possible, A must also be as large as possible for $\mathrm{A}\,r$ to be as small as possible, or, taking the form

$$r = \frac{c + d\,l}{\mathrm{A}^2},$$

since A is to be as large as possible, r must evidently be the smallest external resistance that can be used with the longest arc that can be maintained having the fixed P.D. between the carbons. Thus, when the P.D. between the carbons is fixed, the greatest value of $\dfrac{\mathrm{V}}{\mathrm{E}}$ will be obtained by using

(1) The longest arc that can be maintained with the fixed P.D. between the carbons ;

(2) The largest silent current that will flow with this length of arc ;

(3) The smallest external resistance and E.M.F. of the dynamo with which that length of arc and current can be maintained.

It will be seen that some of the conditions necessary to ensure the greatest value of $\frac{V}{E}$ change considerably according to which of the conditions of the circuit are taken as fixed. The E.M.F. of the generator and the resistance outside the arc have always to be as small as possible, but the length of the arc and the current have sometimes to be as small as possible and sometimes as large. Table XLII. gives a short *resumé* of these conditions, each variable being fixed in turn.

Table XLII.—*Conditions for* $\frac{V}{E}$ *to be Greatest when* E, V, A, *l and r, are Fixed in Turn.*

E.	V.	A.	*l*	*r*
Fixed		Largest	Longest	Smallest
Smallest	*Fixed*	Largest	Longest	Smallest
Smallest		*Fixed*	Shortest	Smallest
Smallest		Largest	*Fixed*	Smallest
Smallest		Smallest	Shortest	*Fixed*

It will not be out of place to give a few examples of the use that may be made of the foregoing methods of securing the largest value of $\frac{V}{E}$ when one or more of the quantities with which we have to deal is fixed by circumstances.

To take first the case of a constant E.M.F., let us consider how to arrange two arcs in series with a dynamo giving a constant E.M.F. of 110 volts, so as to make the ratio of the power expended in the arc to the power developed by the dynamo—that is $\frac{V}{E}$—as large as possible.

There is a constant E.M.F. of 55 volts for each arc and half the resistance in circuit external to the arcs, therefore we must make E equal to 55 in equations (12), (13) and (16), to find what current, length of arc and resistance external to the arc will make $\frac{V}{E}$ as large as possible.

In this way we find that roughly

$$A = 19\cdot6,$$
$$l = 4\cdot7,$$
$$r = 0\cdot16,$$
$$\frac{V}{E} = \frac{E - Ar}{E} = 0\cdot94.$$

These are, of course, the extreme conditions, and if these values were actually used the arc would be very unsteady, for not only would the line representing the resistance be a tangent to the curve connecting P.D. and current, in which case it has been shown that the arc would be unstable (pp. 245–247) but the current would be the largest silent current for the length of arc, and, therefore, the arc would be on the point of hissing. But, although we cannot actually use the current, length of arc, and resistance found from equations (12), (13) and (16), yet these will *serve as guides* for the choice of the actual values to be employed, so that $\frac{V}{E}$ may be as large as it is possible to be when the arc is perfectly steady.

Let us then take an arc of 2·5mm. and a current of 14 amperes, so as to be sure of being well within the limits of steadiness. We have now to put E equal to 55, A equal to 14, and l equal to 2·5 in equation (11), which is the *general* equation connecting E, A, l and r. For, as we are not now using the largest current and longest arc that can be maintained with an E.M.F. of 55 volts, we cannot use equations (12), (13) and (16).

Putting the above values for E, A and l in equation (11), we find that roughly

$$r = 0\cdot6,$$
$$\frac{V}{E} = 0\cdot85.$$

Hence, with a constant E.M.F. of 55 volts, and with the solid carbons I employed, the largest value of $\frac{V}{E}$, when the arc was silent and perfectly steady, would be obtained by having a current of about 14 amperes, an arc of about 2·5mm., and a resistance of about 0·6 ohm in the circuit external to the arc.

With a dynamo having a constant E.M.F. of 110 volts, two such arcs could be maintained in series, and, as each arc would require a resistance of 0·6 ohm in the external circuit, the total resistance in the dynamo, leads and regulating coil would have to be 1·2 ohms, and the ratio of V to E would be the same as when one arc was maintained with a constant E.M.F. of 55 volts— viz., 0·85. Hence, with such an arrangement, 85 per cent. of the power developed by the dynamo would be expended in the arcs.

Next let the current be fixed. This will be the case when there are many arcs in series. We must consider one of these arcs with its proportion of the E.M.F. of the generator, and of the resistance of generator and leads.

Let us take a fixed current of 10 amperes, and let us suppose that the arcs are found to give the most light when they are 3mm. in length.

Then we must put A equal to 10 and l equal to 3 in equations (12) and (13) to find what E.M.F. and resistance to use, so that $\dfrac{V}{E}$ may be as large as possible. Doing this we find that roughly

$$E = 54 \text{ volts},$$
$$r = 0.43 \text{ ohm},$$
$$\frac{V}{E} = 0.92.$$

With these values, however, as before, the arcs would be unsteady. We must use a greater E.M.F. and a higher resistance to maintain a steady arc of 3mm. with a current of 10 amperes. With an E.M.F. of 60 volts for each, the arcs would certainly be steady. If, then, we put E equal to 60, A equal to 10, and l equal to 3 in equation (11), we shall find what resistance must be used outside each arc, so that a steady arc may be maintained. In this way we find

$$r = 1.06,$$
and
$$\frac{V}{E} = 0.82.$$

Thus, if a constant current of 10 amperes were flowing, and if the arcs were kept exactly 3mm. in length, an E.M.F. of about 60 volts and a resistance external to the arc of about 1 ohm for each arc would, with solid carbons of the diameters I employed, make $\dfrac{V}{E}$ as large as it could be for the arcs to be in a perfectly stable condition. As, however, they would not remain exactly 3mm.

in length, the E.M.F. would rise as the arc lengthened, until
the regulator acted and brought the carbons together again.
Thus the *average* E.M.F. of the generator should be 60 volts for
each arc, and this would be sufficient to allow of the arcs being
perfectly stable. In that case we have seen that about 82
per cent. of the power developed by the generator would be
consumed in the arcs.

Next let the arc be at such a distance from the dynamo that
the resistance of the leads, dynamo and regulating coil cannot
be made less than 1 ohm, and let us suppose that under these
circumstances an arc of 4mm. gives the best light with uncored
carbons of the size and hardness employed in these experiments.

Then, putting r equal to 1 and l equal to 4 in equations
(12) and (13), we find that

$$A = 7\cdot3 \text{ amperes,}$$
$$E = 61\cdot8 \text{ volts,}$$
$$\frac{V}{E} = 0\cdot88.$$

In this case the current must be increased to make the arc
steady, for we must *raise* the resistance line (Fig. 79, p. 242)
to make it *cut* the curve instead of *touching* it. Let us then use
a current of 11 amperes. Then, putting A equal to 11, l equal
to 4, and r equal to 1 in equation (6), we find that roughly

$$E = 63 \text{ volts}$$
and
$$\frac{V}{E} = 0\cdot82.$$

Hence, with a resistance of 1 ohm in the circuit external
to the arc, an E.M.F. of 63 volts would maintain a steady arc
of 4mm. with a current of 11 amperes flowing, and 82 per
cent. of the power developed in the generator would be
expended in the arc.

Lastly, let a constant P.D. of 50 volts be maintained between
the carbons. Then from equations (3), (12), (13) and (16),
we find that

$$l = 4$$
$$A = 19$$
$$r = 0\cdot15$$
$$E = 53.$$

T

For perfect stability it will be better to take

$$l = 3 \cdot 5$$
$$r = 0 \cdot 5$$
$$A = 12.$$

Then from the equation $\quad E = V + Ar$
$$E = 56.$$

Therefore $\quad \dfrac{V}{E} = 0 \cdot 89.$

The Smallest Resistance that can be Placed Outside a Silent Arc with Given Carbons.

It has been shown that r is always the least resistance that can be placed outside the arc for any given current to flow through a given length of silent arc when equations (12) and (13) express the relations between the variables. From these equations we get $A^2 r = c + d l$, which shows that when l is fixed and r is the smallest resistance that can be placed outside a silent arc of $l$mm, and when a current of A amperes is flowing, r varies inversely as A^2.

Hence, when the arc is silent and its length is fixed, the smallest external resistance that can be used with any current varies inversely as the square of that current.

From this it follows that with a fixed length of arc the larger the silent current the smaller is the smallest resistance that must be inserted in the external circuit when that current is flowing. Consequently the smallest resistance that can be used in the external circuit at all, when the arc is silent and of the given length, is the smallest with which the largest current will flow silently. Thus the smallest resistance that can be used outside a silent arc of given length is that which is repre. sented by the line which touches the curve connecting P.D. and current for that length of arc at the hissing point.

But the relation between the length of the arc and the current at the hissing point will be shown later to be given by equation (16), and the fact that the resistance line is a tangent to the curve is expressed by equations (12) and (13). Therefore from those three equations we can find the smallest external resistance that can be used with *any* silent arc of fixed length.

Not only this, however, but we can also find from the same three equations the smallest external resistance with which a silent arc can be maintained at all. For, from equations (12), (13) and (16), we have

$$r = \frac{(e+fl)^2}{c+dl},$$

or $$l = \frac{dr - 2ef \pm \sqrt{(dr - 2ef)^2 - 4f^2(e^2 - cr)}}{2f^2}. \quad (17)$$

Since the quantity under the root cannot be negative if l is to have a real value,

$(dr - 2ef)^2$ cannot be less than $4f^2(e^2 - cr)$,

or $d^2r^2 - 4dref + 4e^2f^2$,,   ,,   ,,   ,, $4e^2f^2 - 4cf^2r$,

or $d^2r - 4def + 4cf^2$,,   ,,   ,,   ,,   zero.

Hence $d^2r - 4def + 4cf^2 = 0$

gives the smallest value that r can have, and consequently

$$r = \frac{4f}{d^2}(de - cf)$$

gives the value of the smallest resistance that can be placed in the circuit outside the arc in order that a silent arc may be maintained at all.

Since this value of r depends only on the constants c, d, e and f, it is apparent that—

In order to maintain a silent arc at all the total resistance external to the arc cannot have less than a certain minimum value, which depends solely on the carbons employed.

By putting the numerical values for the constants in the last equation, it will be found that, with the carbons used in my experiments, the smallest resistance that can be placed in the circuit outside the arc, for a silent arc to be maintained, is one of 0·1117 ohms. Further, by substituting 0·1117 for r in equations (12), (13) and (16), it will be found that with that resistance in circuit the length of the arc must be 0·6mm., and the current, which will be the largest that will maintain a silent arc of that length, will be one of 12·7 amperes, while the value of $\dfrac{V}{E}$ will be 0·967.

Equation (17) gives two values of l that make $\dfrac{V}{E}$ a maximum for each value of r, but those obtained from the negative root

T 2

are either negative or so small that practically very little light would escape if the arc were maintained at those lengths. Therefore in practice only the lengths given by the positive root would be used.

SUMMARY.

I. The ratio of the power expended in a silent arc to the power developed by the generator is greater

  (1) The shorter the arc ;
  (2) The larger the current ;
  (3) The more nearly the resistance outside the arc and the E.M.F. of the generator approach to the smallest with which the arc can be maintained.

II. The influence on this ratio of the first of these, viz., the length of the arc, is very trifling.

Taking each of the variables in turn as fixed we have—

III. With a fixed length of arc the above ratio is greater the *larger* the current and the smaller the outside resistance and E.M.F. of the generator.

IV. With a fixed E.M.F. in the generator the ratio is greater the *longer* the arc, the *larger* the current and the *smaller* the outside resistance.

V. With a fixed current the ratio is greater the *shorter* the arc and the *smaller* the E.M.F. of the generator and the external resistance.

VI. With a fixed external resistance the ratio is greater the *shorter* the arc, the *smaller* the current and the *smaller* the E.M.F. of the generator.

VII. With a fixed P.D. between the carbons the ratio is greater the *longer* the arc, the *larger* the current and the *smaller* the external resistance and the E.M.F. of the generator.

VIII. The smallest resistance outside the arc that can be used with any current varies inversely as the square of that current.

IX. There is a certain minimum resistance needed in the circuit outside the arc, which depends only on the nature of the carbons employed.

CHAPTER X.

HISSING ARCS.

The sounds made by the direct current arc are very varied, and in many cases depend upon causes too obscure to have been yet detected. There are, however, three, which accompany well-defined phenomena, and which may, therefore, be easily classified. These are :—(1) The sort of sound like a kettle just about to boil, which belongs to a long arc with a small current, but which is *not* accompanied by a diminution of the P.D. (2) The sharp hiss of an arc of any length when the current sent through it is too great for a silent arc to be maintained ; this, which is accompanied by a fall of P.D. is the hissing usually referred to when hissing arcs are mentioned. (3) A sound rather like the wind blowing through a crack, which comes just before the hissing with big currents begins. It is a sound very difficult to maintain, for the arc has a great tendency at that particular stage to become either hissing or silent ; it is accompanied by a beautiful intense pale green light along the edge of the positive carbon, or, rather, probably coming from the crater itself, but seen just above the edge of the crater in the image on the screen. It was when this condition was reached that Mr. Trotter found that the arc rotated at a speed varying from 50 to 450 revolutions per second, the latter speed being attained just before the arc actually hissed. (*See* p. 69.)

There seems to be some idea that a "hissing arc" proper is necessarily a short arc, an idea which the curves in Figs. 38 and 41 must immediately dispel. Hissing may occur with any length of arc, if the current be increased beyond what the arc can bear silently. The hissing obtained with a 7mm. arc and a current of

21 amperes, for instance, is of exactly the same character as that obtained with a current of some 10 or 11 amperes when the carbons are quite close together. In both cases it simply means that the current is too big for the arc to burn silently with the particular length employed. What happens is this : If, with *any* length of arc, the current be made small at first, and then gradually increased, a point is reached at which, however slightly the outside resistance may be diminished, the current will no longer increase *gradually*, it suddenly leaps up 2 or 3 amperes, the P.D. as suddenly drops about 10 volts, and the arc begins to hiss.

The part in the curves marked "Hissing Unstable State" in Fig. 41 really indicates a condition of the arc such that, with the given conditions in the circuit, no current between the highest for the silent arc and the lowest for the hissing could be sent through the arc at all. For instance, consider the curve for a 3mm. arc. It is there indicated that no current between 17·5 and 21 amperes would pass through the arc. If, when a current of 17·5 amperes was flowing, the resistance in the circuit was lowered by ever so little, the current immediately leaped up to 21 amperes or thereabout, and the P.D. fell from about 46 volts to about 37·5 volts, and, however cautiously and slowly the resistance was diminished or increased, the same result followed, there would be either a silent arc with about $17\frac{1}{2}$ amperes, or a hissing arc with about 21 amperes.

It follows, then, that the smallest current with which continuous hissing first takes place, and the largest current with which a silent arc of given length can be maintained continuously are almost impossible to determine with absolute certainty. Within the unstable part of the curve, before even arriving at the currents which simply refuse to go through the arc at all, one gets currents which act in a very exasperating manner. They will flow silently for several minutes, and when you think the carbons are just formed for that current and length of arc, down goes the pressure, up goes the current, and the arc begins to hiss.

And, *vice versâ*, when the arc is apparently hissing quite steadily, it will suddenly become silent with a reversal of the action of pressure and current ; and, of course, when once either of these changes has really set in, you have to change the outside resistance to bring the arc back to the state you require, and wait while the carbons form again.

In Fig. 80, which is a copy of Fig. 38 with a few additions, all the lines to the left of the curve A B C represent silent arcs, while immediately to the right of this curve are the dotted lines denoting the unstable condition, and still further to the right are the lines representing the hissing arc.

It is important to bear in mind, for reasons which will appear later, that these curves all represent *normal* arcs.

Any discussion of the hissing arc must necessarily deal with it in its three states: (1) when on the point of hissing, (2) when in the unstable condition, and (3) when actually hissing. Hence, in the present chapter, our attention will be entirely directed to that part of Fig. 80 that is to the right of the line representing, say, 12 amperes ; for that part includes all three states for each of the constant lengths.

An examination of these curves shows that with the carbons used, and with the *normal* arc, the following results are met with :—

(1) When the length of the arc is constant and the arc is silent, it may be made to hiss by increasing the current sufficiently.

(2) The largest current that will maintain a *silent* arc is greater the longer the arc.

(3) The hissing point always occurs on the flat part of the curve ; that is to say, the P.D. has always gained a value, which changes very slightly with change of current before hissing begins.

(4) When the current is constant and the arc is silent, *shortening* the arc will make it hiss.

(5) When the arc begins to hiss, the P.D. suddenly falls about 10 volts and the current suddenly rises 2 or 3 amperes.

(6) For the hissing arc the P.D. is constant for a given length of arc, whatever the current.

It was Niaudet,[*] who, in 1881, first observed the fall of about 10 volts in the P.D. between the carbons at the moment hissing began, and in 1886 Messrs. Cross and Shepard[†] made

---

[*] *La Lumière Electrique*, 1881, Vol. III., p. 287.

[†] *Proceedings* of the American Academy of Arts and Sciences, 1886, p. 227.

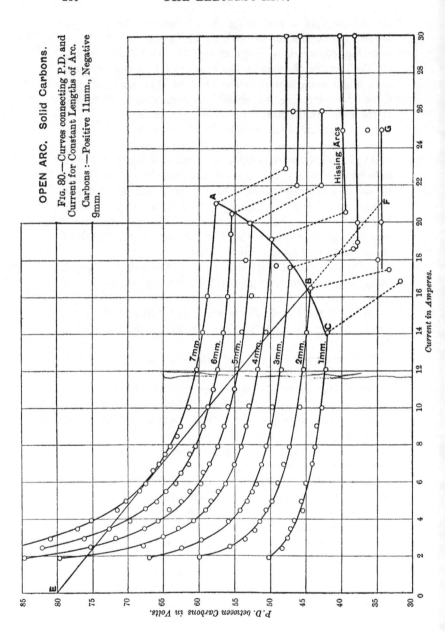

OPEN ARC. Solid Carbons.

FIG. 80.—Curves connecting P.D. and Current for Constant Lengths of Arc. Carbons:—Positive 11mm, Negative 9mm.

some very careful experiments to see whether Edlund's resistance law applied to hissing as well as to silent arcs. They found that it did, and that

if   *r* be the apparent resistance of the arc in ohms,
         l its length in millimetres
and   *a* and *b* constants for given carbons, depending on the current alone, the equation

$$r = a + b\,l$$

applies to the hissing no less than to the silent arc, *a* being smaller however and *b* greater with the first than with the second.

As far back as 1889, also, Luggin* showed that, however long an arc might be, it would still hiss were the current increased sufficiently.

At the Congress at Chicago in 1893, Prof. Ayrton† first drew attention to the region of instability, or rather, the region of blankness corresponding with the impossibity of maintaining any *normal* arc with a particular range of current for each length. At the same time he pointed out in Fig. 41, shown at Chicago, that whether the P.D. was descending as the current increased for, say, a 4mm. arc, or was ascending for, say, a 0·5mm. arc, it became quite constant for wide variations of current with a hissing arc.

Lastly, by a comparison of Fig. 41 with Fig. 39 he brought out the fact that the largest current that would flow silently with any given length of arc was increased by using thicker carbons. For the carbons in Fig. 39 have about twice the diameter of those in Fig. 41, and while the largest silent current for, say, the 2mm. arc in Fig. 41 is 15·5 amperes, that for the same length of arc in Fig. 39 is about 49 amperes, or more than three times as great.

It is plain that the dotted lines in Figs. 80, 39 and 41, divide the curves into two perfectly separate parts, governed by different laws. For to the left of the dotted part the lines are all curved, and curved differently according as solid or cored positive carbons are used, showing that with silent arcs the P.D. varies as the current varies, and that the law of variation is different with solid and cored carbons. To the

---

* *Wien Sitzungsberichte,* 1889, Vol. XLVII., p. 118.
† *The Electrician,* 1895, Vol. XXXIV, pp. 336-7.

right, on the other hand, the lines are all straight, and more or less parallel to the axis of current, whether the positive carbon is solid or cored, showing that with *hissing* arcs the P.D. is the same for a given length of arc and a given pair of carbons, *whatever* current is flowing, and that this law is true whether the carbons be cored or solid. In fact, some complete sudden break-down appears to occur when hissing begins, upsetting all the laws that have governed the arc while it was silent, and bringing the behaviour of cored and solid carbons into accord.

Thus, the subject of the hissing arc divides itself quite naturally into two distinct portions, the one dealing with the arc when the break-down is imminent, but before it has actually occurred—dealing, that is to say, with the points at which the current is the largest that will flow silently—the *hissing points* as I have called them; and the other dealing with the arc after the break-down has occurred, and when, therefore, the arc is really hissing.

An examination of Fig. 80 shows that the hissing points lie well on the curve A B C, the equation to which has been obtained in the following way : The P.D. of each point on the curve was plotted as ordinate, with its corresponding length of arc as abscissa, and the result was found to be a very fair straight line, the equation to which was

$$V = 40 \cdot 05 + 2 \cdot 49 \, l. \qquad . \quad . \quad . \quad (18)$$

which shows that at the hissing points any given increase in the length of the arc causes an increase in the P.D. between the carbons that is simply proportional to the increase of length. That is to say, for every millimetre that is added to the length of the arc, 2·49 volts is added to the P.D. between the carbons at the hissing point.

In Table XLIII. a comparison is made between the observed P.Ds. between the carbons at the hissing points and the P.Ds. calculated from equation (18). It will be seen that the greatest difference between any two corresponding values is only 0·6 volt.

Equating the two values of V given by equations (18) and (3) (p. 184), we get

$$40 \cdot 05 + 2 \cdot 49 \, l = 38 \cdot 88 + 2 \cdot 074 \, l + \frac{11 \cdot 66 + 10 \cdot 54 \, l}{A},$$

or
$$A = \frac{11 \cdot 66 + 10 \cdot 54 \, l}{1 \cdot 17 + 0 \cdot 416 \, l}. \qquad . \quad . \quad . \quad (19)$$

Table XLIII.—*Observed Values of P.D. between Carbons at Hissing Points, Values Calculated from Equation (18), and Differences between the Two.*
Solid Carbons : Positive, 11mm. ; negative, 9mm..

Length of arc in millimetres.	Observed P.D. between carbons in volts.	P.D. between carbons in volts, calculated from equation (18).	Difference in volts.
1	42·2	42·5	− 0·3
2	44·5	45·0	− 0·5
3	47·5	47·5	0
4	49·4	50·0	− 0·6
5	53·0	52·5	+ 0·5
6	55·5	54·9	+ 0·6
7	56·9	57·4	− 0·5

This equation gives, with a very fair degree of accuracy, the current for each of the points on the curve ABC for the different lengths of arc, as will be seen from Table XLIV., which gives the observed values of the currents at the hissing points, the values calculated from equation (19), and the differences between the two.

Table XLIV.—*Observed Values of Currents at Hissing Points, Values Calculated from Equation (19), and Differences between the Two.*
Solid Carbons: Positive, 11mm. ; negative, 9mm.

Length of arc in millimetres.	Observed current in amperes.	Current in amperes, calculated from equation (19).	Difference in amperes.
1	14·06	13·95	+ 0·11
2	16·55	16·37	+ 0·18
3	17·54	17·88	− 0·34
4	19·22	19·02	+ 0·2
5	20·0	19·8	+ 0·2
6	20·5	20·5	0
7	21·0	20·94	+ 0·06

If we put equation (19) in the form

$$l = \frac{1 \cdot 17 \, A - 11 \cdot 66}{10 \cdot 54 - 0 \cdot 416 \, A},$$

it becomes obvious that when

$$10 \cdot 54 - 0 \cdot 416 \, A = 0$$

—that is, when A = 25·3 amperes—*l* is infinite, and hence, if this equation holds for all lengths of arc, and not only for arcs of from 1mm. to 7mm., *there is a maximum current with which a silent arc can be maintained, and any current greater than this will cause the arc to hiss, however long it may be.* With the carbons I used this maximum current is evidently 25·3 amperes.

Equating the two values of *l* obtained from equations (18) and (19), we get

$$V - 40\cdot05 = 2\cdot49 \left(\frac{1\cdot17\,A - 11\cdot66}{10\cdot54 - 0\cdot416\,A} \right),$$

or
$$V = 40\cdot05 + \frac{2\cdot91\,A - 29\cdot02}{10\cdot54 - 0\cdot416\,A}.$$

as the equation to the curve ABC when the same axes and units are used as for all the other curves in Fig. 80, p. 280.

To turn, now, to the arc when hissing has actually begun.

In Figs. 44, 45, 46 and 47 curves were given connecting P.D. with length of *silent* arc for various constant currents with each of the four pairs of carbons. If, however, we proceed to draw such curves for *hissing* arcs, then, in consequence of the curve connecting P.D. with current for each length of *hissing* arc being practically a horizontal straight line, we obtain only *one* curve connecting the P.D. with the length of the arc. For this curve is the same for any current which causes the arc to hiss. In other words, all the curves in any one of Figs. 44, 45, 46, or 47 which refer to *silent* arcs close up into a single curve for *hissing* arcs. Thus, the law connecting the P.D. between the carbons with the length of the arc, when hissing, can be found from Fig. 80, by plotting the mean P.D. between the carbons for each length of arc, when hissing, with the corresponding lengths of arc. In this way we get a straight line, the equation to which is

$$V = 29\cdot25 + 2\cdot75\,l \quad . \quad . \quad . \quad . \quad (20)$$

How far equation (20) really sums up the facts may be seen from Table XLV., which gives the mean value of the observed P.D. between the carbons for each length of hissing arc, the P.D. calculated from equation (20), and the difference between the two.

Table XLV.—*Hissing Arcs. Mean of Observed Values of P.D. between Carbons, P.Ds. calculated from Equation (20) and Differences between the Two. Solid Carbons: Positive, 11mm.; negative, 9mm.*

Length of arc in millimetres.	Mean P.D. between carbons in volts.	P.D. calculated from equation (20).	Difference in volts.
1	32·0	32·0	0
2	34·4	34·75	− 0·35
3	37·8	37·5	+ 0·3
4	40·0	40·25	− 0·25
5	43·0	43·0	0
6	46·5	45·75	+ 0·75
7	48·0	48·5	− 0·5

Equation (20) shows that, *with the hissing as with the silent arc, a straight line law connects the P.D. between the carbons with the length of the arc when both carbons are solid.*

There is, however, this vast difference between the two laws: that for silent arcs the law only holds for constant currents or for the currents at the hissing points, whereas for hissing arcs it holds *whatever the current may be.* Thus, while for silent arcs the constants, which correspond with the terms 29·25 and 2·75 in equation (20), are constant only for each separate current, and change when the current changes, with *hissing* arcs they remain the same *whatever* the value of the current may be. For instance, the equation equivalent to equation (20) for a normal silent arc with a current of 4 amperes may be found from equation (3) (p. 184) to be

$$V = 41·79 + 4·71\, l,$$

and with a current of 12 amperes it is

$$V = 39·85 + 2·95\, l ;$$

but with the hissing arc the equation is

$$V = 29·25 + 2·75\, l,$$

whether the current be one of 20 amperes, or of 50, *and whether the arc be normal or not.*

And here I may explain the reason for the great importance of distinguishing between arcs that are normal and those that are not. We have seen that, with normal arcs of any given length, hissing only starts when all the silent arcs have been used up, as it were ; that is to say, when the current is *greater* than it can be with any silent arc of the same length. But with a *non*-normal arc of 2mm. I have been able to produce hissing with a current of 11 amperes, and to have a silent arc burning with a current of 28 amperes, the same carbons being used in each case. This apparent anomaly will be fully explained later, when we go into the causes that produce hissing.

In 1889, Luggin found, by measuring the fall of potential between each carbon and the arc, that the principal part of the diminution of P.D. caused by hissing took place at the junction of the positive carbon and the arc. Some experiments of the same sort made by myself gave the same result. The carbons used were, as usual, solid Apostle carbons, the positive 11mm. and the negative 9mm. in diameter. The third carbon to place in the arc was rather thick, 2mm. in diameter, but, like the similar carbons mentioned in Chapter VII., p. 210, they burnt well to a point in the arc, and, with the current employed —25 amperes—thinner carbons were consumed too rapidly for good observations to be made.

The P.D. between the positive carbon and the arc was found by placing the third carbon in the arc as close as possible to the positive carbon, and measuring the P.D. between the two with the high resistance voltmeter mentioned in Chapter VII. This was easily done when the arc was hissing, but was impossible when the largest silent current was flowing, for then the mere insertion of the third carbon was sufficient to make the arc hiss. Accordingly, the P.D. between the positive carbon and the arc when the largest *silent* current was flowing has had to be calculated from the formula (given in Chapter VII. p. 222) for calculating that P.D. with *any* silent current, viz. :

$$V = 31 \cdot 28 + \frac{9 + 3 \cdot 1 \; l}{A}.$$

In Table XLVI. two sets of currents are dealt with, viz., the largest silent current for various lengths of arc, and a hissing current of 25 amperes, and for each of these sets of currents and lengths of arc two P.Ds. are given, viz., the P.D. between the main carbons and the P.D. between the positive carbon and the arc itself.

Table **XLVI.**—*P.D. between Carbons, and P.D. between Positive Carbon and Arc with Largest Silent Current and with Hissing Current of 25 Amperes.*
Solid Carbons : Positive 11mm. ; negative 9mm.

Length of arc in millimetres. (1)	Largest silent current.		Hissing current of 25 amperes.	
	P.D. between Carbons in volts. (2)	P.D. between positive carbon and arc in volts (calculated). (3)	P.D. between carbons in volts. (4)	P.D. between positive carbon and arc in volts. (5)
1	42·2	32·1	32·1	24·4
2	44·5	32·2	34·6	25·2
3	47·5	32·3	37·0	25·7
4	49·4	32·4	40·5	25·7
5	53·0	32·5	43·9	27·9
6	55·5	32·6	45·9	27·2

Now in order to compare the change in the P.D. between the main carbons caused by hissing with the corresponding change in the P.D. between the positive carbon and the arc, we must subtract column (4) of Table XLVI. from column (2), and column (5) from column (3), and compare the differences. These differences are given in Table XLVII.

Table **XLVII.**—*Diminution of P.D. between Carbons due to Hissing compared with Corresponding Diminution of P.D. between Positive Carbon and Arc.*
Solid Carbons : Positive 11mm. ; negative 9mm.

Length of arc in millimetres. (1)	Diminution of P.D. between carbons due to hissing. (2)	Diminution of P.D. between positive carbon and arc due to hissing. (3)
1	10·1	7·7
2	9·9	7·0
3	10·5	6·6
4	8·9	6·7
5	9·1	4·6
6	9·6	5·4

Thus, for the lengths of arc dealt with, hissing causes a mean fall of about 9·7 volts in the total P.D. between the carbons, and a mean fall of about 6·3 volts in the P.D. between the positive carbon and the arc. Hence of the whole diminution of the P.D. between the carbons caused by hissing, about two-thirds takes place apparently at the junction of the positive carbon and the arc.

Further, my experiments showed that very little of the remainder of the diminution, if any, was due to a fall of the P.D. between the arc and the negative carbon; therefore, this remaining diminution must be attributed to a lowering of the resistance of the arc itself. We may sum up these results as follows:—

Of the total diminution of the P.D. between the carbons caused by hissing, about two-thirds takes place at the junction of the positive carbon and the arc, and the remaining third seems to be due to a lowering of the resistance of the arc itself.

From equations (18) and (20) we can find the law that connects the change that takes place in the P.D. between the carbons when hissing begins with the length of the arc. For if we call V the P.D. between the carbons at the hissing point with any given length of arc l, and V' the same P.D. when the arc of the same length is actually hissing, then, from these equations, we get

$$V - V' = 10\cdot8 - 0\cdot26l \quad . \quad . \quad . \quad . \quad (21)$$

which shows that *the longer the arc the less the P.D. between the carbons is diminished when its condition changes from silence to hissing.*

From Fig. 80 it might be supposed that, given the length of the arc, the sudden increase of current that occurs when the arc starts hissing was as definite for that length of arc as the diminution in the P.D. This, for a long time, I imagined to be the case, but, while trying to find out what law connected the smallest hissing current for any given length of arc with that length, I saw that the value of that current really depended on the circuit *outside* the arc.

For let E be the E.M.F. in volts of the generator, which we will assume to be constant and independent of the current;

    ,, r ,,   resistance in ohms of the whole circuit *outside* the arc;

    ,, l ,,   length of the arc in millimetres;

Let A be the largest silent current in amperes ;

  ,, V ,,   corresponding P.D. between the carbons in volts ;

  ,, A′ ,,   smallest hissing current in amperes ;

  ,, V′ ,,   corresponding P.D. in volts.

Then $$E = V + A\,r,$$

and, since the change in either E, or r, or both, is infinitely small, when the largest silent current changes to the smallest hissing current

$$E = V′ + A′r,$$
$$\therefore\ A′ - A = \frac{V - V′}{r}.$$

That is, the sudden increase of current when hissing begins is equal to the sudden diminution of P.D. divided by the resistance of the circuit outside the arc.

Again, $$\frac{A′}{A} = \frac{E - V′}{E - V},$$

or $$A′ = \frac{E - V′}{E - V}A.$$

But for a given hissing point V, V′ and A are all constants ; therefore, for such a point, A′ depends simply on the external conditions, and may be calculated in terms of A, V, V′, and either E, the E.M.F. of the generator, or r, the resistance in the circuit outside the arc.

It is now possible to see what the dotted lines in Fig. 80 really mean. For let B be the hissing point for a given length of arc, and F G the line connecting the hissing P.D. with the current for the same length of arc. Let E be a point on the axis of P.D., such that its distance from the axis of current measures the E.M.F. of the dynamo when the hissing point B was found. Draw the line E B, and continue it to meet F G in F. Then the distance of F from the axis of P.D. measures the smallest hissing current possible for the given length of arc with the given E.M.F.

For, as has already been shown on page 241, the slope of the line E B indicates the resistance in circuit outside the arc when the point B was found, and, since this resistance remains practically unchanged when the arc begins to hiss, the point at

which the line E B F meets F G must give the current that will
flow when the arc first begins to hiss—that is, the smallest hissing
current. Hence the slope of the line B F shows the resistance that
was in circuit outside the arc when both the points B and F
were found, and the point at which the continuation of this line
meets the axis of P.D. shows the E.M.F. that the dynamo had
at the time.

Consequently it now appears that the dotted lines in the
unstable region constitute records of the particular E.M.F.'s
the dynamo was made to give and the particular resistances
that were in the circuit outside the arc, on the various days
when the experiments were made with the different lengths of
arc several years ago.

*Hence, when the largest silent current changes to the smallest
hissing current for the same length of arc, the value of that
smallest hissing current depends only on the E.M.F. of the
generator or the resistance in the outer circuit, whichever is chosen
first.* Thus, it is possible, by choosing suitable E.M.F.'s, to
make the sudden smallest hissing current have any value greater
than that of the largest silent current for the same length of
arc, and the larger the E.M.F. the more nearly equal will the
two currents be.

It is evident from Fig. 80 that the smaller the E.M.F. of
the generator, the larger will be the value of the smallest
hissing current, for the lower down will E be on the axis of
P.D., and therefore the farther will the point F be along the
line F G. This explains a circumstance that puzzled me
greatly when it happened, but which is now perfectly com-
prehensible. Some years ago I was using accumulators to maintain an arc, and in order to be able to keep the P.D. between
the carbons as constant as possible, for the experiment described
in Chapter V., p. 171, I was employing as small a number of
cells as possible. I was able to have quite a moderate current
as long as the arc was silent, but as soon as it began to hiss,
the current rushed up to some huge value which would inevit-
ably have ruined the cells, if I had not had a cut-out arranged
to break the circuit. Why the first hissing current should be
so much greater than I was accustomed to find it with the
dynamos I ordinarily used, I could not imagine, but the reason
is now perfectly obvious. The hissing current was so great

simply because the E.M.F. of the cells was so small, and had it been possible to maintain a silent arc without any resistance in the outside circuit except that of the cells, which is what I was trying to accomplish, I might, except for the cut-out coming into operation, have had practically an infinite current when the arc began to hiss.

The change produced in the law connecting the hissing P.D. with the length of the arc, by coring the positive carbon, may be gathered from Table XLVIII., which gives the values of the abscissæ and ordinates of the curve connecting P.D. with length of arc for all hissing currents, when the positive carbon was 9mm. cored and the negative 8mm. solid.

Table **XLVIII.**—*Hissing Arcs. P.D. between Carbons and Length of Arc for Any Current that Causes Hissing. Carbons : Positive, 9mm., cored ; negative, 8mm. solid.*

Length of arc in millimetres.	P.D. between the carbons in volts.	Length of arc in millimetres.	P.D. between the carbons in volts.
0·0	29·8	3·0	37·5
0·5	32·1	4·0	40·1
1·0	34·2	5·0	41·5
2·0	35·8	8·0	49·2

On plotting these numbers we do not obtain a straight line, as with solid carbons, but a curve. Thus, although, when the arc hisses, the connection between P.D. and current for a constant length of arc follows the same law whether the positive carbon be cored or not, yet this is not the case with the connection between P.D. and length of arc. For this connection, which follows a straight line law when both carbons are solid, follows some far more complicated law—probably depending on the relative dimensions, composition and hardness of the core and its case—when the positive carbon is cored.

We now pass from the consideration of the electrical measurements of the arc to the appearance of the crater, arc, and carbons.

Every alteration of the current and of the distance between the carbons naturally produces a corresponding modification of all parts of the arc, but, until the value of the current attains a certain magnitude, which depends only on the length of the arc, with a given pair of carbons, this change is one of degree merely, and not of character. A greater current simply produces a larger crater, a larger arc, and longer points to the carbons. When the special current is reached, however, a change, which is no longer simply one of degree, takes place in the crater. Instead of presenting a uniformly bright surface to the eye, this becomes partly covered with what appear to be alternately bright and dark bands, sometimes placed radially, like the spokes of a wheel, sometimes in one or more sets of concentric circles, moving round different centres in opposite directions. The directions of rotation and whole positions of the images change continually, and the motion grows faster and faster as the current is increased.

When the current is so much increased that the motion becomes too fast for the eye to detect, the arc begins to hum, and then, as Mr. Trotter* first showed in 1894, it rotates at from 50 to 450 revolutions per second. These rapid revolutions, which the unaided eye is incapable of observing, he discovered by the use of a disc having alternate arms and spaces, and kept in rapid rotation. They appear to begin just where the slower oscillations and rotations described above become too quick for the eye to see unaided, and end just as the arc begins to hiss, for he mentions that at 450 revolutions per second the arc breaks into a hiss.

As soon as hissing begins the whole appearance of the crater changes again; a sort of cloud seems to draw in round a part of it, moving from the outer edge inwards as in (*a*) (Fig. 81), and varying continually in shape and position. Sometimes but one bright spot is left, sometimes several, but always the surface is divided into bright and dull parts, giving it a mottled appearance, as is seen in each figure of Fig. 81. After hissing has continued for some time the surface of the crater is pitted with holes separated from one another by ridges as seen in (*c*) Fig. 81. If, then, the current be diminished, so that the arc becomes silent again, the whole surface of the crater grows dark for an instant, (*e*) Fig. 81, then the ridges brighten as in (*f*) Fig. 81, and finally it becomes bright again all over†.

---

* *Proc.* Roy. Soc., Vol. LVI., p. 262.

† The photograph from which Fig. 81 was made was kindly taken for me by Messrs. Fithian and Denny, assisted by Mr. Fawnthorpe, students at the Central Technical College.

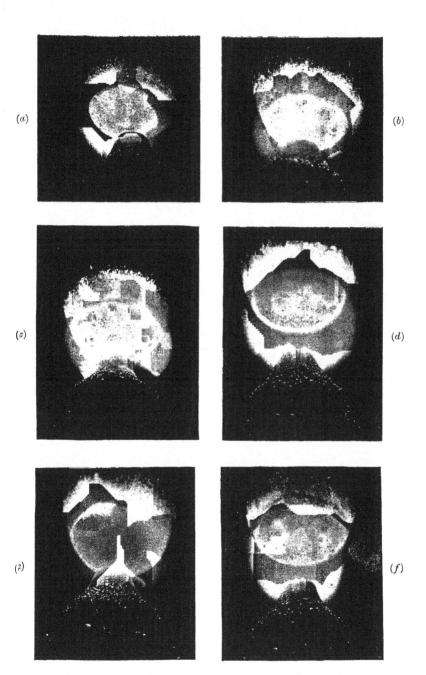

Fig. 81.—Photographs of Arcs (a) immediately after hissing has begun, (b) after it has continued for a very short time, (c) and (d) after it has continued longer, (e) immediately after the arc has become silent again, (f) after it has been silent for a very short period.

The vaporous arc itself undergoes fewer modifications; it preserves the ordinary characteristics of the silent arc while bands of light and darkness hold possession of the crater, but, when humming begins, a green light is seen to issue from the crater, and with hissing this becomes enlarged and intensified, till the whole centre of the purple core is occupied by a brilliant greenish-blue light, as is indicated in Fig. 82. The vapour also becomes apparently less transparent, sometimes even almost opaque enough to hide parts of the crater with a sort of violet mist, as was first mentioned by M. Blondel in 1893.*

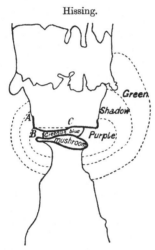

Fig. 82.—Solid Carbons : Positive, 11mm. ; negative, 9mm. Length of Arc, 1·5mm.   Current, 28·5 amperes.

The *shape* of the arc now alters also.   While it is silent or humming, no great difference can be observed in its form. With solid carbons it is rounded or pear-shaped according to its length, and has an appearance of great stability.   But as soon as hissing occurs, the arc seems to dart out suddenly from between the carbons, and to become flattened out.   In Fig. 81 this flattened appearance is well marked, as it is also in Fig. 83 and in (d) Fig. 84; and, indeed, these figures show that every part of the vaporous arc itself is involved in this

* *The Electrician,* 1893, Vol. XXXII., p. 170.

flattening—the purple core, the shadow round it, and the green aureole. In (*b*) and (*d*) Fig. 81, which were taken with isochromatic plates, the flattening has a curious brush-like appearance, especially near the positive carbon. The vertical shadowy lines in most of these figures are noticeable, and want accounting for.

As regards the carbons themselves, the only important modification of the *negative* carbon that appears to be due to hissing is the formation of the well-known "mushroom" at the end of that carbon with a *short* hissing arc. This mushroom, of which a good example is seen in Fig. 82, is well named, not only because of its shape, but also because of the

Silent. Hissing.

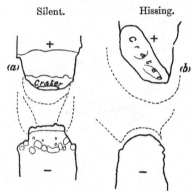

FIG. 83.—Carbons: Positive, 9mm., cored; negative, 8mm., solid.
Length of Arc, (*a*) 5mm., (*b*) 8mm.
Current, (*a*) 3·5 amperes, (*b*) 34 amperes.

rapidity of its growth, which is so great that while it is forming the carbons often have to be *separated*, instead of being *brought together*, to keep the length of the arc constant.

And now we come to the most important of all the changes that take place when the arc begins to hiss, viz., the alteration in the shape of the *positive carbon.*

During the course of his 1889 experiments, Luggin* observed that the arc hissed when the crater filled the whole of the end of the positive carbon. He was thus the first to call attention to the fact that there was a direct connection between hissing

---

* *Wien Sitzungsberichte*, 1889, Vol. XCVIII., p. 1192.

and the relation between the area of the crater and the cross section of the tip of the positive carbon. My own observation

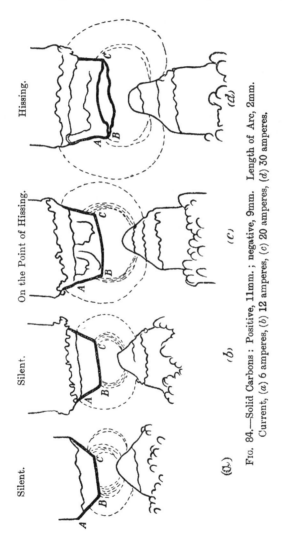

FIG. 84.—Solid Carbons : Positive, 11mm. ; negative, 9mm. Length of Arc, 2mm. Current, (*a*) 6 amperes, (*b*) 12 amperes, (*c*) 20 amperes, (*d*) 30 amperes.

in 1893 led to a conclusion somewhat similar to Luggin's, but yet differing in an important particular. It seemed to me that

with hissing arcs the crater always *more* than covered the end
of the positive carbon—that it overflowed, as it were, along the
side.

How far this is true will be seen from an examination of
Figs. 82, 83, 84 and 85, which show the shaping of the carbons
under various conditions with silent and hissing arcs. These
figures have all been made from tracings of the images of
actual normal arcs, burning between carbons of various sizes.
Fig. 82 is the image of a short hissing arc, for Fig. 83
the diameters of the carbons were the same, but the currents
and lengths of arc were different, for Fig. 84 the carbons

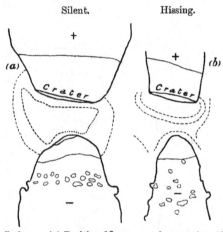

Fig. 85.—Carbons : (*a*) Positive, 18mm., cored ; negative, 15mm., solid.
(*b*) Positive, 9mm., cored ; negative, 8mm., solid.
Length of Arc, 5mm.   Current, 25 amperes.

were all of the same size, and the arcs of the same length, but
the current had four different values, while for Fig. 85 the
current and the length of the arc were the same for both
(*a*) and (*b*), but the diameter of one of the positive carbons was
twice that of the other. The figures were carefully chosen with
special reference to the shaping of the positive carbons ; for,
with normal arcs, the shape of the end of a positive carbon,
even taken quite apart from that of the negative carbon and of
the vaporous arc itself, is capable of revealing almost the whole
of the conditions under which the arc was burning when the

positive was shaped. It is possible, for instance, with a normal arc, to tell, from a mere drawing of the outline of the positive carbon and of its crater, whether the arc with which it was formed had been open or enclosed, short or long, silent or hissing, burning with a large or with a small current for the size of the carbon.

Take, for example, Fig. 83, and note the difference in the shape of the positive carbon with a current of 3·5 amperes as in (a), and with one of 34 amperes, as in (b). In the first case the tip of the positive carbon is rounded, so that the crater lies in its smallest cross-section; in the second, the tip would be practically cylindrical for some distance, but that the crater has burnt away a part of the cylinder, making the tip look as if it had been sheared off obliquely. Comparing now the tips of the positive carbons when the arc is silent and when it is hissing in all the four figures, 82, 83, 84, 85, we find the same difference. With all the silent arcs the tip is more or less rounded, and the crater lies in its smallest cross-section, and consequently is less in area than any but the smallest cross-section. With all the hissing arcs, on the other hand, the tip of the positive carbon is practically cylindrical for a short distance at least, or would be, but that it is sheared away by the crater; consequently the area of the crater is *greater* than the smallest cross-section of the tip, or indeed than the cross-section of the tip for some little distance along its length.

We have now arrived at the real, the *crucial*, distinction between a silent and a hissing arc. When the crater occupies the *end* only of the positive carbon, the arc is *silent*; when it not only covers the end, but also extends up the *side*, the arc *hisses*. Hence, it must be at the *hissing point* when the smallest increase in the area of the crater will make it begin to cover the *side* of the positive carbon, and this can only be when the tip of that carbon has very nearly the same cross-section for some little distance from its end—in other words, when its sides are nearly vertical.

It is thus impossible to doubt that there is some connection between the extension of the crater up the side of the positive carbon and hissing, although, so far, it has not been possible to detect which was cause and which was effect. We shall presently see that the extension of the crater is the *cause* and hissing the *effect* ; that, in fact, *hissing is produced by the crater becoming too large to occupy the end only of the positive carbon, and by its, therefore, extending up its side.*

Before proceeding to prove this, however, it will be interesting to see how the laws for the largest silent currents with normal arcs, which have been already obtained from the electrical measurements on pages 279–284, may be deduced on the above hypothesis from Figs. 84 and 85.

In Fig. 84 we have a series of four normal arcs of the same length, burning between solid carbons of the same diameter, but in (*a*) the current is 6 amperes, in (*b*) 12, in (*c*) 20, and in (*d*) 30 amperes. The roundness of the tip of the positive carbon may be measured by the obtuseness of the angle ABC between its side and end. In (*a*) the tip is very nearly round, and the area of the crater is certainly less than any but its smallest cross-section ; therefore the arc is certainly silent. In (*b*) the tip is less rounded, but the arc is still evidently silent ; in (*c*) the angle ABC is much more nearly a right angle, and it is plain that a very small increase in the area of the crater would cause it to burn up the side of the tip, therefore the arc is near the hissing point. In (*d*) the angle ABC is practically a right angle, the tip of the positive carbon is cylindrical, and the crater has evidently burnt partly up its side, so that the arc is hissing. Thus, keeping the length of the arc constant and gradually increasing, the current must gradually bring us to a hissing point.

Next, I have shown (pp. 13-17), that with, a constant current, the end of the positive carbon becomes rounder, and occupies a larger portion of the entire cross-section of the carbon rod, the more the carbons are separated. Hence, the longer the arc, the greater must be the area of the crater, and consequently the greater must be the current before the crater extends up the side of the positive carbon. Consequently, the longer the arc, the greater is the largest silent current.

Thirdly, it follows that when the current and the length of the arc have been increased to such an extent that the round, tip of the positive carbon occupies the whole cross section of the carbon rod itself, no further increase in the size of the crater is possible, without a part of it extending up the side of the positive carbon. Hence the largest silent current for a positive carbon of a particular diameter cannot exceed a particular value, however long the arc may be made. Lastly, similar reasoning used in conjunction with Fig. 85 tells us that the thicker the positive carbon the greater must be the largest silent current for a particular length of arc.

Consequently, the fact that hissing occurs when the crater covers more than the end surface of the positive carbon and extends up its side, combined with our knowledge of the way in which the positive carbon shapes itself in practice, is sufficient to enable us to deduce *all* the laws given on page 279, which govern the largest current that will flow silently with the *normal* arc under given conditions.

It is also now obvious why, when the arc is *not* normal, it may be made to hiss with small currents and will be silent with quite large ones. For suppose, for instance, the end of the positive carbon were filed to a long fine point, then a very small current would make a crater large enough to extend up the side of the point, and produce a hissing arc. But if, on the contrary, the end were filed flat, so as to have as large a cross section, as possible, quite a considerable current could flow silently, for in that case it would require the current to be very great for the crater to be large enough to fill up the whole of the end of the positive carbon.

We come now to the question, why should the arc hiss when the crater burns up the side of the positive carbon—what is it that happens then that has not happened previously ? In pondering over this question, the possibility occurred to me that as long as the crater occupied only the end surface of the positive carbon it might be protected from direct contact with the air by the carbon vapour surrounding it, but that, when the crater overlapped the side, the air could penetrate to it immediately, thus causing a part at least of its surface to *burn* instead of volatilising. The crater would probably burn more quickly than it would volatilise, and hence, though the burning parts would be at a lower temperature than the remainder, and so look duller, they would consume more rapidly, so that little pits would form, which would deepen while the air continued to get to them. Thus, the darker, spherical parts of the crater shown in Fig. 81 (which you can see deepen by watching the image after the arc has begun to hiss) would be the burning parts, while the brighter ridges would be volatilising.

Many circumstances at once seemed to combine to show that this was the true explanation. The whirling figures, and Mr. Trotter's still faster rotations, how were they caused but by draughts getting into the arc? Then the humming noise, which is so like the wind blowing through a crack, was not this probably caused by the air rushing through a slight breach in the crater already getting near to the critical size? This air, pouring in faster and faster as the breach widened, would cause the arc to rotate faster and faster, sometimes in one direction, sometimes in another, according as the draught was blown from one side or the other. Then finally the air would actually reach the crater, burn in contact with it, and the P.D. would fall and the arc would hiss.

The following is Mr. Trotter's own explanation of the rotation discovered by him. It is taken from a letter on the subject written by him to Prof. Silvanus Thompson, about the end of June, 1894.

"The crater is pouring out a stream of carbon vapour. With a strong stream of vapour and a short arc, the stream may touch the negative; when it does so in sufficient volume, and to exclusion of oxygen, mushrooming occurs. But as a rule most of it, if not practically all of it, ceases to be carbon vapour before it reaches the negative. This, I want to settle by spectroscope.

"There is a combustion of the vapour, and that means an inrush of air. . . . The inrush of air is radial. It is partly due to the oxygen-carbon combustion, partly also, perhaps, to the oxygen-nitrogen combustion. If any accidental cause, such as a spurt of vapour from an impurity in the carbon, cause the inrush to be otherwise than radial, a rotatory motion is started, and persists, as when water running from a wash-basin moves in a vortex. In a washbasin the water can get away, in a tornado also, the air can get upwards and outwards; but in the arc condensation due to chemical combination and lowering of pressure must be looked for as a sink for the vapour stream."

In the open arc, whether silent or hissing, the outer envelope of the vaporous portion is always bright green. With the hissing arc the light issuing from the *crater* is also bright green, or greenish blue. What so likely as that the two green lights should have a common origin, viz., the combination of carbon with air ? For the outer green light is seen just at the junction of the carbons and carbon vapour with the air, and the inner one only appears when air can get direct to the crater.

Again, why does the arc always hiss when it is first struck ? Is it not because a certain amount of air must always cling to both carbons when they are cold, so that when the crater is first made its surface must combine with this air ?

The cloud that draws in round the crater when hissing begins would be a dulness caused by the burning part of the crater being cooler than the parts which were still volatilising. In fact, everything seemed to point to the direct contact of crater and air as being the cause of the diminution in the P.D. between the two carbons which is the important part of the hissing phenomenon.

One easy and obvious method of testing this theory immediately presented itself. If air were the cause of the hissing phenomena, exclude the air and there would be no sudden diminution of the P.D. between the carbons, however great a current might be used. Accordingly I tried maintaining arcs of different lengths in an enclosed vessel, and increasing the current up to some 40 amperes. *No* sudden diminution of the P.D. could be observed with any of the currents or lengths of arc employed, although when the same carbons were used to produce *open* arcs, the sudden diminution of about 10 volts in the P.D. between the carbons occurred with a current as low as 14 amperes for a 1mm. arc. Indeed, so far from there being any sudden diminution in the P.D. when the current through an *enclosed* arc is raised to higher and higher values, the P.D. appears to increase slightly for large currents.

It was, of course, impossible, in these experiments, to avail myself of an ordinary enclosed arc lamp, since a current of some 5 or 8 amperes only is all that is used with such a lamp, whereas to test my theory it was necessary to employ currents up to 40 amperes, although my carbons were of smaller diameter than those fitted in ordinary commercial enclosed arc lamps. Accordingly, I constructed little electric furnaces, some made out of fire-clay crucibles with lids of graphite sealed on, as in Fig. 86, some moulded out of fire-clay with mica windows inserted, so that the image of the arc could be projected on to a screen and its length kept constant; some constructed of iron lined with asbestos; some with tubes inserted in them through which the air could be admitted when required, &c.

It was found that when the vessel was entirely enclosed, the pressure in it was so great when the arc was first started, that occasionally the lid was blown off. Consequently, the space between the positive carbon and the lid was left open till the

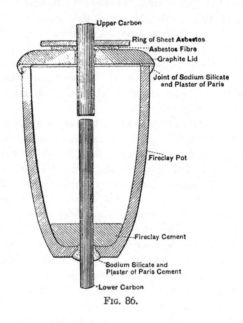

FIG. 86.

arc was well started, and then was tightly closed. This sudden increase of pressure probably took place when the carbons were first *brought into contact,* for Mr. Seaton, while conducting some experiments for Messrs. De la Rue and Müller in 1879 (p. 38) observed that, when the arc was completely enclosed, the increase of pressure when the carbons were first brought into contact was far greater than could be accounted for by the rise of temperature of the gas in the vessel, and that the pressure fell the moment the carbons were separated, almost to what it had been before contact was made. This fact was confirmed by some experiments made by Stenger (p. 44), in 1885. This first great rise of pressure may, of course, be partly caused by the gases occluded in the carbons being expelled on the current being started, but a complete investigation of this phenomenon has not, as far as I am aware, yet been made.

Some curves connecting the P.D. between the carbons with the current, when the arc was completely enclosed in the crucible (Fig. 86) are given in Fig. 87. The carbons were solid, the positive being 11mm. and the negative 9mm. in diameter, similar to those I have used for all my experiments. As this crucible—the first one made—had no window, the length of the arc could not be kept quite constant, but the distance by which the carbons were separated was noted at the beginning of the experiment, and they were then allowed to burn away without being moved till the end, when the distance the positive carbon had to be moved in order to bring it tightly against the negative was noted. Measured in this way, the length of the arc was 1·5mm. at the beginning and 2mm. at the end of the experiment. The current was started at 6 amperes, and gradually increased to 39 amperes; then as gradually diminished to 6 amperes again, increased to 36 amperes, and diminished to 5 amperes, when the arc was extinguished. The P.D. between the carbons for a given current seems to have increased, as the length of time during which the arc had been burning increased; this was undoubtedly partly due to the lengthening of the arc, but was probably also partly due to the whole of the air in the pot having been gradually burnt up, or driven out through the slag wool and the asbestos ring, by the pressure of the carbon vapour.

Many other sets of curves were obtained, but all with the same result, viz., that when once the crucible had been freed from air, no sudden diminution in the P.D. could be observed on increasing the current far beyond the value at which this diminution took place on lifting up the lid and allowing the air to have access to the arc.

The next thing to do was to try if an open arc could be made to hiss, and the P.D. to diminish suddenly, by blowing air at the

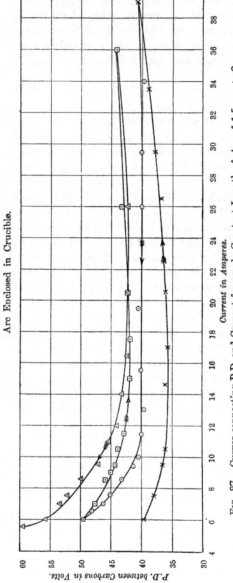

Fig. 87.—Curves connecting P.D. and Current for nearly Constant Length of Arc of 1·5mm. to 2mm. The Arrows show the Direction in which the Current was varied. Solid Carbons: Positive, 11mm.; negative, 9mm.

crater when the current was so small that the crater remained well at the end of the positive carbon—in fact, to bring the air in contact with the crater artificially, when a much smaller current was flowing than would usually produce hissing. I first tried inserting a carbon tube in the arc and blowing through it, but this almost invariably blew the arc out. Then a tubular positive carbon was used and the air was blown down it. This plan answered admirably, for when a current of 10 amperes was flowing with an arc of about 3mm., so that the arc was quite silent, each puff of air blown down through the positive carbon was followed by a hiss and the characteristic diminution of the P.D. between the carbons. With a current of 6 amperes, however, I could get no hiss, but simply blew the arc out with each puff, probably because, with such a small current, the arc was cooled sufficiently to be extinguished before the action could take place.

Oxygen was next tried, still with the open arc, and again each puff produced a hiss and diminution of the P.D., the latter being exactly the same in amount as when air was used, namely, about 10 volts. As my idea was that the diminution of P.D. was due to the chemical combination of air with carbon at the temperature of the crater, the fact of oxygen causing the same diminution of the P.D. as air seemed to show that nitrogen would produce no effect, and that all the effect produced by air was due to the oxygen in it. Accordingly nitrogen was blown down the positive carbon of an open arc, and *no* change in the P.D. followed, if the nitrogen was blown through gently; but beyond a certain pressure, the arc was blown to one side, and thus lengthened, so that the P.D. *rose*, and, if the pressure continued, the arc went out.

This experiment proved two things—firstly, that it is the *oxygen* in the air that causes the diminution in the P.D. with hissing; secondly, that this diminution in the P.D. is not due to cooling, for nitrogen would cool the arc as effectually as oxygen or air.

To make assurance doubly sure on this point, carbon dioxide was blown down the tubular positive carbon, with the same result as when nitrogen was used, viz., no change was produced in the P.D. between the carbons unless the pressure of the gaseous stream were large enough to blow the arc on

x

one side, and then an increase and not a diminution in the P.D. was observed.

If, however, the current was *very near* the value that made an open arc of the particular length used start hissing, blowing either nitrogen or carbon dioxide through the positive carbon sometimes started hissing ; but this was due, *not* to any direct action of the stream of gas on the carbon, but to the arc being deflected by the gaseous stream and burning obliquely up the side of the carbon, and thus allowing the air to come into contact with the crater. The proof of this was that this diminution in the P.D. had the same value as if air had been employed, and that the hissing did not cease on stopping the stream of nitrogen or carbon dioxide.

This was not the case with hydrogen, however. When that gas was blown down the positive carbon in the open air, the arc would start hissing if the current were large enough, *and stop hissing the moment the hydrogen was shut off.* Not only this, but the diminution in the P.D. had a different value from that produced by air, being only about 6·5 volts instead of 10 volts. Table XLIX. gives the current and the P.D. between the carbons just before the hydrogen was turned on, just after it was turned on, just before it was turned off, and just after it was turned off.

Table **XLIX.**—*Effect of Blowing Hydrogen down a Tubular Positive Carbon of an Open Arc.*

Carbons : Positive, 11mm., tubed; negative, 9mm., solid.

Length of arc about 3mm.

P.D. between carbons in volts.				Current in amperes.
Before H was turned on.	After H was turned on.	Before H was turned off.	After H was turned off.	
52	46	46	53	14
52	45	47	52	12
52	45	45	52·2	12
52·5	46	46	53	12
52	45	46	52	9

Thus, the mean diminution of P.D. accompanying the hissing caused by hydrogen being sent down the positive carbon of an arc burning in the air was about 6·6 volts, or about $3\frac{1}{2}$ volts lower than when the hissing was caused by air alone,

In order to exclude all possibility of doubt as to the effect of
the various gases, the experiments were repeated with the arc
entirely enclosed, so that the only gases that could reach it
were those blown down the tubular positive carbon.   The
current was distinctly below the hissing point, being only 10
or 11 amperes, with an arc of from 2mm. to 3mm. long.

When air was blown down the positive carbon, each puff
lowered the P.D. by about 10 volts, and the moment the
puff ceased the P.D. rose again.   Next, oxygen was tried, with
the same result.   Thirdly, nitrogen with *no* result or with
the result that the arc was blown out if the pressure was
too great.   Carbon dioxide had the same effect as nitrogen.
and lastly hydrogen was tried.   This gas gave a totally
different result with the enclosed arc from that already obtained
with the open arc.   For whereas, as has been previously stated,
hydrogen produced a distinct hissing of its own when blown
down the positive carbon in the *open air*, it produced *none* when
used in the same way with the *enclosed* arc.

To prove that, in order to produce the sudden diminution of
P.D. under discussion it was necessary for the active gas to
actually touch the crater, a tubular *negative* carbon was used,
and each gas was blown up through it in turn, gently enough
not to force the gas directly against the crater.

In *no* case was there any sudden diminution of the P.D.,
whatever was the gas blown through the negative carbon, and
whether the arc was open or enclosed.   On the contrary, there
was generally a small increase, probably due to the lengthening
of the arc by its being blown on one side.   If oxygen or air
were blown *very hard* up the negative carbon, they would either
produce hissing, or blow the arc out, or both ; for in that
case some of the gas got to the crater uncombined with the
carbon vapour, and acted exactly as if it had been blown down
the tubular positive carbon.

An interesting proof that the air must be in contact with the
crater to produce hissing is afforded by an experiment carried
out by Cravath, mentioned on page 63.   He tried the effect of
moving one carbon horizontally over the other, while a steady
silent arc was burning.   When the positive carbon was pointed
and the negative flat, the arc burned silently as before, but
when the *negative* carbon was *pointed* and the *positive flat*, each

x 2

change of position caused a hiss. The reason is obvious. Each change of position caused a new crater to form of carbon that had previously been in contact with the air, and, consequently, still had some air clinging to it.

The case, then, stands thus :—

(1) When the arc begins to hiss in the ordinary way, the P.D. between the carbons diminishes by about 10 volts.

(2) If the air is excluded from the arc, this diminution of the P.D. does not take place, even when the current is nearly three times as great as would cause hissing in the air.

(3) If, however, while the air is excluded, puffs of air are sent against the crater, the diminution of the P.D. *does* occur, even with currents much *smaller* than would cause hissing in the air.

(4) If, instead of air, *oxygen* is sent against the crater, the P.D. is diminished to exactly the same extent as when air is used.

(5) If, on the other hand, *nitrogen* is sent against the crater, *no* diminution of the P.D. is observable.

(6) If air or oxygen is gently blown through the *negative* carbon, so that it cannot get direct to the crater, *no* diminution of the P.D. follows.

Thus there can be no shadow of doubt that *the sudden diminution of P.D. that accompanies the hissing of the open arc is due to the oxygen in the air getting directly at the crater and combining with the carbon at its surface.*

It only remains to show how the actual hissing *sound* may be produced by the burning of parts of the surface of the crater. The moment after this burning has begun, a cloud of gas, formed of the products of combustion, must spread over the burning part, protecting it momentarily from the action of the air as effectually as the carbon vapour had hitherto done. When this gas is dispersed the air will again come into contact with that part of the crater, a fresh cloud will form, and the whole action will start *de novo*. Thus a series of rushes and stoppages of the air will take place, setting up an irregular vibration of the very kind to cause a hissing noise. Not only this, however, but, since the crater must cease to burn each time that it is protected by the gas, the diminution of P.D. must also cease to exist at that part, since its cause is removed, and the P.D. will, therefore, rise momentarily. Thus an oscillation of the P.D. between the carbons, and, consequently, of the electric current must be created, corresponding with the oscillation of the *air* current.

That the air current does oscillate when the arc hisses was proved beyond a doubt by the following experiment. One end of a very fine single fibre of asbestos was fastened to the hole of the crucible shown in Fig. 86 through which the positive carbon was moved. Sufficient space was left between the hole and the carbon for the free end of the fibre to stretch out horizontally without touching the latter. While the arc was silent the fibre remained fairly motionless, but as soon as hissing began, instead of being sucked into the crucible, as it would have been with a steady inward current of air, it vibrated rapidly up and down, thus showing the oscillatory character of the current.

The oscillation of the *electric* current was also proved beyond a doubt, by Messrs. Frith and Rodgers,* in 1893 ; and it has recently been shown, graphically, in the most convincing manner by the curves published by Messrs. Duddell and Marchant.† Mr. Duddell has since made much more detailed curves of the same kind which will shortly also be published.

Thus we have seen that not only the sudden diminution of the P.D. between the carbons, but every other phenomenon that attaches to the hissing arc may easily be caused by the oxygen of the air getting directly at the crater, and combining with the carbon at its surface.

SUMMARY.

I. When the length of the arc is constant and the arc is silent, it may be made to hiss by increasing the current sufficiently.

II. The largest current that will maintain a *silent* arc is greater the longer the arc.

III. The *hissing point* always occurs on the flat part of the curve, at a point where the P.D. changes very slightly with change of current.

---

* *Phil. Mag.*, 1896, p. 407.

† *Journal* Inst. Elec. Eng., Vol. XXVIII., pp. 78, 79.

IV. When the current is constant and the arc is silent, *shortening* it will make it hiss.

V. A straight line law connects the P.D. at the hissing point with the length of the arc.

VI. With a given pair of carbons the current cannot have more than a certain maximum value without causing the arc to hiss, however long it may be.

VII. When the arc begins to hiss, the P.D. suddenly falls about 10 volts and the current suddenly rises.

VIII. For the hissing arc the P.D. is constant for a given length of arc, whatever the current, and whether the carbons are cored or solid.

IX. A straight line law connects this constant P.D. between the carbons with the length of the arc, when both carbons are solid, but not when the positive is cored.

X. A straight line law connects the diminution of P.D. that accompanies hissing with the length of the arc.

XI. The longer the arc, the less is the P.D. between the carbons diminished when hissing begins.

XII. About two-thirds of the diminution of P.D., when hissing begins, takes place at the junction of the positive carbon and the arc. The remainder is apparently due to a diminution of the resistance of the arc vapour, and none to any change in the P.D. between this vapour and the negative carbon.

XIII. When the largest silent current changes to the smallest hissing current for the same length of arc, the value of that smallest hissing current depends only on the E.M.F. of the generator or the resistance in the circuit outside the arc, whichever is fixed first.

XIV. When the arc is silent and the current small, the crater presents a uniformly bright appearance, but when the current is increased sufficiently, patches of bright and dark bands appear on it, whirling and oscillating faster and faster as the current is increased.

XV. When the current is so great that the arc is near humming, the speed of revolution is too great to be detected by the eye, and it continues to increase to about 450 revolutions a second, when the arc begins to hiss.

XVI. With humming and hissing, a green light appears in the crater, and with hissing, clouds partially cover the crater; and the carbon vapour becomes flattened out between the carbons.

XVII. With a short hissing arc a mushroom forms on the end of the negative carbon.

XVIII. With a silent arc the end of the positive carbon is rounded, and the crater occupies the smallest cross-section of it. With a hissing arc the end is nearly or quite cylindrical, except where the crater has cut it away obliquely.

XIX. *Hissing is produced by the crater becoming too large to occupy the end only of the positive carbon, and by its therefore extending up the side.*

XX. When the arc is enclosed in such a way that very little or no air can get to it, there is no sudden diminution in the P.D. between the carbons, even with currents three times as great as would produce that diminution with the open arc.

XXI. If, however, whether the air is excluded or not, puffs of air are sent against the crater, the diminution of the P.D. *does* occur, even with currents much *smaller* than would ordinarily cause hissing.

XXII. If, instead of air, *oxygen* is sent against the crater, the P.D. is diminished to exactly the same extent as when the air is used.

XXIII. If, on the other hand, *nitrogen* or *carbon dioxide* is sent against the crater, *no* diminution of the P.D. is observable.

XXIV. If air or any of the other gases are gently blown through the *negative* carbon, so that they cannot get direct to the crater, *no* diminution of the P.D. follows.

XXV. Thus there can be no doubt that *the sudden diminution of P.D. that accompanies the hissing of the open arc is due to the oxygen in the air getting directly at the crater and combining with the carbon at its surface.*

XXVI. Hydrogen, blown against the crater of a silent arc, causes hissing and a diminution of about 6·6 volts in the P.D. between the carbons, when the arc is open to the air. When, however, the arc is enclosed, so that air is excluded, *no such effect* can be observed.

CHAPTER XI.

THE LIGHT EMITTED BY THE ARC. DIFFERENT CANDLE POWER
IN DIFFERENT DIRECTIONS. MEAN SPHERICAL CANDLE
POWER UNDER DIFFERENT CONDITIONS. LUMINOUS EFFI-
CIENCY UNDER VARYING CONDITIONS. HOW TO OBTAIN
THE MAXIMUM LUMINOUS EFFICIENCY UNDER ANY GIVEN
CONDITIONS.

The value of a source of light depends upon two conditions—
(1) the total amount of light that it emits, (2) the distribution
of that light. It is absolutely essential to know both these
factors in order to judge of the utility of the source, for a
large total flux of light is of little use if emitted in the wrong
direction, and a light may be all in the right direction, but so
dim as to be practically valueless. Suppose, for instance, that
the arc would only burn with the positive carbon underneath;
then, however brilliant it might be, it would be useless for
street lighting; for, although the tops of our houses would
be well illuminated, the streets would be left in darkness.
Again, imagine a farthing dip in a ball-room: though every
ray were utilised, it would only suffice to make darkness
visible. What we want to know about a source of light, then,
is the *quantity* of light it emits and the *direction* of the light.

Some sources are so constituted that it is physically impossible to utilise all the light they emit. The very conditions
under which they exist cause the obstruction of some of their
light. Thus, while the light of a candle or a gas jet is practically unobstructed, in paraffin and glow-lamps part of the light
evolved is necessarily absorbed by the glass covering needed—
in the one case to create a sufficient draught of air to completely consume the oil, and in the other to maintain a
vacuum round the filament. In the arc there is a still greater
difference between the quantity of light evolved and the

amount that can be usefully employed, for the negative carbon usually obstructs far more of the light from the principal source—the crater—than would be absorbed by a clear glass covering. Moreover, with every change in the current or the length of the arc, in the diameter or construction of either carbon, the form of the negative carbon changes, and, consequently, the amount of the light that is obstructed changes also. Hence arise many complications in the laws governing the light of the arc, which vanish more or less completely when the amount of light evolved and the quantity that escapes and becomes perceptible to the eye are studied separately.

The sources of light in the arc are (1) the crater, (2) the remainder of the hot end of the positive carbon, (3) the white-hot spot on the negative carbon, " the white spot," as I have called it, (4) the remainder of the hot end of the negative carbon, (5) the arc vapour. I shall call the light of the whole five sources together the light of the arc, and shall speak of the light emitted by the arc proper as the vapour light, or the light of the vapour.

Not only the quantity but the proportion of the whole light emitted by each of the sources probably varies with each current and length of arc, as well as with the construction and thickness of either carbon. But in all cases by far the larger part of the light is due to the crater, the next greatest source being the white spot ; and, last of all, the hot sides of the carbons and the vapour, which, even when the arc is long enough to " flame," give comparatively little of the light. What the exact proportions are under any given set of conditions, and how they change when the conditions change, has never yet been accurately determined ; nor, indeed, has much attention been paid, as far as separate photometric measurements of intensity are concerned, to the light of any other part of the arc but the crater. Sir William Abney discovered, for instance, as long ago as 1881,[*] that the quantity of light emitted per square millimetre of crater was practically a constant for a given quality of carbon, however the current and the length of the arc might be varied ; but whether the intrinsic brilliancy of the white spot, or of the vapour, is also a

---

* *Phil. Trans.*, 1881, Vol. CLXXII., p. 890.

constant, has never been determined. Similarly, it is known that with a given length of arc the area of the crater increases as the current increases, and I have shown (p. 154) that the area of the crater increases also, as the length of the arc is increased, with a given current; but no attention whatever has been paid to the variations in the area of the white spot, which I have, nevertheless, found to depend as definitely on the current, though not on the length of the arc, as the area of the crater itself. The reason of this neglect is obvious. In the ordinary vertical arc with the positive carbon on top (which alone we are now considering) the light from the white spot must principally escape upwards, and can thus be of little use in the region far below the arc for which the light is needed. Thus, though this white spot is nearly, if not quite, as brilliant as the crater (though far inferior to it in extent), the light from it is, in a sense, unimportant. The light emitted by the arc vapour is also small compared with that of the crater, so that it also has received very little attention. In determining the light emitted by the arc then, the *important* points to consider are (1) the quantity of light given out by the crater, and (2) the extent to which this light is obstructed by the negative carbon.

It must always have been noticed from the first that the negative carbon cuts off more or less of the brilliant light emitted by the crater; but how much, under any given conditions, was never made clear till Mr. Trotter* put the whole matter in a nutshell by enunciating and proving experimentally the delightfully simple theorem that the great difference observable in the candle power of the arc in different directions is due solely to the different amounts of crater visible in those directions. He showed that in directions in which the view of the crater was entirely unobstructed, the candle power varied directly as the apparent area of the mouth of the crater, and that where it was obstructed by the negative carbon the candle power diminished in proportion to the increase of the obstruction. He found also that the light emitted by all parts of the arc and carbons except the crater was practically the same in all directions, with given carbons, current, and length of arc, as, of course, it would have to be for his theorem to be correct.

---

* *The Electrician,* 1892, Vol. XXVIII., p. 687, and Vol. XXIX., p. 11.

The Light received from the Crater in Different Directions.

Let us consider the light and the obstruction—the crater and the negative carbon—separately. Mr. Trotter began by correcting the erroneous impression frequently held, that the hollowing of the crater caused the light to be concentrated and cast downwards. He pointed out that the same amount of light was received from the crater in any direction as would be received in that direction from a disc of equal brightness fitting into the mouth of the crater. This is only absolutely true when none of the light is absorbed in its passage from the surface of the crater to its mouth. It will be seen later that it is more than probable that this condition is not entirely fulfilled; but the error thus introduced is in most cases so small that we may neglect it, and consider the crater as a luminous disc of area equal to its mouth. Mr. Trotter's theorem is, then, that with the exception of a comparatively small quantity of light, which is constant for all directions, the candle power of the arc in any direction is directly proportional to the apparent area of the crater as seen from that direction.

We may talk about the apparent area of the crater as looked at in a *direction* instead of from a point, because the diameter of the crater is always so small compared with the distance of the eye or the photometer screen from it that its apparent area is the same from all points in any one direction. Let AB, for instance (Fig. 88), be the diameter of the crater, and EC a direction in which it is viewed, then if the eye is at C, the apparent area of the crater will depend upon the angle B C A, and if it is at D it will depend upon the angle B D A. If D and C are both very far from A B, these two angles will be practically equal, and so the apparent areas of the crater, as seen from these two points, will be equal also. For the same reason—the smallness of the crater compared with its distance from the eye—the line joining the point from which it is viewed to *any* point on the crater may be called the *direction* in which it is viewed, for clearly when C is far away, C A, C B and C E are all parallel.

Mr. Trotter pointed out that, when there is no obstruction the apparent area of the crater varies directly as the cosine of the inclination, that is, as the cosine of the angle between the plane through the mouth of the crater and a plane perpendicular to the line joining the eye to the crater.   In other words, the apparent area of the crater is proportional to the cosine of the angle between the direction from which it is viewed

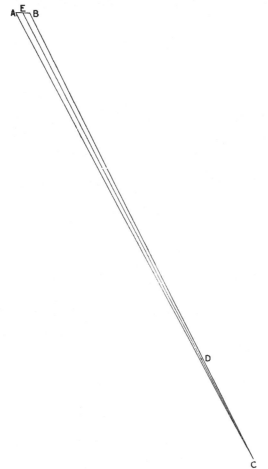

Fig. 88.—Disc Viewed from a Great Distance.

and the perpendicular to the mouth of the crater.   This also is, of course, only strictly true when the diameter of the crater is small compared with its distance from the eye.   A complete mathematical proof of it is given in the Appendix (page 441).

Mr. Trotter's experiments were very simple, but they were quite conclusive for the cases he tried. The apparent area of the crater, seen from different directions, was measured, and the candle-power of the arc in the same directions taken, and it was found that the two sets of values both varied as the

(Trotter).

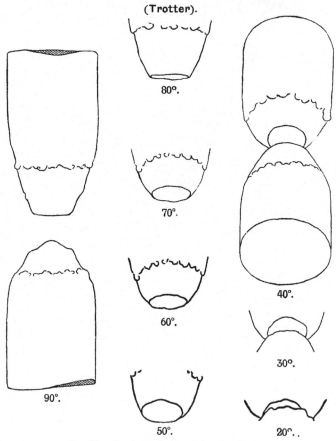

Fig. 89.—Tracings of normal arc $1\frac{2}{3}$ full size.

cosine of the inclination, for all directions in which the view of the crater was unobstructed.

Mr. Trotter pointed out that when the radius vector of a polar curve is proportional to the cosine of the angle between it and the fixed line, the curve is a circle, of which the pole is one point. Thus, he argued, a polar curve with a line proportional to the apparent area of the crater as radius vector, and the inclination of the crater as the angle between the radius vector and the fixed line, must form a part of a circle. Hence, the candle-power of the crater, plotted as a polar curve, must also form a part of a circle for those directions in which

(Trotter).

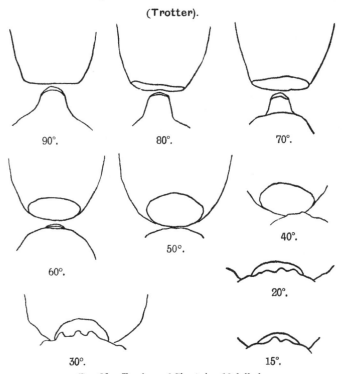

90°.   80°.   70°.

60°.   50°.   40°.

30°.   20°.   15°.

Fig. 90.—Tracings of Short Arc 1⅔ full size.

the view of the crater is unobstructed, if this candle-power varies directly as the amount of crater visible.

The tracings of the crater and the negative carbon that Mr. Trotter used to test his theory are given in Figs. 89 and 90. These show that with an inclination of 90°, that is, when the eye was in a horizontal line with the plane of the crater, the crater could not be seen at all, as one would expect. With

inclinations of from 90° to between 50° and 40°, the crater was
entirely unobstructed by the negative carbon, and with smaller
inclinations the obstruction was greater the less the inclination.
The apparent areas of the crater, and the candle-power of the
arc, taken from the same points, are both plotted in Figs. 91
and 92, the angle between the radius vector and the fixed line
being made equal to the inclination of the crater in each case.
Triangles represent areas of crater, and crosses candle-power.
It will be seen that the observations of areas and of candle-

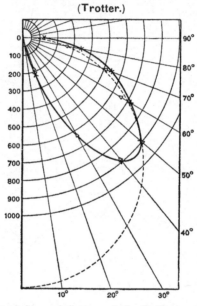

(Trotter.)

Fig. 91.—Curve of Areas of Crater and Candle-Power of Normal Arc.
  △ = area and × = candle-power.   The scale of radii is arbitrary.

powers coincide in a very remarkable manner with one another,
and with a part of the circle which is the polar curve connecting the cosine of the inclination with the inclination.

Fig. 93 shows the connection between candle-power and
apparent area of crater, from different directions, with rectangular co-ordinates, ordinates representing candle-power and
abscissæ apparent areas of crater.   The points lie very fairly
well in a straight line which cuts the axis of candle-power at
about 100.   This, therefore, must have been the candle-power

of the white spot, the glowing ends of the carbons and the vapour, and the curve shows that the part of the candle-power of the arc that is due to these is practically the same in all directions, with a given current and length of arc and a given pair of carbons.

Mr. Trotter's results were qualitative rather than quantitative; his candle-powers were, in most cases, relative, and not absolute, but there can be no doubt, nevertheless, that he proved his point, and that the variable part of the candle-

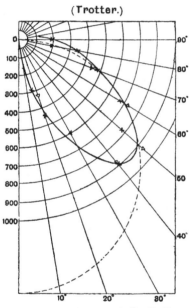

Fig. 92.—Curve of Areas of Crater and Candle-Power of Short Arc. $\triangle$ = area and $\times$ = candle-power. The scale of radii is arbitrary.

power of the arc was very fairly proportional to the apparent area of the crater in the cases he tried. This carries with it, as he stated, the necessity for the light being *uniformly distributed* over the crater also; that is, his experiments show that, roughly, the quantity of light emitted per unit surface of the crater is constant over the whole surface, for otherwise the variable part of the candle-power would depend upon *which part* of the crater was visible from any point as well as upon *how much.*

Y

Although Mr. Trotter's experiments show that the light received from the crater is fairly uniform, his own later experiments and mine prove that it cannot be entirely so. For when a part of the crater is covered with the swiftly whirling figures that he discovered in 1894, and the more slowly moving ones that I showed at the Institution of Electrical Engineers in 1899, it is quite evident that more light per

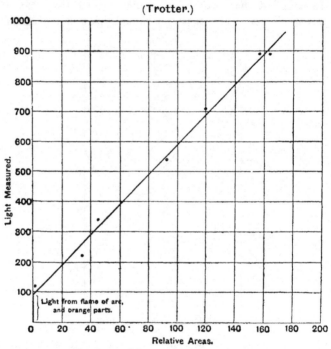

(Trotter.)

Fig. 93.—Areas of Crater, and Light of Normal Arc.

square millimetre must be received from the entirely bright parts of the crater than from the parts where there are dark bands.

Quantity of Light Obstructed by Negative Carbon.

We have next to consider how far the light emitted by the crater is obstructed by the negative carbon. If this carbon were of constant shape for all currents and lengths of arc, clearly the amount of light obstructed by it would depend simply on the length of the arc, for the farther the negative carbon was from the crater the less crater light it would obstruct. This is by no means the case, however. As has already been mentioned, the shape of the negative carbon alters with every change in the current and the length of the arc, and alters more, for a given change, the shorter the arc and the smaller the current. With short arcs and large currents it is very sharply pointed, and the point becomes blunter the longer the arc and the smaller the current. Take, for instance, the diagrams for arcs of 3mm. in Fig. 8 (p. 10). There is considerably more difference in form in the negative carbons for currents of 6 and of 10 amperes than in those for currents of 16 and 21 amperes. Or, take the 6mm. arc : the difference between the negative carbons for 6 and 16 amperes is decidedly greater than between those for 16 and 30 amperes. Again, in the diagrams for 10 amperes in the same figure, the change in the negative carbon caused by lengthening the arc from 1mm. to 2mm., is greater than that caused by lengthening it from 2mm. to 3mm., and with a current of 6 amperes the difference between the negative carbons for 1mm. and 2mm. is greater than that caused by lengthening the arc from 2mm. to as much as 6mm. A comparison of the diagrams in Figs. 7, 8 and 9 (pp. 9, 10 and 12) shows that the shape of the negative carbon depends on the diameters of the carbons as well as on the current and the length of the arc.

In order to study with greater ease the changes in the shape of the negative carbon, some of the diagrams in Figs. 7, 8 and 9 have been enlarged. In Fig. 94 the carbons and the current are constant, or nearly so, but the length of the arc is 1mm., 2mm. and 6mm., going from left to right.

There is a certain similarity in the forms of all three negative carbons : each has two shoulders—a small one, of which the diameter is about the same as that of the crater, and a larger one, where the burning away of the sides of the carbon ceases. This larger shoulder is usually of rather greater diameter than the unburnt part of the carbon, because of the ragged fringe of frayed carbon that sticks out from its sides. The differences in the forms of the various negative carbons depend chiefly upon the relative diameters of the two shoulders and the distance between them, and on the shape and height of the point that rises above the smaller. Let us examine more closely the way in which the negative carbon interferes with the light of the crater. As the arc is practically an axially symmetrical source of light, we may, for theoretical purposes, examine the light in one plane passing through the axis of the two carbons, and whatever is true of the light in that one plane will be true of the light in all other planes passing through the axis, except in so far as any error is introduced by the carbons not being strictly in line, or the arc burning on one side.

Let A B and A C represent, the one a diameter of the crater, and the other a line in the same plane, drawn from the end

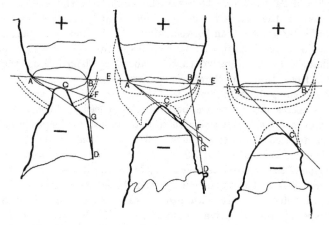

Fig. 94.—Shape of Negative Carbon, with same Current but Different Lengths of Arc.

of that diameter to touch the negative carbon, The angle between these two lines includes all the directions on the right-hand side of the arc from which an eye moving in the same plane as the two lines could see the crater unobstructed by the negative carbon. Hence, the angle B A C, or the mean of the two corresponding angles on either side of the arc, in the same plane, roughly measures that part of the crater light that is unobstructed by the negative carbon.

Next, if B D be drawn to meet the lower shoulder of the negative carbon in D, B D is the nearest direction to the vertical in which any of the crater can be seen at all. That is to say, the angle E B D is the largest angle that the line joining the eye to the crater can make with the horizontal for any portion of the crater to be visible. We now have in the angle B A C a rough measure of the total quantity of light received from the crater in one plane in directions in which it is unobstructed by the negative carbon, and a similar rough measure of the quantity received in directions in which it is partially obstructed in the angle B F A, which is equal to the angle E B D minus the angle B A C. It will be observed that the second is only a very rough measure indeed, for the amount of light obstructed depends on the shape of the negative carbon between the points C and D, even more than on the angle B F A. Take, for instance, the 1mm. and 2mm. arcs in Fig. 94, and compare the angle F A G in the one with the same angle in the other. This angle, which lies between the line touching the *point* of the negative carbon and the line touching the smaller shoulder, includes all the directions from which only a very little of the crater is hidden. Now, in the 1mm. arc this angle is more than four times as great as in the 2mm. arc, and in the 6mm. arc it does not exist at all, for the lines A C and A G coincide. Hence, although the unobstructed crater light is much less in the 1mm. arc than in the 2mm. arc, that which is only slightly obstructed is much greater in the 1mm. arc, and it is thus possible that the total amount of light received from the crater of the 1mm. arc may be as great as, or even greater than, that received from the crater of the 2mm. arc.

The Light emitted by a very short Arc is greater than when the Arc is longer, with a large Constant Current.

It is, of course, only the *proportion* of the crater light that gets out, compared with the whole light emitted by the crater that can be estimated in this way; but when the areas of two craters are equal, or nearly so, a very good rough comparison of the relative mean spherical candle-powers of the two arcs can be made by comparing the sizes of the angles B A C, C A G, and E B D in each. Now the 1mm. and 2mm. arcs in Fig. 94 have craters of very much the same area; for although (*see*

p. 154) this area does increase as the arc is lengthened, even
when the current is constant, yet the increase in this case is so
small that I have calculated it to be only about a half per cent.
Thus the relative candle-powers of the two arcs may well be
estimated by the above method, which seems to show that,
in this one instance at any rate, the mean spherical candle
power of the 1mm. arc was at least equal to that of the
2mm. arc.   This conclusion, so contrary to the generally
accepted ideas, and arrived at by a mere examination of the
shape of the negative carbon, could not, however, be accepted
without reference to actual candle-power experiments.   Let us
see what these say.

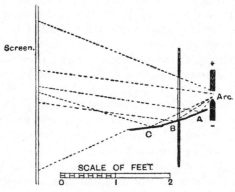

Fig. 95.—Side View of Three Mirrors, A, B and C, throwing the Light of
the Arc through a Slit on to the Screen.

Table L. shows the results of the experiments on the mean
spherical candle-power of the arc made for the Paper read by
Prof. Ayrton at the Electrical Congress at Chicago in 1893.
They were obtained by taking what I may call a sample of
the light with each current and length of arc.   The arc was
enclosed in a light-tight box (of which one side is seen in
elevation in Fig. 95), and a wedge of light was allowed to
escape, through a slit of constant width, long enough to
allow the light in all directions in one vertical plane to pass.
This wedge of light was received on a screen of white
blotting paper (Fig. 96), and in order that the whole of the
light in the one vertical plane in which it was being measured

Fig. 96.—General Plan of Apparatus for taking Mean Spherical Candle-power of Arc. A B C, three mirrors at different angles (*see* Fig. 95); L, projecting lens; M, mirror; Photometer bar about 12ft. long.

should be collected on the screen, the lower part of the light was reflected from the three plane mirrors A, B and C (Fig. 95) on to the screen. From the blotting paper the light was reflected on to the fixed photometer screen, and was compared with that of a 2 c.p. Methven Standard, which was movable along a bar graduated directly in mean spherical candles. Corrections were made for the light absorbed by the mirrors and the blotting paper, and thus the mean spherical candle-power for each arc was obtained at one reading, or rather in one series of readings, for the mean of from 6 to 12 readings was taken in each case. The carbons employed were 13mm. cored positive and 11mm. solid negative.

For currents of 4, 7 and 10 amperes, the candle-power of the 2mm. arc is greater than that of the 1mm. arc, but for currents of 15 and 20 amperes it is considerably less, and with a current of 32 amperes the 1mm. arc gives more light than the 2mm. arc does with a current of 32·5 amperes. Thus with currents that are fairly small for the sizes of the carbons, the longer arc gives the larger amount of light, but for currents great enough for the negative carbon to be sharply pointed, as it was in Fig. 94, the shorter arc does actually give the larger amount of light. These experiments, therefore, confirm the conclusions gathered from a comparison of the shapes of the negative carbons, and show that with short arcs and large currents the candle-power of the arc may first diminish and then increase again as the arc is lengthened.

Table L.—*Mean Spherical Candle-power of Arcs of Different Lengths with various Constant Currents.*

Carbons: Positive, 13mm., cored; negative, 11mm., solid.

Current in Amperes.	Mean Spherical Candle-power.				
	1mm.	2mm.	3mm.	4mm.	6mm.
4	70	103	78	105	...
6	...	238	...	...	247
7	270	280	353	326	...
10	560	700	...	736	900
11	...	.–	730	...	...
15	1,220	904	1,087	1,480	1,180
20	1,754	1,586	1,714	2,300	1,914
23·5	...	1,874	...	...	...
23·75	...	...	...	3,204	...
24	...	...	2,486	...	...
25	...	...	...	...	3,332
32	2,880	...	...	...	3,600
32·5	...	2,600	...	...	...
34	...	...	3,870	...	...
35	...	...	...	4,732	...

It seems probable that the length of arc with which the candle-power is a minimum diminishes as the current diminishes, with carbons of any given size. For the smaller the current the shorter must be the arc in order that the negative carbon should have a sharp point Thus it is most likely that if Prof. Ayrton's experiments had been carried on with arcs of less than 1mm., a minimum value for the candle-power with a current of 10 amperes would have been found for some length of arc between 0mm. and 1mm. When the current is very small for the sizes of the carbons, however, the negative carbon *never* becomes sharply pointed, however short the arc may be, and in that case the light obstructed by the negative carbon must steadily increase, as the arc is shortened, till the carbons touch. The curves in Fig. 97, which are plotted from Table L., show very clearly how the mean spherical candle-power decreases and then increases again with the short arcs and large currents.

Among the very complete and beautiful series of experiments on the total flux of light emitted by the direct current arc,

made by M. Blondel,* and published in 1897, were several
in which the current was kept constant and the arc was
lengthened from 0 to many millimetres.   M. Blondel plotted
curves connecting the flux of light with the P.D. between the
carbons instead of with the length of the arc, and hence the
phenomenon under discussion escaped his observation.   Nevertheless, indications of it are not wanting in his experiments,

(Ayrton.)

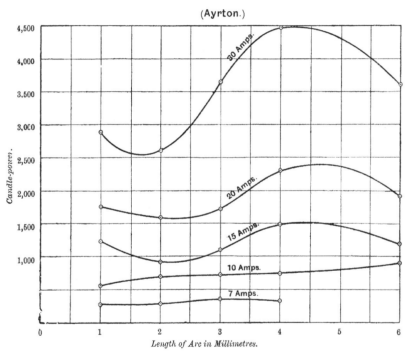

FIG. 97.—Curves connecting Mean Spherical Candle Power with Length of Arc.
Carbons : Positive, 3mm., cored ; Negative, 13mm., solid.

notably in the two curves in Fig. 98, which are plotted from
the numbers given by M. Blondel for Curves I. and III. in
Table III., p. 296, of his articles.

M. Blondel's method of measurement was the same in
principle as Prof. Ayrton's.   He placed the arc in the centre
of an opaque sphere, on opposite sides of which were two

* L'Éclairge Électrique, 1897, Vol. X., pp. 289, 496, 539.

vertical openings of 18° each (Fig. 99). The whole of the light that escaped from these two slits was caught by an elliptic mirror M (Fig. 100), and reflected on to a screen of white blotting paper, and the illumination of this screen was measured by means of the " Universal Photometer," invented by M. Blondel himself.

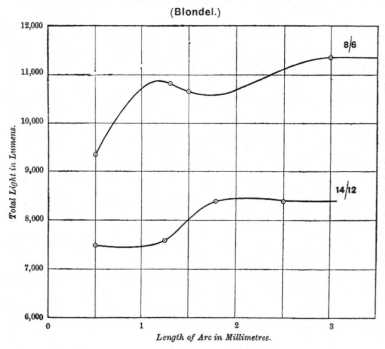

Fig. 98.—Curves connecting Total Light emitted with Length of Arc. Positive Carbon cored ; negative solid.

M. Blondel preferred to measure the total flux of light emitted by the arc rather than the mean spherical candle power, and the unit he employed was the "lumen," or the total flux produced by a source having a uniform intensity of one decimal candle in a solid angle, cutting off one square millimetre of surface from a sphere of radius 1mm. As the total flux of light emitted by a source is numerically equal to

4π times the mean spherical candle power, M. Blondel's numbers have only to be divided by 4π to give the mean spherical candle power of the arc in *decimal candles*, a unit

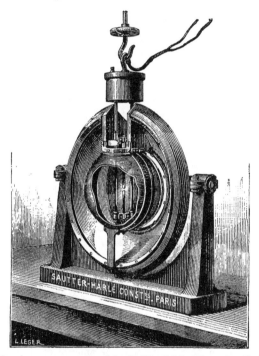

Fig. 99.—Apparatus employed by M. Blondel in Measuring the Total Light emitted by the Arc.

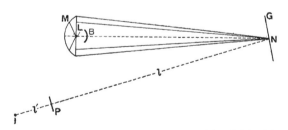

Fig. 100.—Arrangement of Apparatus used by M. Blondel.

L = Arc, M = Lumenometer, G = Diffusing Screen of White Blotting Paper, P = Photometer Screen.

which is one-twentieth part of the Violle unit. As this unit
is not universally adopted, however, it will be better to leave
M. Blondel's results in lumens.

For the upper curve in Fig. 98, the carbons employed were
"Nanterre," 8mm. cored positive and 6mm. solid negative.
For the lower curve they were both 10mm., the positive cored
and the negative solid. The constant current was 10 amperes
in both cases. Both curves show a tendency on the part of
the light emitted to *diminish* or at most to remain constant
as the length of the arc is increased from 1mm. to 2mm. in
the one case, and from 0·5mm. to 1mm. in the other.* Thus
M. Blondel's experiments also confirm the conclusion reached
by a simple examination of the diagrams of arcs and carbons,
in Fig. 94, viz. : that with certain currents the lighting power
of the arc, after increasing at first as the carbons are separated, diminishes or remains stationary as the arc is further
lengthened, and then increases again.

With a Constant Current the Illuminating Power of the Arc
Increases to a Maximum as the Arc is Lengthened, and then
Diminishes again.

To return to Fig. 94. When once the arc is long enough
for the point of the negative carbon to have become quite
blunt and round, it is plain that lengthening the arc can only
enlarge both the angles, B A C and E B D, on which the
amount of crater light that escapes depends. We should, therefore, gather from this figure that the amount of light received
from the crater must increase continually as the arc is
lengthened beyond about 2mm. ; and if Mr. Trotter's theorem
is correct in all cases, it follows that the candle power of the
arc must also increase continuously as the arc is lengthened.
To see whether this is so, we must turn again to Fig. 97,
taken from Prof. Ayrton's experiments. These curves do not
seem to bear out the deduction made from the shape of the
negative carbon, for they prove that, far from increasing continually as the arc is lengthened, *the illuminating power of*

---

* The lengths of arc corresponding with P.Ds. of 43·5 volts in the
upper curve and 48·7 volts in the lower were not given by M. Blondel,
and were therefore found by plotting the curves connecting the other
P.Ds. with the corresponding lengths of arc in each case.

the arc increases only till it is of a certain length, and then diminishes again as it is further lengthened. This most inte-

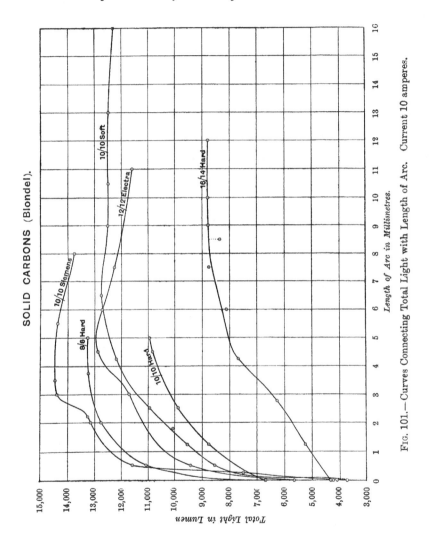

SOLID CARBONS (Blondel).

Fig. 101.—Curves Connecting Total Light with Length of Arc. Current 10 amperes.

resting and important point was first noticed by Prof. Ayrton, who announced it in the Paper he read before the Electrical

Congress at Chicago in 1893. At the same time Prof. Carhart mentioned that he had found *the luminous intensity of the arc to be a maximum with a certain definite P.D. between the carbons,*

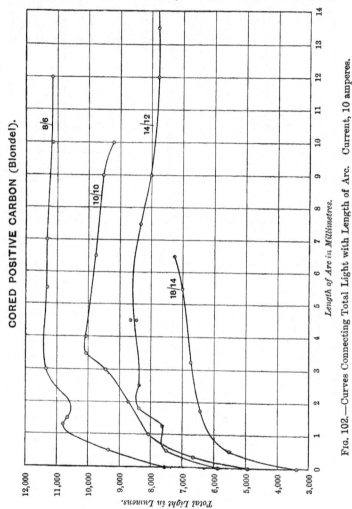

FIG. 102.—Curves Connecting Total Light with Length of Arc. Current, 10 amperes.

the P.D. depending on the nature and size of the carbons. As the P.D. between the carbons, for any particular current, is determined by the length of the arc, it is plain that the two discoveries were identical, though I shall show later that Prof. Ayrton's was the more correct way of putting it, and that the maximum illuminating power depends directly on the length of the arc, and therefore only indirectly on the P.D. between the carbons.

M. Blondel, as I have mentioned before, drew no curves connecting the *length of the arc* with any of the other variables, but, from the tables he gave of the results of his experiments, I have drawn the curves in Figs. 101 and 102, connecting the total light emitted by the arc with its length. For Fig. 101 both carbons were solid, and their sizes varied from 8mm. and 6mm. to 16mm. and 14mm., while for Fig. 102 the positive carbon was cored and the negative solid, and their sizes varied from 8/6 to 18/14. (This is a very convenient way, adopted by M. Blondel, of denoting the sizes of carbons. The left-hand figure always gives the diameter of the positive carbon in mm., and the right-hand that of the negative.) The current was 10 amperes in all cases. In both sets of curves, in every case where the arc was made sufficiently long, the light flux increased to a maximum and either remained stationary or diminished as the arc was further lengthened. Thus both Prof. Ayrton's and M. Blondel's experiments contradict the evidence of the diagrams of the arc and carbons, and, in order to find out where the error arises, it will be well to inquire a little more in detail into the manner in which the most important part of the light—the crater light—varies when the current is kept constant and the arc is lengthened.

For this purpose the diagrams in Fig. 103 will be found useful. They were taken from arcs of $\frac{1}{2}$mm., 1·1mm., 2mm., 3·2mm. and 6·6mm., burning between 18mm. cored and 15mm. solid carbons, with a constant current of 20 amperes flowing. On the assumption that Mr. Trotter's theorem was true for the arcs from which these diagrams were taken, it will be possible, with the help of Rousseau's method of finding the total light received from an axially symmetrical source, to construct figures of which the areas will be proportional to the total quantity of light received from the craters of the arcs. Then, taking the number of square millimetres in each area as abscissa, and the corresponding length of arc as ordinate, we shall be able to draw the curve connecting the total quantity of light

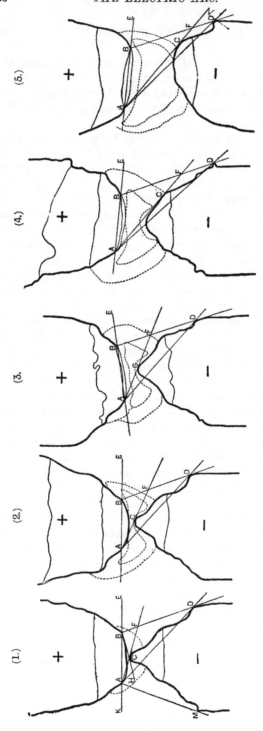

Fig. 103.—Diagrams of Arcs of Different Lengths.

Carbons : Positive, 18mm., cored ; negative, 15mm., solid.   Current, 20 amperes.

received from the crater with the length of the arc, for the constant current of 20 amperes. If, then, to each ordinate we add a length equivalent to the quantity of light received from the remaining four sources in the arc of corresponding length, we should, if our premises and measurements have been correct, get a curve of the same general form as those in Figs. 97 and 102.

Rousseau's figures* can only be drawn for axially symmetrical sources of light, and although the arc is this theoretically, yet it is so in theory *only*, for the carbons are rarely perfectly in line, and consequently the arc is seldom quite central. To minimise this source of error, we shall use only measurements which are the *mean* of the measurements made on either side of the diagram in each case. For instance, instead of taking the angle B A C (Fig. 103) as the angle containing all the directions from which the whole crater can be seen, we shall take the mean of the two angles B A C and A B H ; and so on with all the other measurements.

To Find, from a Diagram of the Arc and Carbons, a Figure proportional to the Total Amount of Light received from the Crater.

Let us first consider the figure that would represent the total light received from the crater if none of it were cut off by the negative carbon. Let A (Fig. 104) be the centre of the mouth of the crater. The distance between the crater and any point at which the light from it is measured is always so great, compared with the diameter of the crater, that the light may all be considered to come from one point. Thus, by Trotter's theorem, we have to find the total light received from a point A, when the quantity of light received in any direction is proportional to the cosine of the angle that that direction makes with the vertical, the unit quantity of light being the light emitted by one square millimetre of crater.

With centre A and unit radius describe the circle B C D, cutting the horizontal at B and D and the vertical at C. Draw the tangents B E and C E, and produce C E to F, making

---

* A detailed description of the manner in which these figures are drawn to represent the total light received from an axially symmetrical source is given in the Appendix (p. 448).

E F proportional to the light that would be received from the crater in a vertical direction, *i.e.*, proportional to the area of the crater. Let A G be another direction from which the light is measured, and draw G H K parallel to C F, making H K proportional to the light received in the direction A G. Join F K and K B, and draw G M parallel to A C. Then by Trotter's theorem

$$\frac{H K}{E F} = \frac{\cos G A C}{\cos 0}$$

$$= \frac{G M}{A C}$$

$$= \frac{B H}{B E}.$$

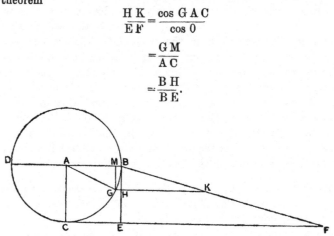

Fig. 104.—Figure used in finding Total Light received from Crater if none were obstructed by Negative Carbon.

Therefore B, K, and F are in one straight line, and B F must be the locus of the ends of all lines drawn in a manner similar to H K, each proportional to the light received from the crater in some given direction between A B and A C. But the figure representing the total light received from the source in one plane is that which is included between B E, E F, and this locus. Hence the triangle B E F must be the figure of which the area is proportional to the whole light that would be received from the crater in one plane if none of it were cut off by the negative carbon, and $2\pi \times$ area B E F is the total quantity of light that would be received from the crater in *all* directions if none were cut off by the negative carbon. We have thus the means of drawing a series of triangles, of which the areas would represent the total light that would be received from the craters of the arcs of which diagrams are given in Fig. 103, if none of the light were cut off by the negative carbon. We have only to draw triangles having their bases proportional to the areas of the craters in the diagrams and their heights all equal to B E.

In order to compare the quantities of light actually received from the craters, however, this is not sufficient, we must know what part of the light *is* cut off by the negative carbon. To find this, let B A G (Fig. 105) be the mean of the two angles B A C and A B H in 1, Fig. 103, and let B A N be the mean of the two angles E B D and K A M in the same figure. Then the triangle B H K represents the whole light received from the crater in directions in which the whole crater can be seen, and no light will be received from it at all in any direction nearer to the vertical than A N. Therefore the

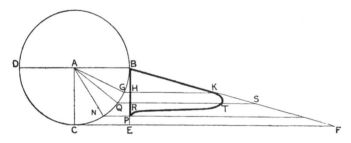

Fig. 105.—Figure proportional to Total Light received from Crater.

figure representing the total light received from the crater when part of it is cut off by the negative carbon must be bounded by P B, B K, and a curve which starts at K and ends at P. One point on this curve, if well chosen, will be sufficient to show what the shape of the curve must be, for we know that the occultation produced by the negative carbon must both increase and diminish quite gradually, so that the curve must join both B K and B P quite smoothly. The best point to take is that belonging to the direction in which the occultation begins to be serious, the point corresponding with the direction that just touches the larger shoulder of the negative carbon. Let A Q (Fig. 105) be this direction, then R S is the line representing the total light that would be received from the crater in the direction A Q if none of it were

z 2

cut off by the negative carbon. To find how much of the light
in this direction is cut off, we must turn to Fig. 106.

Let A B be the direction in which the light is being received,
and let C D be the tangent to the negative carbon that is
parallel to A B. Then, since both eye and photometer screen
are far from the crater, all the rays of light that reach the eye
from the crater will be parallel to A B, and all rays parallel to
A B that enter the negative carbon will be cut off from the eye.
The outermost of these rays are those that just touch the
negative carbon, so that to determine what region of the crater
is obscured by the negative carbon from an eye looking at it

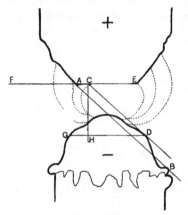

FIG. 106.—To find the Quantity of Light obscured by the Negative
Carbon in any one direction.

in the direction B A, we must find what is the area of the
crater that is cut off by these tangent rays. The true area of
this part is that which would be cut off from the crater by the
horizontal section of the negative carbon through D, if it were
moved parallel to itself along the line D C. The *apparent area*,
to which the light cut off is proportional, is the true area
multiplied by cos D C H, where C H is the vertical. As all
horizontal sections of the negative carbon are practically
circular, this true area would be that cut off from the crater
by a circle in the same plane, having C F = D G for its
diameter. To find this area, let A E, A C, C F (Fig. 107) be

equal to the corresponding lengths in Fig. 106, and draw the circle E G H, representing the area of the mouth of the crater, and having B for its centre, and draw F G H equal to the area of the cross-section of the negative carbon at D, and having D for its centre. Then A G C H is the area of that part of the crater from which the rays in the direction A B (Fig. 106) are cut off by the negative carbon.

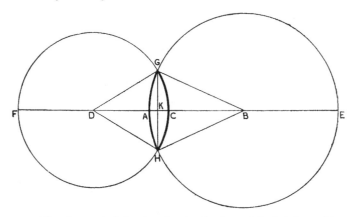

Fig. 107.—Geometrical Construction for the Area of Crater obscured by the Negative Carbon in any one direction.

To find this area, draw G H meeting A C in K, and join B G, B H, D G, and D H.

Let r = A B, the radius of the mouth of the crater.

r' = C D ,, ,, horizontal section through D.

x = A C

y = G K or K H.

Let a be the angle G B H in degrees, and

a' ,, ,, G D H ,,

Then the area A G H is $\dfrac{a\pi}{180}r^2 -$ G K, K B ;

the area C G H is $\dfrac{a'\pi}{180}r'^2 -$ G K, D K ;

∴ the area A G C H = $\dfrac{a\pi}{180}r^2 + \dfrac{a'\pi}{180}r'^2 -$ G K (B K + K D)

$$= \frac{\pi}{180}\left(ar^2 + a'r'^2\right) - y\left(r + r' - x\right).$$

Thus the required area has been found in terms of quantities, all of which are easily obtainable for any given arc by drawing such a figure as Fig. 107 from a diagram such as those in Figs. 94 and 106.

The ratio of the above area to the area of the crater is the ratio of the part of the crater light cut off in the direction A B (Fig. 106) to the whole light emitted in that direction. If, then, A Q (Fig. 105) represents the direction A B (Fig. 106) *i.e.*, if the angle B A Q (Fig. 105) = the angle E A B (Fig. 106), then, to find what part of the line R S represents the light received from the crater in the direction A B we must take a point T such that S R : S T :: area of crater (Fig. 106) : area G A H C (Fig. 107), and the curve bounding the figure representing the total light received from the crater must pass through T. We now have four points, B, K, T and P, through which the curved part of the figure must pass, and thus we can draw the figure B K T P R H, of which the area is proportional to the total light received from the crater in one plane.

The numerical value of the total light received from the crater in *all* directions is $2\pi \times$ area B K T P R H, when the unit of light is the quantity of light emitted by a square millimetre of crater. If, however, we take as our unit of light the quantity of light emitted by a square millimetre of crater multiplied by 2π, the area B K T P H (Fig. 105) will then represent the total light received from the crater in *all* directions in those units.

Curves deduced from Diagrams in Fig. 103 connecting Total Light received from Crater and Total Light from All Sources with Length of Arc.

From the diagrams in Fig. 103 we can thus find figures similar to Fig. 107, proportional to the total light received from the crater for arcs of 0·5mm., 1·1mm., 2mm., 3·2mm. and 6·6mm. when the positive carbon was 18mm. cored, and the negative 15mm. solid, and the current was 20 amperes. Then, by taking distances proportional to these as ordinates, and the corresponding lengths of arc as abscissæ, we can draw the curve connecting the total light received from the crater with the length of the arc *on the assumption that the quantity of light emitted by the crater per square millimetre of surface is constant, with a constant current, whatever the length of the arc.*

This curve is given in A B C (Fig. 108), and it is obvious from it that it is no diminution in the light received from the

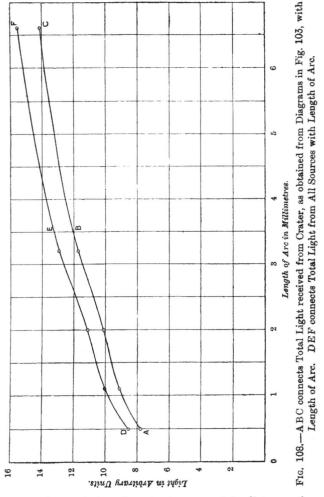

FIG. 108.—A B C connects Total Light received from Crater, as obtained from Diagrams in Fig. 103, with Length of Arc. D E F connects Total Light from All Sources with Length of Arc.

crater, as the arc is lengthened after a certain distance, that causes the total light received from the arc to diminish. On the contrary, the total light received from the crater *increases* continuously as the arc is lengthened with these particular carbons and this current. The curve does not pass through the zero point, because length of arc 0 does not mean that the carbons are touching, but that the tip of the negative carbon is in the same plane as the mouth of the crater. And when the carbons are so close the first is always smaller than the second, so that a space is left between the two through which a certain amount of the light emitted by the crater can escape.

It is very important to bear well in mind the different meanings that attach to (1) the ordinary polar curve of candle-power, (2) the Rousseau figure for the same arc, and (3) a curve such as A B C (Fig. 108). Each *radius vector* of the polar curve is proportional to the whole light emitted by *crater, arc and carbons, in one direction only*; but the area of this polar curve is *not* proportional to the whole light emitted by the source, as is frequently erroneously stated (*see* Appendix, p. 450). It is the area of the Rousseau figure drawn from this polar curve that is proportional to the whole light emitted by the source, or to its mean spherical candle power. Each Rousseau's figure of crater light obtained from the diagrams in Fig. 103 represents the *total* light received from the crater. In the curve A B C (Fig. 108) each ordinate is proportional to the whole area of the Rousseau figure of crater light, for the length of arc represented by the corresponding abscissa. Thus, the curve A B C shows the connection between the total light received from the crater and the length of the arc.

Fig. 108 shows, then, without a doubt, that it is not the light received from the crater that increases to a maximum and then diminishes again as the arc is lengthened. It remains to be seen whether the light from any of the other sources—the vapour, the white spot, or the glowing carbon ends—has a maximum value for a given length of arc. First, as regards the arc vapour. Both the length and the cross-section of this increase as the arc is lengthened, as will be seen from a comparison of Figs. 3 and 4,* which are pictures of the arc and carbons with the same carbons and the same current, but different lengths of arc, in which the shape of the vapour has been very carefully noted. The emitting area of the vapour increases, therefore, as the arc is lengthened ; and,

---

* Between pp. 6 and 7.

since we may take it that its intrinsic brilliancy is fairly constant, it follows that the total light emitted by the vapour is *greater* the longer the arc.

The white spot is, I find, practically constant in area with a constant current, and as, like the crater, it must reveal more of its light the longer the arc is made, the light received from this also must *increase* continuously as the arc is lengthened.

The glowing end of the negative carbon gives so little light that we may certainly neglect it; but that of the positive carbon must be taken into account. The length of this glowing part *increases*, I find, as the arc is lengthened, so that if we assume that the intrinsic brilliancy of this also is fairly constant with a constant current, it is clear that the light emitted by it increases as the arc is lengthened. Thus the light from each of the minor sources in the arc increases as the arc is lengthened, with a constant current. If we take this light as being always about 10 per cent. of the total light received from the crater, the curve connecting the total light received from the arc with the length of the arc will be of the form D E F (Fig. 103). From this curve we should say that the total light received from the arc with a constant current increases very rapidly as the arc is first lengthened from zero, and continues to increase, though more and more slowly, till the arc is infinitely long. Thus, a careful examination of the light received from all the sources in the arc fails to show how there can be a maximum amount of light received with a given length of arc. Nevertheless, there can be no doubt that this maximum does exist—experiment has proved it again and again. Such able experimenters as Prof. Ayrton, Prof. Carhart and M. Blondel have all come to the same conclusion.

Why Light-Length-of-Arc Curves deduced from Diagrams Differ from those found by Experiment. Absorption of Crater Light by Arc.

In seeking an explanation of this curious discrepancy between two sets of facts connected with the same phenomenon, we must turn again to a consideration of the light emitted by the crater. Mr. Trotter's experiments, while showing that the light of the crater is practically uniform over the whole surface, give no indication as to whether the

light received from the crater is constant for all currents and
lengths of arc or not.　It is plain that the two conditions—
uniformity over the whole surface and constancy for all
currents and lengths of arc—are quite distinct.　The first
might easily exist without the second, though it is more
difficult to conceive of the second without the first.　Each
square millimetre of a 10 ampere 2mm. arc might easily give
the light of 100 candles, while each of a 100 ampere 10mm.
arc gave 200 candles, but it is difficult to believe that the
average amount of light per square millimetre emitted by two
craters of very different areas could be the same, without the
light being uniformly distributed over the whole surface of
each.　There is a third condition, however, which is actually
known to exist in the arc, and which necessitates the existence
of both the other conditions, viz., the constancy of the
temperature of the crater with a given quality of carbon.

Although there was a good deal of vague surmise about the
volatilisation of carbon at the crater from a very early period
in the history of the arc, it was Sir William Abney* who, in
1881, first clearly enunciated the theory that the temperature
of the crater was that of the volatilisation of carbon, and must,
therefore, be constant.　He then remarked, "Whether the crater
be one-eighth or half an inch in diameter, the brightness remains
constant, *being apparently due to the temperature at which
carbon is vaporised.*" (The italics are mine).　This theory, which
marks as distinct an epoch in the history of the arc as Edlund's
discovery of the law of the apparent resistance, has been
attributed at different times to several different observers, but
Sir William Abney was undoubtedly its originator, and, in
formulating it, he made a very great advance in our ideas
concerning the physics of the arc.　He was led to it by finding
that with given carbons both the intensity and the colour of
the light emitted by the crater appeared to be quite unaltered
by any change in the current or in any of the other conditions
of the arc.

It would, perhaps, be more correct, as several observers have
suggested, to speak of the *sublimation* of the carbon at the
surface of the crater than of its *volatilisation*; for there is no

---

* *Phil. Trans.*, 1881, Vol. CLXXII., p. 890.

evidence to prove that the carbon liquefies before volatilising. Indeed, I have tried pressing the two ends hard together, when a large current was flowing, and then turning off the current suddenly before separating them, without finding any signs of their having stuck together, or even moulded one another. On the other hand, a small carbon placed in the arc near the positive pole will sometimes bend right over, so that the two parts make a very distinct angle with one another, and Prof. Elihu Thomson[*] has succeeded in bending a carbon by merely sending a current through it large enough to make it white hot.

Prof. Silvanus Thompson has expressed an opinion[†] that the temperature of the crater is that of the *boiling* point of carbon. "My present view of the physical state of the arc crater is that the solid carbon below is covered with a layer or film of liquid carbon just boiling or evaporating off." But there seems to be no evidence to support this view, for even if the thin film of liquid carbon exists, which is very doubtful, it probably evaporates long before it has reached the *boiling* point.

The question has been attacked from another side by various investigators, of whom the principal are Rossetti,[‡] and Violle,[§] who found the temperature constant under varying conditions, Rossetti estimating it at 3,900°C., and Violle at 3,500°. M. Violle's experiments covered a very wide range of current— 10 to 1,000 amperes—and were made both with arcs open to the air and with arcs enclosed in furnaces ; but they all led to the same conclusion, viz., that the temperature of the crater, with given carbons, was absolutely independent of the current and of the P.D. between the carbons. Such testimony as this must, I think, negative a suggestion made by M. Blondel, that the diminution in the light emitted by the arc, after it had reached a certain length, was caused by a diminution in the intrinsic brilliancy of the crater owing to cooling.

Since the temperature of the crater with given carbons is constant, it must follow that its intrinsic brilliancy—the

---

[*] *Electrical Engineer* of New York, March 27, 1891, p. 322.

[†] *The Electrical Review*, 1895, Vol. XXXVII., pp. 571.

[‡] *La Lumière Electrique*, 1879, Vol. I., p. 235.

[§] *Comptes Rendus*, 1892, Vol. CXV., p. 1, 273 ; 1894, Vol. CXIX., p. 949. *Journal de Physique*, 1893, Vol. II., p. 545.

amount of light emitted per square millimetre—must also be both uniform and constant; for a greater supply of energy or a smaller withdrawal through cooling, would not raise the temperature of the crater, but would simply cause a larger quantity of carbon to be volatilised, and so increase the area or depth of the crater, or both. Similarly an increased withdrawal of energy, or a diminution in the supply would only result in the formation of a smaller crater.

Because, however, the temperature of the crater is constant, and the light *emitted* by it must therefore be uniform and constant, it by no means follows that the light we *receive* from it is the same. There may be some reason why part of it is lost in transmission, and, if this is so, the evidence gathered from photometric measurements of the light of the crater will disagree with that obtained from measurements of its temperature. Such a disagreement does actually exist, for not only, as I have mentioned (p. 322), is the light received from the crater not always *uniform*, but M. Blondel* has shown it to be more than doubtful whether it is *constant* for all currents. While pointing out that such experiments must be received with caution, owing to the way in which the crater moves about, M. Blondel mentioned, in the admirable series of articles from which we have already gathered so much information, that the maximum brilliancy of the crater had varied in one series of experiments from 163 decimal candles per square millimetre with a current of 5 amperes to 210 candles with one of 25 amperes. Thus, although the intrinsic brilliancy of the crater may be, and probably is, constant in the main, yet it must be acknowledged that there are circumstances under which the light *received* from the crater is neither uniform over its whole surface nor constant for all currents.

What follows, then? That the emissivity should vary, when the substance is the same and the temperature constant, is inconceivable. Before it reaches the eye, however, the light of the crater has to pass through a region filled with what I have hitherto called the arc vapour. Our eyes tell us that this vapour must consist of at least three layers, an inner purple kernel, a dark envelope, and a green outer aureole, and

---

* *L'Éclairage Électrique*, 1897, Vol. X., p. 500.

Prof. Dewar has given us some information about its chemical constitution ; but as to its average density, or the density of the various layers, or as to whether these densities change when the current and length of arc are altered, we are entirely ignorant.

It has hitherto been taken for granted that this complex vapour (although it is known to be full of solid carbon particles) absorbs none of the light of the crater, or a quantity so small as to be negligible. But what if the amount of light absorbed, though very small when the quantity of vapour to be traversed is small, as it is when the arc is short, becomes quite noticeable when the quantity is considerably increased, as it is by lengthening the arc ? What if the quantity absorbed depends not only on the amount of vapour, but also on its density, its constitution, or the arrangement of its layers ? Then the light *received* from the crater, far from being uniform and constant, as is the light *emitted* by it, will depend upon the current, the length of the arc, the motions of the various layers—upon everything, in fact, that can cause a change in the density, the constitution, the number, or the relative positions of the layers of vapour. Absorption such as this would account for all the anomalies that have been observed in the light of the arc, all the discrepancies that appear to exist between the evidence gathered from direct measurements of the light of the crater and that obtained by measuring its temperature. Let us see what evidence there is of the existence of such absorbent power in what is usually called the arc vapour.

In 1822 Silliman,[*] the editor of *Silliman's Journal* first observed that carbon particles were shot out from the positive pole of an arc on to the negative. The observation has since been confirmed by experimenters too numerous to mention, but Herzfeld established the existence of solid carbon particles in the arc beyond doubt by the beautiful experiment described on page 85, in which he attracted the particles out of the arc on to a highly charged insulated plate placed near it. The existence of these particles in the arc at once disposes of the theory that it is composed of pure carbon vapour, or even of

---

[*] *Silliman's Journal,* 1822, Vol. V., p. 108.

such vapour mixed with gases, and places what I shall for the future call the *arc mist* in the same category as the flame of a candle, which is also known to contain solid carbon particles.

Having arrived at this conclusion, it occurred to me that some experiments with a candle flame might prove useful in suggesting the simplest method of testing the light-absorbing power of the arc mist.   The first I made was to place a piece of printed paper behind the flame and try to read through it. I found that all the unshaded part of the flame in Fig. 109 completely hid the print, but that I could see through the

FIG. 109.—Photograph of Candle Flame.

remainder as through a mist, the most transparent part being the dark region all ronnd the wick.   When the paper was placed so close to the candle that it was in danger of burning, it could be read through any part of the flame, but when it was even only an inch away from the flame the words immediately behind the bright part were completely blotted out. If, however, black marks were made on a piece of *transparent* paper, and, when they were quite invisible through the flame, another candle was brought up *behind* the paper, they reappeared as soon as the light from the second flame shone through the paper. If, also, the light from an incandescent gas burner shone full on the opaque printed paper from the front, the words on it could be easily read through the flame of a candle placed even as much as six inches away. Thus, if the paper were only brightly enough lighted, either from behind or in front, the words on it could be easily read through the candle flame. This proved that it was want of light, and not excess of light, *i.e.*, dazzling, that prevented the paper from being read, for in the latter case throwing more light on it would have increased the difficulty.

On the other hand, the want of light was not necessarily all, or even principally due to absorption. Some part of the loss—probably a large part—was due to reflection, internal reflection among the solid particles, and reflection back on to the paper. A very pretty proof of the latter was discovered by Mr. Burch in 1885.* Concentrating a strong beam of sunlight on the flame of a candle by means of a lens, he found that enough of the light was reflected by the flame to enable him to analyse it in a spectroscope and to prove that it was indeed sunlight and not candle light that was analysed. He also found that the reflected light was polarised in directions at right angles to the incident rays, showing that the reflection was due to minute solid particles in the flame.

Sir George Stokes,† without having heard of Mr. Burch's experiments, re-discovered the power of a candle flame to reflect sunlight later. He sent a highly condensed beam of sunlight through the flame of a candle, and noticed two bright spots of white light at the parts of the surface of the flame where the cone of rays entered and left it. He could not trace the passage of the beam *through* the flame, and hence concluded that the layer of solid particles was extremely thin and situated only at the surface. I have tried the same experiment, but it seemed to me that in some parts of the flame a faint haze joined the two bright splashes of blue light. It is extremely difficult to be sure of this, however, because the faintest flicker of the flame causes a *ring* of bluish light to encircle it, instead

---

* *Nature*, 1885, Vol. XXXI., p. 272.

† *Nature*, 1891, Vol. XLV., p. 133.

of the two bright spots only, and the apparent faint haze may
be only the inner coating of the farther side of this ring.

Owing to dark weather I was unable to send a beam of
sunlight on to the arc, but I tried whether objects were hidden
by the arc mist as they are by candle flames. I looked through
a slit that protected the eyes from all light but that of the
mist, which was much brighter than might have been expected,
but not so strong as to completely dazzle me. The result was
just what I anticipated, as regards the purple core—it behaved
exactly like the candle flame in blotting out printed letters
that were looked at through it. The green aureole, however,
was much more transparent, allowing the letters to be read
easily through it. Thus, in this one way at least, the arc
behaves to light transmitted through it exactly as the flame
of a candle does.

The Shadow of the Arc.

The success of this experiment made it seem worth while to
try whether the arc, like the candle, would cast a shadow. To
do this I thought at first of sending the light of one arc, through
another, on to a screen from which all the direct light from the
crater and carbons of the second arc was cut off, so that the
shadow, if there were one, might be as sharp as possible.
Mr. Mather, however, who very kindly superintended the first
experiments for me, contrived a far simpler and better method
involving the use of one arc only. The arrangement is shown
in section in Fig. 110.

A was a plane mirror placed so as to receive a large amount
of the light from the crater and to reflect it on to a screen
of white cartridge paper B C. The screen was placed high
above the tip of the positive carbon, and was tilted so that it
should catch none of the direct light from the crater, and
as little as possible from the carbons. The mirror A was
wider than the arc, so that only a part of the light it reflected
went back through the arc, and a part on either side went
straight on to the screen. Thus, if any of the light reflected
by the mirror were absorbed in the arc, or even if it were
simply reflected back again by the arc mist on to the mirror,
a shadow of the arc would appear on the screen. This is just
what happened. Deep shadows of the carbons appeared, with

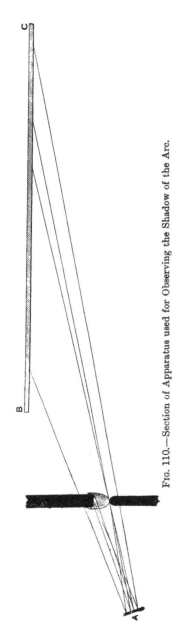

FIG. 110.—Section of Apparatus used for Observing the Shadow of the Arc.

a fainter shadow between. This faint shadow deepened when the arc was lengthened.

This seemed, at first, conclusive proof that a comparatively large part of the light sent through the arc was either absorbed by it or reflected back on to the mirror. The affair was not so simple, however. Mr. Mather noticed a rim of light round the shadow of the arc, which was brighter than any other part of the screen. He pointed out that this could only be caused by the light sent through the arc mist being *refracted*, on account of the difference of density between the mist and the surrounding air. The arc, in fact, acts as a lens towards light sent through it. Now, a *concave* lens which is denser than air throws a shadow that has a rim of light round it—just such a shadow as the arc throws. But the arc is distinctly *convex*, and the peculiar light-rimmed shadow that it throws proves, therefore, that it is less dense than the surrounding air. It acts, in fact, as a double-convex *negative* lens, one having a refractive index less than unity.

This lens effect is not confined to the arc; it is even stronger in candle and gas flames, as I found by sending

A A

the light of the arc through them. An account of some interesting results obtained in this way will be found in the
Appendix (p. 452).

If the arc shadow is produced simply by refraction, it does
not, of course, necessarily involve any *loss* of crater light ; the
distribution only may be altered, and not the quantity that is
received outside the arc. It is possible, however, that even a
re-distribution of the crater light might involve a certain
amount of loss or gain owing to the peculiar way in which the
arc lies between the carbons, for the part of the vapour lens
which is near the negative carbon is so shaped that it must
bend back into that carbon many rays from the crater that
would otherwise pass it, and bend others away from it that
would otherwise enter it.

It is clearly only the refraction of rays whose paths,
without it, would lie within a certain distance from the
surface of the negative carbon, both internally and externally,
that can have any effect on the total amount of crater light
received outside the arc ; and this distance must be determined by the refractive index of the mist. What this is,
under any given circumstances, and whether it alters with the
current and the length of the arc, is not known. Indeed, the
whole question of the refraction is fraught with difficulties
and surprises. Who could have foretold, for instance, that all
the three parts of the arc, the purple core, the shadow, and
the green envelope would have the same refractive index ?
Yet they must have, for the shadow they throw on the screen
is apparently of uniform depth, and surrounded by a single
rim of light, and is of a size and shape which show that it
belongs to all three portions.

Again, the refractive power of the mist must depend on its
density relatively to the surrounding air, and that again on
the relative temperatures of the two. Now it is true that
M. Violle* found the temperature of the arc higher than that
of the crater, but this was with an arc enclosed in a furnace.
Such an enclosure alters all the conditions of temperature, and
results thus obtained can give no clue to the relative
temperatures of the arc and carbons in the open air. No very

---

* *Journal de Physique*, 1893, 3rd Series, Vol. II., p. 545. *Comptes Rendus*,
1894, Vol. CXIX., p. 949.

definite experiments have yet been made on this point with the open arc, but the following considerations point to the conclusion that the average temperature of the arc mist must be lower with a long arc than with a short one, when the same current is flowing in both.

Since Sir William Abney first announced that the crater was at the temperature of volatilisation of carbon it has never been doubted that the stuff of which the arc was composed consisted chiefly of the vapour thus volatilised. But how can this be so ? That it leaves the positive carbon as vapour there can be but little doubt, but that by the cooling action of the air around it its temperature must be lowered, and that it must therefore condense, at a very short distance from the crater, there can be as little doubt. My own belief is that the vapour, in leaving the crater, acts just as steam does when issuing from the mouth of a kettle. Through a distance small enough for its temperature to continue unaltered it still remains vapour ; at greater distances it is condensed into carbon fog or mist. The true vapour is probably invisible, just as water vapour is (the space that is always seen between the arc and the crater confirms this view), but the mist is visible. The resistance of true vapour is very great, and consequently the resistance of the thin layer of vapour that lies over the crater is so great compared with that of the remainder of the arc that it is usually supposed not to be a resistance at all but a back E.M.F. The heat evolved by the passage of the current through this great resistance is sufficient to volatilise the surface of a part of the positive carbon, and thus to keep up the supply of vapour. The part of the surface that is thus volatilised becomes hollow with short arcs in the way explained in Chapter XII. (p. 393), and thus the crater is formed by the action of heat supplied to it by the thin layer of vapour that spreads over its surface.

According to this theory of the arc, the temperature of any horizontal section of the mist must depend upon (1) the temperature at which it left the crater, (2) the constant supply of heat conveyed to it, by radiation from the crater, (3) the heat evolved by the passage of the current through it, and (4) the cooling effect of the surrounding air. If we take a section at, say, 1 mm. from the tip of the negative carbon, the supply of heat received from the crater will be less the longer the arc, and the cooling effect of the surrounding air will be greater the longer the arc. On both accounts the *average* temperature of the mist must be lowered by lengthening the arc, and consequently its average density must be increased.

The reasons for considering that the arc mist absorbs an appreciable amount of the light emitted by the crater are, then—

(1) That this mist shares with candle and gas flames the power to hide anything placed behind it, as if it were opaque.

(2) The acknowledged and proved existence of solid particles in it.

(3) Its casting a shadow, which can hardly be due merely to refraction.

Change of Colour of Arc Light as the Arc is Lengthened.

The strongest proof that the arc, when it is long, absorbs quite a considerable portion of the light of the crater is, however, this :—Sir W. Abney has shown [*] that crater light is very like sunlight, but has a slight excess of orange and green rays and a slight deficiency of blue. He measured the light in such a direction that he was, as far as possible, getting rays from the crater *only*, that is to say, in a direction in which as little of the arc as possible was interposed between the spectroscope and the crater. The crater light seen through a small quantity of mist, then, is *yellower* than sunlight. But this very light, when it has penetrated through the mist of a long arc, is bright purple in colour, as may be deduced from the colour of the opalescent globes surrounding arc lamps, when the arc is long. And this colour has nothing to do with the fact that our being accustomed to the yellow light of gas and incandescent lamps after dark makes a pure white light appear blue, for even in broad daylight the globes surrounding arc lamps, in which the arc is long, appear bright purple. Indeed, I have seen the inside of a screen surrounding a long arc flooded with purple light as if lighted by a stained glass window, and retaining this colour *even when daylight was let*

---

[*] *Report* on the Action of Light on Water Colours, c. 5,453, 1888, pp. 23 and 69,

in on it. There can, therefore, be no doubt that the light of the crater becomes tinged with violet or purple as it passes through the arc, and that the tint deepens as the arc lengthens. If light became coloured in this way by being passed through coloured glass, we should say it was because the glass absorbed rays of certain colours and allowed other rays to pass. Why, then, should not this explanation apply to the arc mist?

What probably happens is this. The arc, except a thin layer quite close to the crater, consists of a mist of solid carbon particles, which are continually forming and falling, surrounded by burning gases. The vapour and gases must, of course, absorb a minute—possibly an inappreciable—portion of the light that issues from the crater. If this were all, there would probably be no maximum of light with a given length of arc; but the solid carbon particles have to be reckoned with. If the light simply passed through each of these that it encountered, and suffered only the small amount of absorption that would naturally take place, the whole quantity of light absorbed might still be too small to notice. But a ray of light encountering a white hot particle is not only refracted—some of it is reflected, so that each ray may be reflected from particle to particle, and so may traverse the mist hundreds of times before it finally emerges. At each reflection and refraction part of the light that the particle is capable of absorbing is absorbed, and a ray that has suffered much internal reflection must emerge in a very different state from that in which it left the crater.

Suppose, now, that the carbon particles were capable of absorbing the orange light and a certain amount of the green, but allowed all the violet light to pass. Then, after each successive reflection or refraction, the light would become more violet, and that which had encountered many particles would be entirely violet. No incandescent gases alone give a dazzlingly brilliant light, so that when one looked at the arc mist alone, screening off the whole direct light from both carbons, the part of the crater light that was transmitted to the eye from the solid particles would entirely swamp the feeble light emitted by the gases, and one would only perceive a brilliant violet or purple light. This would account for

the light of the arc alone being so much more brilliant than
one would expect.

A very simple experiment will suffice to show that the light
emitted by the arc *mist* is violet, while that emitted by the
crater and the white hot spot on the negative carbon is white.
If a thin metal plate, containing a horizontal slit *a* (Fig. 111)
about $\frac{1}{16}$in. in width, be held vertically near an arc so that the
slit is about equidistant from the ends of the two carbons,
and if the light from the arc that passes through the slit be
received upon a vertical white screen *cd*, a foot or two away,
this light will form three horizontal bands on the screen, the
upper and lower ones being white and the middle one of a

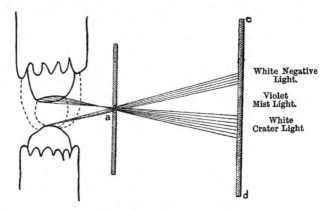

FIG. 111.—Light from Crater, Mist, and White Spot passing through a
Narrow Slit on to a White Screen.

bright violet.   The slit is, of course, simply a pinhole horizontally elongated, so that the upper white light must proceed
from the negative carbon, as indicated by the lines in Fig. 111,
the lower from the crater, and the middle violet band must
be lighted by the arc mist.

A still simpler experiment is that of shading the upper
carbon and part of the arc with any opaque body, one's hand,
for instance.   The shadow on the screen will then be found
to be edged with a broad band of reddish violet light, this
band being the portion of the screen that is illuminated by
the mist and the negative carbon alone (the red-hot part of

this carbon gives the rosy tinge). Below the band is the part illuminated by all three sources—crater, mist and white hot spot—and this naturally looks quite white when contrasted with the violet band. A diagrammatic illustration of this experiment is given in Fig. 112; *ab* is a metal plate, *cd* the white screen on which the light falls. The portion above is entirely in shadow, that between *c* and *d* is lighted by mist and negative carbon alone, and that below *d* is illuminated by the crater as well as by these.

This absorption by internal reflection would also explain a curious anomaly pointed out by Mr. Swinburne in the discussion on the Paper by Mr. Trotter,* already quoted.

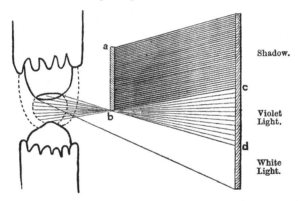

Fig. 112.—Band of Violet Light bordering the Shadow of a Metal Plate which is cutting off the Light from the Crater of an Arc.

Mr. Trotter exhibited the curves, published by Sir W. Abney, which showed that crater light only differed from sunlight in having a slight excess of orange and green rays and a slight deficiency of blue rays. Mr. Swinburne pointed out that as crater light and sunlight were so very nearly of the same colour, the sun and the crater must also be nearly of the same temperature; but that, as the sun gave above ten times as much light per square millimetre as the crater, the emissivity of the sun must be ten times as great as that of the crater, which was incredible. This objection

---

* Journal of the Institution of Electrical Engineers, 1892, Vol. XXI., p. 381.

is perfectly valid if sunlight and the light *emitted* by the crater are really so nearly of the same colour. But Sir W Abney was only able to measure the light *received* from the crater *after it had passed through a certain quantity of the mist*, for it is impossible for the crater light to get out without passing through *some* mist. Suppose, then, that the light *emitted* by the crater were very much yellower than the sun, but that even in passing through this small quantity of mist it was robbed of a considerable portion of its orange and green rays : then when it reached the photometer its colour would by this means have been brought much nearer to that of sunlight than it had been when it left the crater; it would have been rendered *bluer* in the course of transmission, and would appear, therefore, to have been emitted by a surface of much higher temperature than the surface of the crater actually is. Thus, if the mist does really absorb any considerable portion of the orange and green rays, Mr. Swinburne's objection no longer holds, for the crater light is probably far yellower when it is *emitted* than when it is *measured*.

Now take the effect of lengthening the arc on the light emitted by the crater. The arc mist would, on the whole, be cooler, there would be more solid particles, and each ray would, therefore, run a greater chance of encountering one or more of these particles before emerging. Thus more of the light on the whole would be absorbed, and more of the rays would have been robbed of all the light the particles were capable of absorbing before emerging, so that the light would, on the whole, be more violet than with a shorter arc—as it actually is.

Absorption of Crater Light in Successive Layers of the Arc.

So far, all the evidence appears to be in favour of the theory of the absorption of crater light in the carbon mist, but it still remains to be seen how far this theory will account for the curious phenomenon that started the inquiry—the existence of maximum points in the curves connecting the total light of the arc with its length, when the current is constant. For this purpose we will take the curve constructed in Fig. 108, connecting the whole light emitted by the crater, arc, and carbons, with the length of the arc, with a current of 20 amperes, and see whether, by subtracting a fraction of the crater light of the same magnitude as the light absorbed would probably be, from the light of each length of arc, we obtain a curve resembling those found by Profs. Ayrton and Blondel from actual measurement of the light.

We must first find what fraction of the crater light would be absorbed in any given length of arc when it is assumed that some fixed fraction of the whole light that enters any given layer of mist is absorbed in that layer. Let us divide the arc mist into layers of the kind shown by the dotted lines in Fig. 113, each half a millimetre thick at its thickest part, and let us suppose that each layer absorbs one nth part of the crater light that enters it. The law is, of course, really far

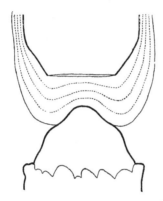

Fig. 113.—Arc with Mist divided into Layers of Equal Thickness.

more complicated than this, but we cannot hope, in any case, to obtain more than a very rough estimate of the light absorbed without more exact data than are obtainable at present. The amount of light that escapes from the first layer will be $L - \dfrac{L}{n}$ or $\dfrac{n-1}{n} L$. After it has passed through the second layer the quantity of light will be

$$\frac{n-1}{n} L - \frac{n-1}{n^2} L, \text{ or } \left(\frac{n-1}{n}\right)^2 L,$$

and after passing through the lth layer it will be

$$\left(\frac{n-1}{n}\right)^l L.$$

The quantity of crater light that escapes from an arc of l mm. will therefore be

$$\left(\frac{n-1}{n}\right)^{2l} \text{L}.$$

Now we have L for each length of arc corresponding with a point on the curve A B C (Fig. 108). Let us take $n = 40$, which is equivalent to saying that two and a-half per cent. of the whole light that enters each half millimetre layer of the arc mist is absorbed by that layer. Table LI. gives the total crater light before absorption, the same after absorption, and this latter crater light plus 10 per cent., to allow for the light emitted by the other sources, for each length of arc.

Table LI.—*Crater Light that would Escape Without and With Absorption, and Total Light Emitted by Arc if there is Absorption, in Arcs used for Curve A B C, Fig.* 108.

Length of Arc in mm.	Crater Light Without Absorption.	Crater Light With Absorption	Crater Light With Absorption, plus 10 per cent. of same.
0·5	7·8	7·6	8·36
1·1	9·12	8·63	9·49
2·0	10·12	9·14	10·05
3·2	11·76	9·90	10·89
6·6	14·15	10·13	11·14

Light-Length-of-Arc Curves drawn from Diagrams, allowing for Absorption of Crater Light in Arc.

If, now, a curve be drawn connecting the total light emitted by the arc (column 4, Table LI.) with its length (column 1) this curve should, if my assumption concerning the absorption of the crater light is correct, resemble the curves in Figs. 97, 98, 101 and 102. In Fig. 114 this curve is given, and it will be seen that it resembles, in every particular, the curves drawn from actual measurements of the light in Figs. 99 and 100, even to having a dip between 1mm. and 3mm., which is practically imperceptible in D E F (Fig. 108), which is the curve that would connect the total light emitted by the arc with its length if there were no absorption of the crater light. The hollow is not so deep as to form an actual minimum point, because 20 amperes is not a large enough current with carbons of

18mm. and 15mm. for the negative carbon to be *very* pointed, even with a short arc.

The real interest of Fig. 114, however, lies in the fact that the curve has a maximum point somewhere between 3mm. and

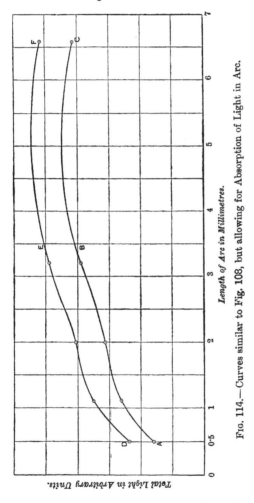

Fig. 114.—Curves similar to Fig. 108, but allowing for Absorption of Light in Arc.

6mm. For this shows that if there is an absorption of anything like $2\frac{1}{2}$ per cent. of the crater light in every half millimetre of the arc mist, the total light emitted by the arc, with 18/15mm.

carbons and a current of 20 amperes, will rise to a maximum with an arc of about 4·5mm, and then diminish. That is to say, on the assumption of this very possible rough law of absorption in the mist, the light curve obtained from measurements of diagrams of arcs and carbons resembles that obtained by actual measurement of the light. This completes the chain of evidence in favour of an appreciable portion of the crater light being absorbed in the arc mist ; and, while each separate piece of evidence may possibly be open to some other interpretation than that given to it, yet the whole together seems to afford an overwhelming presumption in favour of the theory.

Effect of Variation of Current on Total Light emitted by Arc.

Having examined the effect on the total light received from the arc of varying its length, while the current is kept *constant*, the next question to consider is how the light is affected by *varying* the current. This question may be approached in two ways : either the length of the arc or the P.D. between the carbons may be kept constant while the current is varied. Prof. Ayrton chose the first way, M. Blondel the second. The first has more scientific, the second more commercial interest ; for, while the amount of crater light that can get past the negative carbon and the quantity that is absorbed by the arc mist, both depend immediately on the length of the arc, the makers of arc lamps do not concern themselves with the exact length of the arc, but only with the current and the P.D. between the carbons, or rather, between the terminals of the lamp.

The diagrams in Fig. 115 show how the shapes of the carbons and arc change as the current is increased, while the length of the arc remains constant. For the upper row currents of 10, 25 and 35 amperes were used with a 4mm. arc ; and, for the lower, the currents were 6, 20, and 30 amperes, and the length of the arc was 0·5mm. The first thing that strikes the attention in these diagrams is the way in which the crater enlarges as the current is increased. For instance, in the upper row the diameters of the craters for 10 and 25 amperes were 5·4mm. and 8·4mm. and the areas of their mouths were, therefore, 22·9 and 55·4 sq. mm., *i.e.*, the light-giving surface of the 25-ampere crater was more than double that of the

10-ampere crater. Thus the light received in each direction from which the crater could be seen must have been more than doubled by increasing the current from 10 to 25 amperes. It does not necessarily follow, however, that the total amount

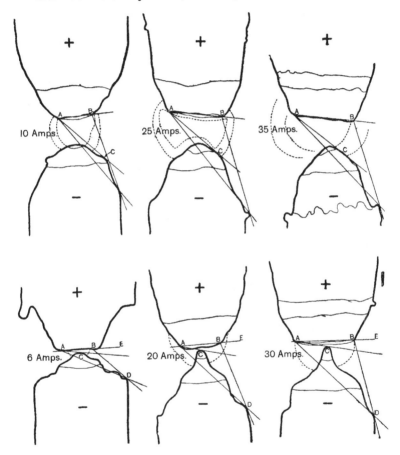

FIG. 115.—Effect of Variation of Current on Shapes of Carbons.

of light received outside the arc was thus largely increased, for the angles through which the whole crater could be seen and through which any part of it could be seen might be proportionately diminished. There is no such diminution, however, as will be seen from a comparison of the angles A B C and E B D for the three currents in each row. In all cases,

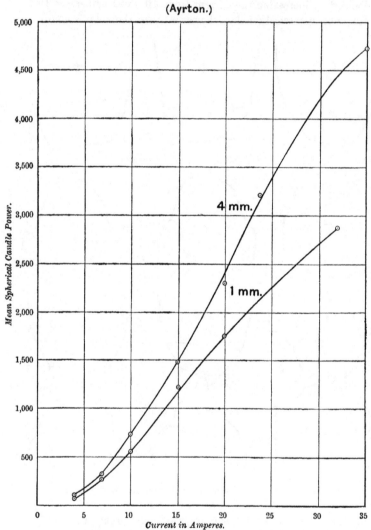

(Ayrton.)

Fɪɢ. 116.—Curves connecting Mean Spherical Candle-power with Current for Constant Lengths of Arc of 1mm. and 4mm.
Carbons : Positive 13mm. cored ; negative, 11mm. solid.

therefore, if none of the crater light were absorbed in the arc mist, the light received from the crater, and therefore the whole light received from all sources would increase as the current was increased with a constant length of arc.

(Blondel.)

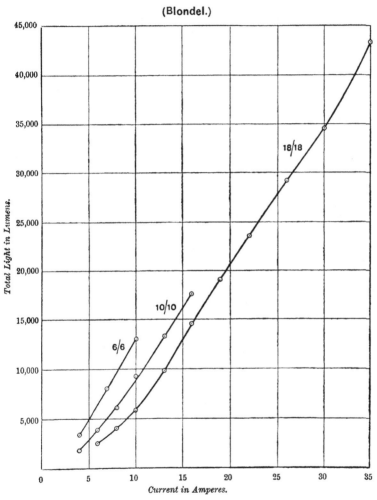

There is, probably, however, absorption, and this absorption must also increase with the current, for the cross section of the arc certainly increases with the current. Whether the extra amount of light absorbed ever balances the extra amount emitted through an increase of current, so that the total light received outside the arc rises to a maximum with some particular current and then diminishes can only be determined by experiment. To answer this question, therefore, we must turn to the actual measurements of the light with constant lengths of arc and different currents.

The curves that Prof. Ayrton found connecting the mean spherical candle-power of the arc with the current for constant lengths of arc of 1mm. and 4mm., are given in Fig. 116. They are plotted from Table L. (p. 328), so that the carbons employed were 13mm. cored and 11mm. solid. These curves show by their form that with small currents the light increases more rapidly than the current, while with large currents the reverse is the case, but there appears to be no maximum of light with any particular current. Thus we are led to the conclusion that with a given length of arc the light continues to increase as the current increases, till hissing occurs and changes all the conditions.

M. Blondel[*] found exactly the same result with a *constant P.D.* and varying current, as may be seen from Fig. 117, which gives some of his curves. The light increases much more rapidly with the current in these curves than in Prof. Ayrton's, because M. Blondel had to increase the length of his arc with the larger currents in order to keep the P.D. constant. The net result is the same, however : there is no maximum of light with any particular current—the light continues to increase as long as the current does so.

THE LIGHT EFFICIENCY OF THE ARC.

So far, our whole attention has been devoted to the *quantity* of light emitted by the arc, without any consideration for the amount of energy expended in producing that light—the *efficiency* of the arrangement. We shall now proceed to examine the conditions under which the greatest efficiency

---

[*] *L'Éclairage Électrique*, 1897, Vol. X., p. 297.

can be obtained. This has already been done in Chapter IX. (p. 264) as far as the ratio of the power developed in the dynamo to the power expended in the arc is concerned; but it still remains to be seen what arrangements enable the arc to turn a given amount of energy supplied to it into the greatest quantity of light.

It is surprising to find how small a part of the energy absorbed by even such an efficient source of light as the arc is utilised in producing light. Some tests made by Mr. Hatsuné Nakano * of the ratio of light-giving radiation to the total radiation in the arc, with various kinds and sizes of carbons, showed that this ratio varied between 0·018 and 0·198, or, roughly, he found that the light-giving radiation was from 2 per cent. to 20 per cent. of the whole. The 45 tests he made gave an average of 10 per cent. for the light-giving radiation.

Now, the most economical way of feeding an arc is, of course, to use a dynamo driven by a steam engine. Leaving out of account all losses in the dynamo and mains, the engine can only transmit to the dynamo about 10 per cent. of the energy of the coal; and it follows that, since only 10 per cent. of this is utilised in the arc to produce light, an average of only *one per cent.* of the energy of the coal is transformed into light energy, and the other 99 per cent. is wasted. Truly there is a wide field for improvement in our ways of producing light!

Mr. Nakano's experiments consisted of four distinct operations. First, he measured the total radiant energy of the arc by means of a thermopile attached to a sensitive galvanometer. Next, he placed an alum bath between the arc and the thermopile, to screen off the " dark heat," and measured the remaining energy. As, however, the alum bath allowed a certain quantity of "dark heat" to pass, and screened off a certain portion of the light, a correction for each of these was needed. The first was made by placing a cell containing an opaque solution of metallic iodine in bi-sulphide of carbon between the lamp and the alum bath. This stopped all the light rays, but allowed all, or a very large proportion, of the dark heat rays to pass. Any deflection of the thermopile

---

* Paper read before the American Institute of Electrical Engineers, New York, May 22nd 1889.

must now be caused by the dark heat rays that passed through the alum bath. These were so few that they were found to be imperceptible in most cases. Fourthly, the quantity of light that was absorbed by the alum bath was measured photometrically, and was found to be about 26 per cent., on an average.

Until Mr. Nakano's experiments were made, we had no idea how much of the energy given to the arc was utilised in producing light : it might have been 10 per cent., it might have been 90 per cent. His measurements are, therefore, of immense scientific value, and of great value also in showing the direction that improvements in our method of producing light should take. For practical purposes, however, something more is required ; light energy needs to be translated into candle power. The light consumer wants to know, not how much of the power he supplies to the arc will be utilised in producing light, but how much actual light he will get for a given power supplied. By making measurements, connecting candle power with luminous energy, it might be possible to translate the ratio $\dfrac{\text{luminous energy}}{\text{total energy}}$ into the ratio $\dfrac{\text{light emittted}}{\text{power supplied}}$, but there are many objections to such a method of finding the latter ratio, not the least being its circuitousness. No, this ratio is best found by measuring directly either the total light emitted and the power absorbed by the arc, as M. Blondel does, or the mean spherical candle power and the power absorbed, as most other experimenters do ; and for either of these measurements the eye and not the thermopile must be the instrument.

Distribution of the Power Supplied to the Arc between the Carbon Ends and the Mist.

Before turning to these experiments it will be interesting to see how the power absorbed in the arc, under given conditions, is distributed among the five sources of light enumerated on p. 314, and how that distribution changes with a change in the conditions. The power absorbed by the hot ends of the carbons, including, of course, the crater and the white spot, may be measured by multiplying by the current the P.Ds. between those carbons and the arc mist. The power given

to the arc mist may be obtained by subtracting the power
absorbed by the two hot ends of the carbons (including
crater and white spot) from the total power supplied to
the arc. Now it seems most likely that the end of the
positive carbon, exclusive of the crater, is only imperceptibly
heated by the passage of the current, and that it is kept hot
by heat supplied to it by the crater. Similarly, the end of
the negative carbon is kept hot by the white spot. We may,
therefore, consider that all the electric energy directly supplied
to the hot ends is consumed in the crater and the white spot,
and is therefore usefully employed in giving light, while that
supplied to the arc mist may be considered as practically
wasted, since so small a part of the total light is emitted by
the mist.

In Chapter VII., p. 231, it was shown that, of the

$$\left(38 \cdot 88 + 2 \cdot 07l + \frac{11 \cdot 66 + 10 \cdot 54l}{A}\right) \text{ volts}$$

always needed to maintain an arc of l mm., with a current of
A amperes, with solid carbons 11/9,

$$\left(38 \cdot 88 + \frac{11 \cdot 66}{A} + \frac{3 \cdot 1l}{A}\right) \text{ volts},$$

at most, were expended at the junctions of the carbons with
the arc mist, and

$$\left(2 \cdot 07l + \frac{7 \cdot 44l}{A}\right) \text{ volts},$$

at least, were used in sending the current through the mist.
Multiplying each of these expressions by A we get

$$(38 \cdot 88A + 11 \cdot 66 + 3 \cdot 1l) \text{ watts}$$

as the power supplied to the carbon ends, and

$$(2 \cdot 07Al + 7 \cdot 44l) \text{ watts}$$

as the power used up in the mist. Thus, every increase of 1mm.
in the length of the arc is accompanied by an increase of 3·1 watts,
at most, in the power supplied to the carbon ends, and of
$(2 \cdot 07A + 7 \cdot 44)$ watts in the power used in the mist. But the
mist in itself gives out so little light that this latter amount
of power may be considered wasted. Thus, *almost the whole
of the increased power that has to be supplied to the arc when it
is lengthened is swallowed up by the mist and is practically wasted*

B B 2

Moreover, since the increased power used in the carbon ends does not depend on the current, and that used in the mist does, it follows that the waste is greater the greater the current. For instance, with a current of 5 amperes each increase of 1mm in the length of the arc entails an increased waste of 17·8 watts in the mist and an increased use of 3·1 watts at the carbon ends. With a current of 20 amperes the power used in the carbon ends is still 3·1 watts per mm., but the waste in the mist is 48·8 watts per mm.

When $l = 0$, $(2 \cdot 07A + 7 \cdot 44) l = 0$ also, that is, the number expressing the power supplied to the mist is zero, when the length of the arc is 0, even although length of arc 0 does not mean that the carbons are touching, but only that the tip of the negative carbon and the mouth of the crater are in the same plane. This can only mean that the power supplied to the mist that lies in the hollow of the crater is always so small as to be inappreciable.

The least fraction of the whole power that is always practically wasted in the mist is

$$\frac{(2 \cdot 07A + 7 \cdot 44) l}{38 \cdot 88A + 11 \cdot 66 + (2 \cdot 07A + 10 \cdot 54) l}.$$

With a 3mm. 10-ampere arc (a very ordinary one for the size of the carbons) this fraction is 0·17, showing that, with such an arc about one-sixth of the whole power consumed is practically wasted in the mist, quite apart from the fact that not all the light created can get out and be made use of. Figs. 118 and 119 bring out with great clearness the way in which power is wasted by lengthening the arc, as far as the amount of light *created* is concerned.

The upper line in Fig. 119 shows the connection between the whole power given to the arc and the length of the arc, with a constant current of 10 amperes. The lower curve shows the power wasted in the mist in each length of arc. The distance between the corresponding points on each line shows the part of the power that is used at the carbon ends, and that is therefore usefully employed. The very small difference in the distance between the corresponding points on the two lines for 0mm, and 10mm. shows how very little the quantity of light *created* in the arc is increased by lengthening it from 0mm. to 10mm.

Fig. 119 shows the fraction of the whole power that is wasted as the arc is lengthened from 0mm. to 10mm. under

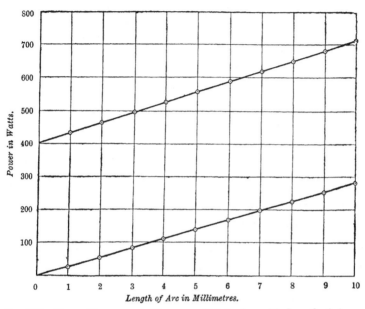

Fɪɢ. 118.—Upper Line connects Power supplied to Arc with Length of Arc ;
Lower Line connects Power wasted in Mist with Length of Arc.
Constant Current, 10 amperes.

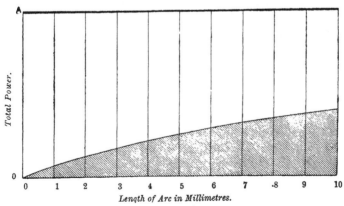

Fɪɢ. 119.—Curve showing the Proportions of the whole Power that
is wasted in the Mist with each Length of Arc.
Constant Current, 10 amperes.

the same conditions as governed Fig. 118.   The distance O A
is taken to represent the *whole* power supplied to the arc,
whatever that power may be, and the part of this distance
that is shaded represents the fraction of the power that is
wasted in the mist in the arc of corresponding length.   The
way in which the shaded portion grows in width gives a very
good idea of the manner in which the power is wasted, as far
as the *creation* of light is concerned, by lengthening the arc.

Condition necessary for the Arc between Given Carbons to emit
the Maximum of Light for a given Power developed by the
Generator.

Not only is the ratio of the light created to the power
consumed in the arc greatest in the shortest arc, but it has
been shown in Chap. IX. (p. 260) that the ratio of the power
consumed in the arc to the power generated in the dynamo is
also a maximum with the shortest arc ; and, further, the
quantity of crater light that is absorbed by the arc mist is
also least in the shortest arc.   Everything, therefore, tends to
make the arc the more efficient the shorter it is.   This leads
us to the conclusion that, *but for the negative carbon stopping*
some of the light, the ideal condition for an arc would be to have
the carbons nearly touching.

Influence of Cross Sections of Carbons on Lighting Power of Arc.

The next question to be considered is the influence of the sizes
of the carbons on the lighting power of the arc.   Schreihage*
attacked this problem in 1888, in the following way.   He took
five pairs of cored positive and solid negative carbons, the
positives varying in diameter from 7·12mm. to 18mm. and the
negatives from 5·75mm. to 16·1mm.   The ratio of the cross
section of the positive carbon to that of the negative was as
constant as possible, varying between the limits 1·54 : 1 and
1·24 : 1, as is shown by the diagrams of them published in the
account of the experiments.   Using the same current, 6·29
amperes, and the same mean P.D. between the carbons, 43·9
volts with each pair, Schreihage gathered that the mean hemi-
spherical candle power varied inversely as the *diameter of the*

---

* *Centralblatt für Elektrotechnik*, 1888, Vol. X., p. 591,

positive carbon within about 9 per cent. on either side of the mean.

This is a very simple and beautiful relation, but it obviously cannot be applied universally without further investigation, if only because there is nothing in the experiments to show whether it is the size of the positive carbon or of the negative, or of both, that determines the lighting power of the arc. The ratio of the diameters of the two carbons varied only between the limits 1·12 : 1 and 1·24 : 1, so that the quantity L *d*, *i.e.*, mean spherical candlepower multiplied by diameters of carbon would be equally constant whether the *d* was interpreted to mean the diameter of the positive carbon or of the negative carbon, or the mean of the two—as long as the same interpretation was taken for the whole five pairs of carbons. What Schreihage really showed, and it was a great advance at the time, was that there was some sort of inverse proportionality between the lighting power of the arc and the diameters of the carbons, when the current and P.D. were constant. He also called attention to the increasing bluntness of the positive carbon, and to the way in which the length of the red-hot part of that carbon diminished as its diameter increased.

The effect on the lighting power of the arc of varying the diameter of *each of the carbons separately* was first tried by M. Blondel* in 1897, as part of the exhaustive series of experiments to which I have made such frequent reference. M. Blondel found that the light could be increased with the same current and P.D. between the carbons, either by reducing the size of the negative carbon with a constant positive, or by reducing the size of a cored positive carbon with a constant negative. He found, for instance, that with a current of 10 amperes, a P.D. of 40 volts and a positive carbon of 16mm., the light could be increased from 5,019 lumens with a negative carbon of 16mm. to 6,575 lumens with one of 6mm.; and with the same P.D. and current, and a negative carbon of 6mm. the light could be increased from 6,575 lumens with a positive carbon of 16mm. to 10,462 with one of 6mm. The relative importance of the sizes of the two carbons in regulating the

---

* *L'Éclairage Électrique*, 1897, Vol.X., p. 497.

total quantity of light emitted, cannot, of course, be determined
from the above numbers, because the size of the constant carbon
was not the same in both cases, but they appear to show that
the diameters of the two carbons are at least equally important
in determining the lighting power of the arc. This appearance
is, however, to a certain extent fallacious, and indeed, while
we know perfectly well that increasing the diameter of the
negative carbon must diminish the light of the arc by
obstructing that of the crater, it is difficult to see how
increasing the diameter of the positive carbon can have an
effect anything like as great. It is true that quite apart from

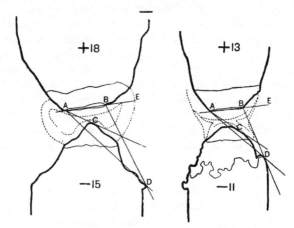

Fig. 120.—Diagrams of Arc with the same Current (10 amperes) and
Length (2mm.), but with Different sized Carbons.
Positive Carbon cored ; negative solid.

the shortening of the red hot part of the positive carbon
mentioned by Schreihage, there are two other ways in which
the light of the arc must be diminished by increasing the
diameter of the positive carbon. The light of the white spot
must be more screened, and the mist being obstructed in its
upward flow must be wider and so absorb more light. But it
is quite inconceivable that all these three causes together
should be sufficient to account for the immense difference in
the light observed by M. Blondel. That the mist has actually
a larger cross section, and the obvious reason of it, will be seen

from Fig. 120, the diagrams in which are diminished copies of images of the arc and carbons taken with a current of 10 amperes, a P.D. of 43 volts and an arc of 2mm. For the diagram on the left the carbons were 18mm. cored and 15mm. solid, and for that on the right they were 13mm. cored and 11mm. solid. It is evident that the wider positive carbon has caused the arc mist to have a much larger cross section, and that therefore the crater light must lose more by absorption when this larger carbon is used.

This, however, is not the most important reason for the great change in the light flux noticed by M. Blondel when he increased the size of the positive carbon. The reason seems, rather, to be that while the negative carbons used for the experiments were *solid*, the positive carbons were *cored*, and as they were ordinary commercial carbons—" Nanterre " was the brand—and not made specially for experimental purposes, the core was certainly made to have some sort of proportionality to the diameter of the carbon : it would be larger in the larger carbon than in the smaller one. Now, it will be presently shown (p. 386) that the surface of the core gives a much less brilliant light than the remainder of the surface of the crater, and M. Blondel himself has shown that the arc gives less light the thicker the core. Thus, of two craters of nearly the same size, the one with the larger core would give the less light ; so that the 18mm. carbon with a core probably 4mm. in diameter would naturally give less light, all other conditions being equal, than the 6mm. carbon with a core of, say, 1·5mm. It was not, therefore, the increase in the size of the positive *carbon* that was the principal cause of the diminution in the amount of light emitted by the arc in M. Blondel's experiments, but the larger core that that increase entailed. Clearly, to see the effect produced by size alone, it would be necessary to repeat the experiments, using a *solid* instead of a cored positive carbon.

The whole question is, however, fraught with difficulties and complications, and should, I think, be attacked in a somewhat different manner. For instance, we have seen how much more directly the light depends on the length of the arc than on the P.D. between the carbons. We have also seen (p. 161) that the P.D. between the carbons is influenced by the diameters of the carbons with some currents, and not with others, and that it always depends to some extent on the size of the core. It seems highly probable, therefore, that by choosing the current and the carbons properly we could almost make the light vary in any way we chose, while getting an effect apparently due to a change in the size of the positive carbon alone. In order, then, to see how a simple increase in the size of the positive carbon alone affects the lighting power of the arc, the following precautions must be taken :—

(1) *Both* carbons must be solid and of the same make and hardness for all the experiments.

(2) The current and length of arc must be chosen in such a way that the length of arc, as well as the current and the P.D. between the carbons, can be constant for all the experiments.

(3) The negative carbon must have a constant diameter, as it had in M Blondel's experiments.

With these precautions which, though difficult it would not be impossible to take, we should really be able to judge of the extent to which a given increase in the size of the positive carbon diminished the lighting power of the arc. That it must diminish it to a certain extent has been shown from *a priori* considerations, and hence, without making such very accurate experiments, it is quite safe to take it for granted that in order to insure the greatest light efficiency in the arc the positive carbon must be as small as it conveniently can. Now there is a limit to the thinness of the positive carbon that does not exist in the case of the negative. I allude to the possibility of hissing, which has been shown (p. 299) to depend on the size of the positive carbon with a fixed current and length of arc. The larger the current the greater must be the positive carbon in order that the arc may remain silent. It comes to this, then ; *with a fixed current, length of arc and P.D. between the carbons, and a given negative carbon, the arc will give most light when the positive carbon is as small as it can be without causing the arc to be so near the hissing point as to be unsteady.*

The diagrams in Fig. 121 show the sort of change that takes place in the shapes of the carbons when the sizes of these are altered, and when the current and the length of the arc are kept constant. The current employed was 6 amperes, and the length of the arc 1mm. The carbons were 18/15, 13/11, and 9/8, the positive being cored and the negative solid in each case. Enlarging the carbons seems to slightly enlarge the crater, which would cause an *increase* in the amount of light emitted, were it not that the significant angles B A D and E B D both *diminish* when the carbons are enlarged. Also when the light is received in any given direction between A D and B D, it is easy to see that a large negative carbon would hide far more of the crater than a small one.

Thus this figure shows, as M. Blondel's experiments did, that *with a fixed current, length of arc, and P.D., between the carbons, the arc gives more light the smaller the negative carbon.*

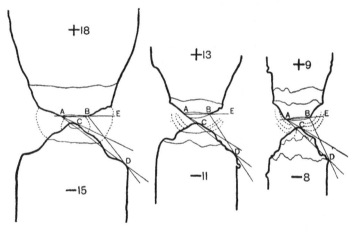

Fig. 121.—Diagrams of Arcs with the same Current (6 amperes) and Length (1mm.), but with Different sized Carbons. Positive Carbon cored ; negative solid.

In Chap. IX. (p. 260) it was shown that the supply of power to the arc was most efficiently conducted, *i.e.*, the ratio $\dfrac{\text{power supplied to arc}}{\text{power developed by dynamos}}$ was greatest, when the current was the largest that would certainly give a silent arc. This is only another way of saying that with a given current the arrangement for the supply of power was most efficient when the positive carbon was the smallest with which there would be no danger of hissing with the given current. The law of the smallest carbons, therefore, applies not only to the light efficiency, but also to the power efficiency of the arc. In other words both the ratio $\dfrac{\text{power used in arc}}{\text{power developed by dynamo}}$ and the ratio $\dfrac{\text{light generated in arc}}{\text{power supplied to arc}}$ are greatest when the carbons are the smallest. Hence the ratio $\dfrac{\text{light generated in arc}}{\text{power developed in dynamo}}$ is also greatest under the same conditions. In every way possible, therefore, *the arc burns most economically when the carbons are the very smallest that can carry the current*, without danger of hissing, in the case of the positive, and without burning too fast, in the case of the negative.

The prime necessity in order to ensure a highly efficient arc is, therefore, to have both carbons, especially the negative, as thin as the current will allow, and then, as the negative carbon cannot be infinitely thin, it must be a matter for experiment to see with what length of arc the disadvantage of the extra power wasted in that length is exactly counterbalanced by the advantage due to the extra amount of light let out ; for this will be the length that will give the maximum efficiency under the given conditions. One would naturally suppose that this length of arc would be shorter the smaller the carbons, and that it is so, at least when both carbons are solid, is proved by the curves in Fig. 122, which are taken from M. Blondel's * results.

These curves give the connection between the light efficiency $\left(\dfrac{\text{total light emitted}}{\text{power consumed}}\right)$ of the arc, and its length, with a constant current of 10 amperes, for three different sets of solid " Nanterre " carbons, 8/6, 10/10, and 16/14. They show at a glance what M. Blondel deduced from the same results plotted with the P.D. between the carbons, instead of with the length of the arc, viz., that whatever the length of the arc (except zero in one case) the light efficiency is *always* greatest with the smallest carbons, and that the *maximum* efficiency for a given pair of carbons is also much the greatest with the smallest carbons, ranging as it does from about 26·6 lumens per watt with the 8/6 carbons to only about 14·6 lumens per watt with the 16/14 carbons. In other words, the maximum.

---

* *L'Éclairage Électrique*, 1897, Vol. X., p. 296.

efficiency in this case is nearly doubled by halving the diameter
of the positive carbon and by taking that of the negative carbon
rather less than half. Lastly, these curves show that with solid
carbons the *maximum efficiency of the arc is reached with a
shorter arc the smaller the carbons.* For with the 8/6 carbons
this maximum is attained with a 1mm. arc, with the 10/10
with a 3·5 mm. arc, and with the 16/14 carbons the arc has
to be 5·5mm. in length before the maximum efficiency is
gained. Thus practical experience fully corroborates the
conclusions drawn from theoretical considerations as to the
need of using small carbons and short arcs in order to obtain
the maximum light efficiency with solid carbons.

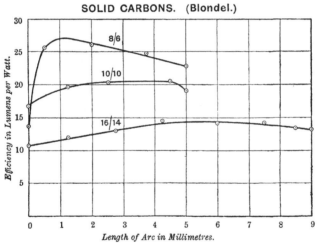

SOLID CARBONS. (Blondel.)

FIG. 122.—Curves connecting Light Efficiency with Length of Arc.
Current, 10 amperes.

When the positive carbon is cored the conclusions are not
all quite so definite, as will be seen from Fig. 123, the curves
in which, like those in Fig. 122, are drawn from M. Blondel's*
researches. The current employed was again 10 amperes, and
the carbons were 8/6, 10/10, 14/12, and 18/14. It may be
gathered from these curves that the efficiency for any particular length of arc is always greater, and the maximum
efficiency is higher, the smaller the carbons, when the positive

---

* *L'Éclairage Électrique,* 1897, Vol. X., p. 296.

carbon is cored as well as when both carbons are solid.   What is not quite so definite with the cored positive carbon is the length of arc with which the maximum efficiency is attained. This seems to vary between 0·5mm. and 1·5mm. with all the carbons except the 10/10, which have very much the same efficiency for any length of arc between about 1·7mm. and 4mm.

On the whole, however, Figs. 122 and 123 both corroborate what has already been said on p. 380, that in order to have the greatest possible amount of light for the power supplied to the arc, with a given current, the positive carbon must be as small as it can be without fear of hissing ; the negative carbon must

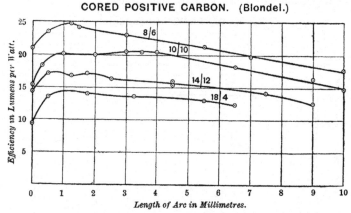

CORED POSITIVE CARBON.   (Blondel.)

Fɪɢ. 123.—Curves connecting Light Efficiency with Length of Arc. Current, 10 amperes.

be the smallest that will carry the current without burning away too fast; and the length of arc must be that which is found by experiment to give the maximum efficiency with the given current and carbons.   With solid carbons this length is less the smaller the carbons, because the smaller the negative carbon the more nearly do the quantity of light *created* and the quantity that escapes from between the carbons approximate to one another, and if the negative carbon could be infinitely small all the light created would escape, and then the maximum efficiency would be obtained with length of arc 0.   With a cored positive carbon the core seems to interfere, and to make the length of arc with which the maximum efficiency is obtained depend less definitely on the sizes of the carbons.

Low Efficiency of Commercial Arc Lamps due to Thickness of Carbons employed.

A reference to Table LII., the data for which were collected by Prof. Ayrton about two years ago, will show how little the need of thin carbons for the arc to burn economically was then realised, and very little improvement in this direction has, I believe, taken place since. This table gives a list of carbons of different makes with the currents intended to be used with them.

Table LII.—*Sizes and Lengths of Carbons employed in Commercial Arc Lamps, with Currents used, Average P.D. between Carbons, and Time of Burning.*

Carbons: Positive cored ; negative solid.

Diam. of Pos. Carb. in mm.	Diam. of Neg. Carb. in mm.	Length of each carbon in inches.	Current in Amperes.	P.D. between Carbons in Volts.	Time of Burning in Hours.
20	13	9·5	10	44	10-12
18	13	9·0	10	43	10
18	12	8·0	10	...	...

Now suppose that these arcs were all burned under such conditions that the maximum efficiency possible with such carbons and such a current was obtained. This maximum efficiency would probably be very much the same as that produced with the 18/14 carbons used by M. Blondel, about 14 lumens per watt, or, since the negative carbon in the lamps was less than 14mm., let us put it at 16 lumens per watt, which is a liberal allowance. If the size of the carbons had been 8/6 instead of what they were, the current and P.D. between the carbons being unaltered, the quantity of light emitted per watt would, according to Fig. 123, have been *24* lumens, or half as much again, so that by using smaller carbons the same amount of light could be obtained with the expenditure of two-thirds of the power, *i.e.*, at two-thirds of the cost for power. If still smaller carbons than 8/6 could be used, the light efficiency would be still greater, and probably might be doubled, for the light efficiency increases more rapidly than the sizes of the carbons diminish. The reply to this is, of course, that smaller carbons cannot be manufactured to burn anything like 10 hours with a 10-ampere current flowing. This is probably true at the present time, but it only shows the imperative necessity of making carbons that shall be both thin and slow burning, and when this is done the enclosed arc will no longer be able to hold its own against the open arc; for there can be no possible doubt of the superior efficiency of the open arc when burnt under proper conditions. The sole superiority of the enclosed arc lies in the labour and expense it saves by the slow burning of its carbons—a great superiority indeed, and one which counts for even more in America than in England—but one that should yield to the ingenuity and patience of the inventor.

Variation of Light Efficiency with Current.

In observing the way in which the efficiency of the arc varies under different conditions, we have hitherto dealt with a constant current of 10 amperes; it is necessary, however, to know also how the efficiency varies with the current. Fig. 124, taken, like so many others, from M. Blondel's Papers,[*] shows the variation in the efficiency produced by changing the current, while a constant P.D. was maintained between the carbons. The carbons employed were ordinary commercial cored positive and solid negative carbons for the first three curves, and Siemens' solid carbons for IV. The sizes were 6/6 for I., 10/10 for II., 18/18 for III., and 10/10 for IV. For the first three curves the P.D. was kept constant at 45 volts, and for the fourth at 50 volts. The efficiency of the arc appears to increase very rapidly at first, as the current is increased, and then much more slowly, while it even has a maximum point, and then diminishes with the solid Siemens' carbons. This maximum point is, I think, explained by the fact that as the P.D. was kept constant and the current increased, the length of the arc had to be increased also, for with a constant *length of arc* the P.D. always *diminishes* as the current

---

[*] *L'Éclairage Électrique*, Vol. X., p. 498.

increases with solid carbons. Now the length of the arc rose from 2·8 to 3·3mm. as the current was increased from 12 to 15 amperes; and we have seen that there are two ways in which the efficiency is diminished by lengthening the arc: Firstly, more of the power is wasted in the mist, and secondly, the longer mist column absorbs more of the light of the crater. At the same time, lengthening the arc enables more of the crater light to escape. It is thus quite probable that with

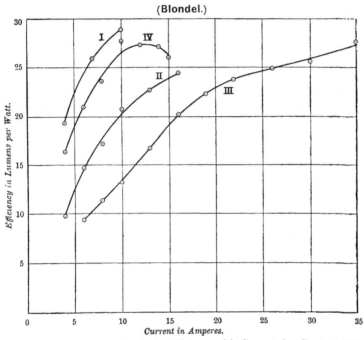

(Blondel.)

FIG. 124.—Curves connecting Light-Efficiency with Current for Constant P.D. Carbons; I., 6/6, II., 10/10, III., 18/18, 45 volts; IV., 10/10, 50 volts. .I., II., III., Nanterre, positive cored, negative solid. IV., Siemens, both solid.

arcs as long as from 2·8 to 3·3mm. the diminution of efficiency due to the first two causes may more than counterbalance the increase due to the third and to the increase of the current. It is not, however, easy to assign phenomena to their right causes when cored carbons are used, seeing how the core complicates everything in the arc—the P.D. necessary to send a given current through a given length of arc; the amount of absorption due to the two sorts of mist that must be generated, that of the core and that of the hard outer carbon; the difference of intrinsic brilliancy between the two parts of the crater surface. All these introduce complications that it is next to impossible to unravel, and hence it is really only upon observations made with solid carbons that it is possible to theorise, except in so far as one can show *why* the results obtained with cored carbons often differ so widely from those gained with solid ones.

Effect of Composition of Carbons on the Lighting Power of the Arc.

The lighting power of the arc depends just as much on the composition of the carbons as on their sizes, or on the current and length of arc. Carbons may be hard or soft, solid or cored. Dr. Louis Marks* has made a very careful study of the lighting power and life of particular sorts of carbons. He finds that soft carbons, rich in lampblack, give a better light for the same amount of power than harder ones, containing more graphite. M. Blondel attributes this to the quicker burning away of the softer carbons, their pointing themselves better, and having a more distinct crater. The American "Electra" carbons, he finds, give a great deal of light for the power they consume, but burn away with extraordinary rapidity. M. Blondel finds also that when both carbons are solid more light is given out, for the same power applied, than when the positive is cored, and that the diminution of the light is greater the larger the diameter of the core. This, as he pointed out in 1893,† is because the core has a lower degree of incandescence, which must, of course, mean a lower temperature. Every one who has seen an image of the crater must have observed this. If the current is large enough for the crater to more than cover the core, the cored part looks quite dim compared with the ring of solid carbon round it.

M. Blondel has found that there are yet other ways in which the make of the carbons influences the light-producing power of the arc. He finds that it is very important that they should be as pure as possible, pure carbons giving a far better

---

* American Institute of Electrical Engineers, July, 1890.

† *Report* of the Electrical Congress at Chicago, 1893.

light than those less carefully prepared. He also lays great
stress on the necessity for their being well-baked, and attributes
the superiority of the Siemens " A " carbons to their being twice
baked, the double baking rendering them purer, he believes.

Arc Lamps in Series.

One great difficulty to be overcome in the arrangement of
arc lamps to-day is the high P.D. that it has lately become
usual to maintain between the mains—200 and sometimes even
230 volts. This necessitates the use of several lamps in
series, and the substitution of resistances for any lamps of the
series that do not happen to be required. The waste thus
involved, whenever it happens that all the lamps are not
always required at the same time, is enormous. The best way
to overcome the difficulty, would probably be to use very small
currents (and consequently very small carbons), and to treat
each series of arcs as a single source of light. Suppose, for
instance, it were possible to have a good steady arc with a
current of, say, 2·5 amperes only, the carbons being so small
that the maximum efficiency of the arc would be obtained with
quite a short arc, 0·2mm., say. Then the P.D. between each
pair of carbons would probably not have to be higher than
about 47 volts, and with 200-volt mains four arcs in series
could be employed, using 188 volts, and leaving 12 volts to be
used up in the mains and the regulating resistances. The
resistance of all these together would be $12 \div 2\cdot5$ ohms, or
4·8 ohms. The power supplied to the groups of arcs would
be 500 watts, and of that 30 watts would be wasted in the
resistances &c., or 94 per cent. would be used and 6 per cent.
would be wasted. The arcs could be grouped so as to let out
the largest possible amount of light, and they would use the
same amount of power as a single 50-volt lamp burning
10 amperes. It seems quite clear that, if a not too complicated
arc lamp could be devised that would hold and regulate
the whole four arcs at once, it would command a very large
sale for use on high voltage mains ; for the convenience of
being able to use the whole large P.D. between the mains
to supply each single group of lights would more than
counterbalance any small waste that might ensue from the
arrangement being less efficient than the ordinary large current
arcs now in use.

c c 2

The Only Fair Method of Comparing the Efficiency of two Sources of Light.

One word, in conclusion, as to the right way of comparing two sources of light. As yet there is no definite standard method which is adopted by the scientific and the practical man alike. Various people test various things and call them candle power. Some take the candle power in the direction in which it is greatest and call that the candle power of the source. In this way we get nominal 2,000 c.p. arc lamps, which are really a bare 600 c.p. Others take the mean spherical candle power of each source, it is true, but under conditions which, while they may be the best for one of the sources, may be the worst for the other. This, again, is unfair.

There is, indeed, only one fair and right method of comparison, and that is to arrange the conditions of each source so as to enable it to do its best, and then to measure the mean spherical candle power, or total quantity of light (it does not matter which), emitted by each; to measure at the same time the power supplied to each, and to divide the first by the second to obtain the *light efficiency* of each source. This is the only fair way of comparing their relative values. For what does it matter to a user of light that one lamp will give him 2,000 c.p. and another only 500? He can always arrange to have four of the latter instead of one of the former, *provided always that the four do not cost him more than the one.* What he wants to know about his lamp is how much light he gets from it for the money he expends, that is, for the power it uses, after the initial cost has been overcome. Doubtless there are many other things to be taken into account in choosing a lamp besides its efficiency—the initial cost, cost of repairs, cost of up-keep (labour in replacing new carbons, &c.), &c.; but in comparing the *light* received from lamps, the only true test of their value is the amount of light per watt that they give when each is in the conditions best suited to it.

SUMMARY.

I. The light emitted by the arc in any given direction is roughly proportional to the apparent area of the crater as seen from that direction.

II, On the whole, the light emitted by the arc, with a constant current, increases, up to a certain point, as the arc is lengthened, because the greater distance between the carbons allows more of the crater light to escape.

III. As, however, the negative carbon becomes very finely pointed when the arc is short, with large currents, the light emitted by a very short arc is greater than that of a longer one, when the current is large.

IV. The illuminating power of the arc does not increase indefinitely as the arc is lengthened, with a contant current, but has a maximum value with a comparatively short length of arc, some 4mm. or so.

V. This is probably due to the absorption of the light of the crater by the carbon mist, which it traverses before escaping from between the carbons.

VI. The fact that the arc does absorb light can be proved in various ways. (1) It casts a shadow; (2) it hides things placed behind it; (3) the light becomes more purple as the arc is lengthened, as if more of the yellow and green rays of the crater light were absorbed, the longer the arc.

VII. When either the length of the arc or the P.D. between the carbons is constant, the illuminating power increases with the current.

VIII. Only about 10 per cent. of the energy supplied to the arc is utilised in producing light.

IX. Of the power supplied to the arc only that which is utilised at the ends of the carbons in the crater and the white spot is useful for light giving purposes, that absorbed by the arc itself being practically wasted.

X. When the arc is lengthened, while the current is kept constant, almost the whole of the extra power supplied is swallowed up by the carbon mist, and is practically wasted as far as the creation of light is concerned. Consequently, more light is *created*, for a given power supplied, the shorter the arc.

XI. Soft carbons give more light than hard for a given amount of power, but they burn away much faster.

XII. The most efficient arc would be obtained with infinitely thin carbons and an infinitely short arc.

XIII. This ideal condition is most nearly reached when the positive carbon is so thin for the current that the arc is as near as it can be to hissing without being unsteady, when the negative carbon is as small as it can be, to burn the requisite number of hours, and when the length of the arc is such that the ratio $\dfrac{\text{light emitted}}{\text{power developed in generator}}$ is a maximum for the particular carbons.

XIV. When the length of the arc is constant the light-efficiency increases with the current.

XV. The only fair method of comparing two sources of light is to *put each into the conditions best suited to it*, and then to measure the ratio $\dfrac{\text{total light emitted}}{\text{total power supplied}}$ in each.

CHAPTER XII.*

———

THE MECHANISM OF THE ARC. ITS TRUE RESISTANCE.
INQUIRY AS TO THE NECESSITY OF A BACK E.M.F. TO
EXPLAIN ITS BEHAVIOUR. WHY CORED CARBONS PRODUCE
DIFFERENT RESULTS FROM SOLID CARBONS.

In the following chapter I propose to see how far the peculiar
behaviour of the arc might have been logically predicted from
the known conditions of its existence, viz., that it is a gap in a
circuit furnishing its own conductor by the evaporation of its
own material; and to show that it is quite unnecessary to
invoke the aid of a negative resistance, or even of a large back
E.M.F. to account for this behaviour.

What happens on making the Gap.

The usual explanation given for the formation of a spark or
flash on opening an electric circuit is that it is caused by self-
induction. The interesting question therefore arises, could an
arc be struck and maintained if there were no self-induction
whatever in the circuit? I think it could. For the surfaces
of all solids are irregular, and therefore all parts of the carbons
cannot be separated at the same instant. The parts that
remain in contact will still conduct the current, but the fewer
of them that remain the greater will be their resistance. The
heat caused by this resistance must at last be great enough to
volatilise the carbon at the remaining points of contact, and, by
the time that no part of one carbon is touching any part of the
other, the small gap will be full of carbon vapour.

To explain the further formation of the arc, we must
remember that when the separation between the carbons is still
greater, all the material in the gap, as I have explained in
Chapter XI. (p. 355),† cannot retain its high temperature. The

---

* The larger part of the following Chapter was embodied in a Paper on
"The Mechanism of the Electric Arc," read before the Royal Society,
June 20, 1901.

† *See also* p. 88 (Herzfeld).

access of the cold air must turn some of it into carbon mist or fog, and I have suggested that the purple interior portion of the image of the arc is composed of such mist, while there is an indication of a space between the mist and the positive carbon which is occupied, I believe, by a thin film of true carbon vapour.

Next, the dissimilar action of the poles, met with in so many electric phenomena, begins. Instead of *both* poles volatilising, so that there is a thin layer of carbon *vapour* over each with a mass of carbon *mist* between them, the positive pole alone volatilises, while the negative appears simply to burn away.

Besides the film of vapour and the bulb of mist, other volatile materials go to make up the whole substance of the arc. The surrounding air not only cools the carbon vapour, but it unites chemically with a certain thickness of the mist, thus forming a sheath of burning gases surrounding both vapour and mist, and even portions of the solid carbons themselves. This sheath of gases is the brilliant green flame shown in Fig. 1, while the shadow between it and the mist (Figs. 3 to 6) probably indicates where the two mingle. There are three sorts of material in the gap therefore, marking the three stages through which the vapour is continually passing.

(1) It starts as a thin film of carbon *vapour* spread over the end of the positive carbon.

(2) It then changes into the *mist* that lies between this vapour film and the negative carbon.

(3) Finally, it burns and forms a sheath of burning gases which encloses not only the fresh vapour and mist, but also the ends of the solid carbons themselves.

The Conducting Power of the Vapour, the Mist and the Flame.

The specific resistances of true vapours are known to be high, therefore, I conclude that the film over the end of the positive carbon has a high resistance, even though it be very thin. The mist, on the contrary, is probably composed, as I have already suggested, of minute solid particles of carbon, and must, therefore, I think, have a lower specific resistance. My experiments on the flame have shown, on the other hand, that *its* specific resistance is so high, compared with that of the inner purple mist, that it is relatively an insulator—a result confirming that obtained by Luggin* in 1889. The current, therefore, flows through the vapour and the mist, but practically not at all through the sheath of burning gases.

The production of the High Temperature at the Crater.

To explain the great production of heat at the end of the positive carbon, as well as the sudden change of potential that is known to exist there, it has been supposed that a back E.M.F. of some 35 to 40 volts existed at the junction of the crater and the arc. But if, as I suggest, there be a high resisting vapour film in contact with the crater, the current passing through this must generate much heat, and this heat is utilised mainly in continuously forming fresh carbon vapour, to be itself turned into mist and then into flame. Hence, it seems probable that the high and constant temperature of the crater is kept up, not by the current flowing against a back E.M.F., but through the resistance of a thin vapour film at the surface of the crater. In other words, *it is not the crater itself that is the source of the heat of the arc but a thin film of carbon vapour in intimate contact with it.*

Why the End of the Positive Carbon has its Particular Shape.

As only the part of the positive carbon that is in actual contact with the vapour film can be at the temperature of volatilisation, evaporation can only take place at that surface; and hence, unless the vapour film is as large as the whole cross section of the positive carbon it must dig down into the carbon and leave the surrounding parts unvolatilised—*i.e.*, the part of the positive carbon against which the film rests must become concave. These surrounding parts, however, are heated sufficiently by conduction from the evaporating surface and by the hot gases surrounding them to burn away, and so there must be a race between volatilisation of the centre portion and burning away of the edges, which must in all cases determine the shape of the surface of volatilisation. When, all other things being equal, the gap between the carbons is small, so that the end surface of each carbon is well protected from the air, volatilisation will gain over burning, and the pit may become very deep. When, on the other hand, the gap is

---

* *Wien Sitzungsberichte*, Vol. XCVIII., Part I., Division II., p. 1,233.

large, so that the air can easily reach all parts of the carbon, except that actually covered by vapour, these parts may burn away as fast as, or even faster than the inner portion is volatilised, and in that case the surface of volatilisation will be flat or even slightly convex.   It is evident, therefore, from the very nature of things, that this surface cannot help being concave when the distance between the carbons is short, and flat, or convex when it is long.   And this is true, whether the volatilisation is due solely to a large back E.M.F., as some have supposed, or to the resistance of a thin film of carbon vapour, as I have suggested, or partly to one and partly to the other.

When only a small bit of the end of the positive carbon is being volatilised, the outer edge of the carbon will not be made hot enough to burn, and the tip will remain relatively blunt, as it does with small currents in Figs. 7, 8 and 9. When, on the contrary, the area of voltilisation is large, the edge of the carbon must burn away, and a long tapering end will be found, terminating in the surface of volatilisation. The shorter the arc the less will the heat be able to escape between the carbons, and, consequently, the longer must be the tapering part, as it is seen to be in Figs. 7, 8 and 9 (pp. 9, 10 and 12).

Why the End of the Negative Carbon assumes its Particular Shape.

It is acknowledged that volatilisation takes place at the end of the positive carbon only, therefore the negative carbon must be shaped entirely by burning away, the heat that raises it to burning temperature being furnished partly by the mist that touches it, and partly by radiation from the vapour film lying against the positive carbon.   The part that the mist rests on is protected by it from the action of the air, and does not, therefore, burn away as fast as the rest.   At the same time it must be hotter than the remainder of the carbon, and so the portion of the carbon near it must burn away *more* readily than the rest, leaving a mist-covered tip which will be longer and slenderer, because its sides will be hotter and burn away more easily, the larger the crater and the shorter the arc.

Hence, with a small crater and a long arc the negative carbon would remain fairly flat (*a*, Fig. 125), whereas, as the crater became larger, its action alone would shape the negative carbon as dotted in (*b*), and the extra heating due to the mist would render it as shown in the full line. With a short arc, on the contrary, a small crater alone would produce an end as dotted in (*c*), while the combined effect of the crater and mist would produce the end outlined by the full line. Finally, with a large crater and a short arc the crater alone would produce an end as dotted in (*d*), while the crater and mist together

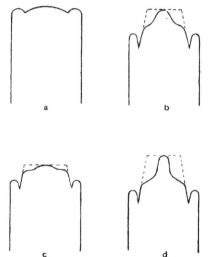

Fɪɢ. 125.—The Shaping of the Negative Carbon with Large and Small Craters and with Long and Short Arcs.

would shape the negative carbon as given by the full line in (*d*). The shapes in (*a*), (*b*), (*c*) and (*d*) are those which as shown by Figs. 7, 8 and 9 (pp. 9, 10 and 12) are actually acquired by the negative carbon under the given conditions.

The Ratio of the Volume of the Arc to its Cross Section Depends on the Shapes of the Carbon Tips as well as on the Distance between them.

The relation between the volume and the cross section of the arc must depend primarily on the quantity of vapour produced per second and on the shapes of the carbon ends between and round which the resulting mist has to dispose itself. If these ends are short and thick, the mist will be flattened out between them, and its mean cross section will be great compared with its volume. If the ends are long and slender, the mist will extend itself along these ends and will have a small mean cross section compared with its volume. The thickness of the envelope of burning gases, on the other hand, does not depend primarily on the quantity of carbon volatilised per second, but simply on the ease with which the air can get at the mist to unite with it. When the carbon ends are thick they protect the mist that is between them from the action of the air better than when they are slender, and when they are near together than when they are far apart. Thus, indirectly, the thickness of the gaseous envelope depends also on the distance between the carbons.

Why the Area of the Crater is not *Directly Proportional to the Current but depends also on the Length of the Arc.*

Suppose that the current and the distance between the ends of the carbons have been kept constant long enough for the arc to have become normal, and that the resistance in the outside circuit is then suddenly diminished. At the first instant the P.D. between the carbons must be increased, a larger current will have to flow through a vapour film of the old dimensions, and, consequently, the heat developed in it per second will increase. The temperature of the film cannot rise, because there is no increase of pressure; consequently, it must expand and spread over a larger area of solid carbon. The moment the film had expanded in the slightest degree it would begin volatilising carbon from a part of the surface hitherto inactive, and thus a larger quantity of vapour per second would be volatilised. At the next instant, therefore, the quantity of carbon volatilised per second would have increased, and the resistance of the vapour film would have become lower, and its tendency to expand would therefore be diminished on both accounts. Thus, at each instant after the change of current the volatilising surface would increase, but more and more slowly, till its area was such that the heat developed per second in the vapour film only just sufficed, after all losses from conduction, &c., to keep up the volatilisation. After that, the vapour film would cease to expand, and the surface of volatilisation would have reached its maximum area for the new current.

The vapour film, besides radiating heat in all directions from its free surface, must lose a certain extra amount of heat all round its edges by conduction through the part of the solid carbon that it does not actually touch. The heat thus lost must be subtracted from the edges of the part that it does touch, and this part will therefore be just *below* the temperature of volatilisation as will also a small ring of the solid carbon outside the vapour film. Suppose, for instance, that the full line in Fig. 126 is the part of the positive carbon that is in contact with the vapour film, then the inner dotted line will enclose the area that is actually volatilising fresh carbon, and the space between the two dotted circles will be at a temperature just below that of volatilisation, because the conduction of heat from the edges of the vapour film will bring the outer circle up and the inner circle down to a temperature a little below that of the vapour film itself.

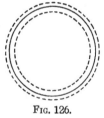

Fig. 126.

The slightly lower temperature of the space between the dotted circles would make it rather less brilliant than the volatilising surface, but it would still be very much more brilliant than the remainder of the positive carbon, so that it must form the outer circle of what we are accustomed to call the crater, viz., the most brilliantly white part of that carbon. The area of the crater is thus rather larger than the cross-section of the vapour film, while the actively volatilising surface is slightly smaller.

When the carbon vapour proceeds from a given area the cross section of the vapour film will be greater the more it is protected from the cold outer air by the end of the positive carbon. If, for instance, AB, Fig. 127, were the diameter of the volatilising surface, the cross section of the vapour film would be greater if the end of the carbon were CD than if it were C′D′, or, since the end of the positive carbon is thicker the longer the arc, *the cross section of the vapour film is greater the longer the arc.* This film will also be able to keep a larger

ring of solid carbon at a temperature just below that of volatilisation when the end of the carbon is CD than when it is
C'D', therefore the whole space that is just below the temperature of volatilisation, *i.e.* that is included between the dotted
circles in Fig. 126, will be greater with a long arc than with a
short one, when the surface of volatilisation is the same size in
each case. In other words, the area of the crater increases with
the length of the arc with a given surface of volatilisation.
Now, I shall show presently that in the normal state of the arc
the area of the volatilising surface is directly proportional to
the current, but is independent of the length of the arc; it
follows, therefore, that with a given constant current the area
of the crater must increase with the length of the arc, as I
have already shown it to do by actual measurement (p. 154).

FIG. 127.—Positive Carbons having the same Area of Volatilisation—to the
left with a Long Arc, to the right with a Short one.

The area of the crater is, then, a function of (1) the *time* after
a change has occurred in either the current or the length (till
the arc becomes normal again), (2) the *current*, and (3) the
length of the arc.

The Film of Vapour in contact with the Positive Carbon acts like a Back E.M.F.

Let a be that area of the film that uses its heat in volatilising fresh carbon, and let x be the part of which the heat is lost
by conduction, radiation, &c. Then the whole area of the film
is $a + x$, and its resistance is $\dfrac{p}{a + x}$, where p is constant, if we
consider the thickness of the film to be constant. The heat
generated per second in the film is proportional to $\dfrac{pA^2}{a + x}$, and, of
this, only $\dfrac{a}{a + x}$ is used in volatilisation. The quantity of carbon volatilised per second is, therefore, proportional to

$$\frac{a}{a + x} \cdot \frac{pA^2}{a + x} = \frac{apA^2}{(a + x)^2}.$$

But the quantity of carbon volatilised per second must be proportional to the surface from which it is volatilised, *i.e.*, to *a*.

$$\therefore \quad qa = \frac{pa\mathrm{A}^2}{(a+x)^2}, \text{ where } q \text{ is constant,}$$

or

$$\frac{a+x}{p} = \frac{\mathrm{A}^2}{(a+x)q}.$$

But

$$\frac{a+x}{p} = \frac{1}{f},$$

where *f* is the resistance of the film,

$$\therefore \quad f = \frac{q(a+x)}{\mathrm{A}^2}. \quad \ldots \quad \ldots \quad (22)$$

Now the heat developed per second in a film of *area a* by a current A is proportional to $\dfrac{\mathrm{A}^2}{a}$; and, when all this heat is employed in evaporating carbon from the surface *a*, the quantity of carbon evaporated per second is proportional to the area *a*, and also to the heat employed in evaporating it, *i.e.*, to $\dfrac{\mathrm{A}^2}{a}$.

We have, therefore,

$$a = \frac{m\mathrm{A}^2}{a}, \text{ where } m \text{ is constant,}$$

or $\qquad\qquad a = n\mathrm{A}, \text{ where } n \text{ is constant.}$

Thus the area of the evaporating surface of the crater is proportional to the current. Substituting this value for *a* in equation (22) we have

$$f = \frac{q(n\mathrm{A}+x)}{\mathrm{A}^2}.$$

Now we have seen (p. 398) that *x*, which is the part of the vapour film that loses heat by conduction at its edge, must increase as the arc lengthens. Also, since $l = 0$ does not mean that the carbons are touching, but that the mouth of the crater and the tips of the negative carbon are in the same plane, conduction at the edge of the film would take place, even with no length of arc, so that *x* must have some value, even when $l = 0$. Thus all we know about *x* and *l* is in accordance with the equation

$$x = v + wl,$$

where *v* and *w* are constants.

Let us assume that this equation is true of x when $l = 0$, then we have

$$f = \frac{q(n\mathrm{A} + v + wl)}{\mathrm{A}^2},$$

or

$$f = \frac{h\mathrm{A} + k + ml}{\mathrm{A}^2},$$

or

$$f = \frac{h}{\mathrm{A}} + \frac{k + ml}{\mathrm{A}^2}, \quad \cdots \cdots \quad (23)$$

which is an equation of exactly the same form as that obtained by direct measurement. For equation (6) (p. 222) was found by measuring the P.D. between the positive carbon and points in the mist quite close to the crater; and dividing this equation throughout by A we get

$$f = \frac{\mathrm{V}}{\mathrm{A}} = \frac{31 \cdot 28}{\mathrm{A}} + \frac{9 + 3 \cdot 1 l}{\mathrm{A}^2}.$$

The identity of the two equations shows that not only is no back E.M.F. at the crater necessary to account for the great fall of potential between it and the arc, but that the film of high resistance vapour, whose existence I have suggested, could cause the P.D. between the positive carbon and the arc to vary exactly as experiment proves it to vary.

When A in equation (23) is changed to 2A, f becomes less than $\frac{f}{2}$, which shows that f diminishes faster than A increases, or, *in other words, the resistance of the vapour film diminishes faster than the current increases.*

The Apparent Negative Resistance of the Arc is caused by the true Positive Resistance Diminishing More Rapidly than the Current Increases.

It has been mentioned (p. 392) that the specific resistance of the green flame is so high as to make it, to all intents and purposes, an insulator, so that nearly the whole of the current flows through the mist. Consequently, it follows that the resistance of an arc of given length must depend (apart from the resistance of the vapour film) simply on the cross-section of the carbon mist, which, as it appears purple in the image of the arc, can easily be measured. To see how this cross-section varies when the current is increased while the length of the arc is kept constant, I have drawn, in Fig. 128, diagrams of images of the normal arc, taking great care to trace the exact limits of the purple mist and the green flame as accurately as possible.

The resistance of the carbon mist (as distinct from that of the vapour film) may be defined practically as being the resistance of that portion of the mist that lies between the two parallel planes that pass, the one through the mouth of the crater, and the other through the tip of the negative carbon.

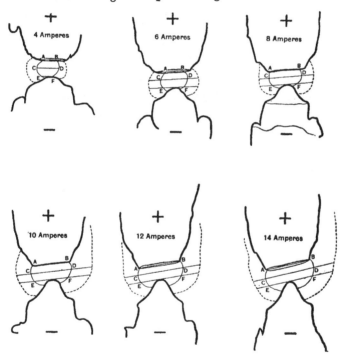

Fig. 128.—Diagrams of Arcs and Carbons with Outlines of Mist and Flame very carefully noted.

The mean cross-section of the mist D^2 given in column (3) of Table LIII. has been obtained by taking the means of the squares of the three lengths AB, CD and EF in Fig. 128. The next column, giving the ratio of D^2 to the current A, shows that the cross-section of the mist increases more rapidly than the current. Column 5 gives numbers proportional to the resistance of the mist, while columns 6 and 7 contain numbers proportional to the power spent in the mist, as obtained by experiment, and from the equation to be subsequently referred to.

The mist carries practically the whole of the current, and, since D^2 increases more rapidly than A (column 4), it follows that in the normal arc the resistance of the mist diminishes more rapidly than the current increases. But I have shown (p. 400) that the resistance of the vapour film also diminishes faster than the current increases. Hence, *the whole resistance of the normal arc diminishes more rapidly than the current increases,* and for this reason the P.D. required to send the

Table LIII. *Mean Squares of Diameters of Mist with corresponding Currents and P.Ds. between the Carbons. Numbers proportional to Resistance of Mist and to Power expended in Mist.*

Normal Arc. Constant Length of Arc, 2mm.
Solid Carbons: Positive, 11mm. ; negative, 9mm.

A	V	D^2 sq. mm.	$\dfrac{D^2}{A}$	$\dfrac{1}{D^2}$	$\dfrac{A^2}{D^2}$ experiment.	$\dfrac{A^2}{D^2}$ from equation.
4	51·7	4·8	1·20	0·208	3·33	3·4
6	49·0	9·8	1·63	0·102	3·67	3·68
8	48·0	16·2	2·02	0·061	3·95	3·95
10	47·0	23·4	2·34	0·043	4·27	4·22
12	45·7	34·9	2·91	0·029	4·13	4·49
14	45·1	41·2	2·94	0·024	4·76	4·76

current through this resistance must also diminish as the current increases. Thus, if, in the normal arc, δA be an added increment of current and δV the accompanying increment of P.D., $\dfrac{\delta V}{\delta A}$ must have a negative value, even although the resistance of the arc is positive, simply because that resistance diminishes faster than the current increases.

There is Nothing to Show that the P.D. between the Carbons Divided by the Current is not the True Resistance of the Arc.

Fig. 129 shows that the curve connecting the values of $\dfrac{A^2}{D^2}$, given in the sixth column of Table LIII., with those of the current, in the first column, is a straight line, having the equation

$$\frac{A^2}{D^2} = 0 \cdot 136A + 2 \cdot 86 \; ;$$

so that, for a normal arc of given length, the power expended in the carbon mist is proportional to a constant plus a term which varies directly with the current. Dividing by A^2 we obtain m, the resistance of the *mist*,

$$m = \frac{0 \cdot 136}{A} + \frac{2 \cdot 86}{A^2}.$$

Now, let us suppose that, for a given current, the resistance of the mist is directly proportional to the length, l, of the arc, then the equation for m must take the form :—

$$m = \left(\frac{a}{A} + \frac{\beta}{A^2} \right)l.$$

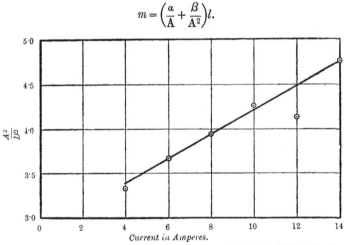

Fɪɢ. 129.—Curve connecting the Power expended in the Mist with the Current, for a Constant Length of Arc of 2mm.

Solid Carbons : Positive, 11mm. ; negative, 9mm.

Combining this with f, the resistance of the vapour film, in equation (23), we have for the total resistance of the normal arc an expression of the form

$$f + m = \frac{p + ql}{A} + \frac{s + tl}{A^2},$$

a form of equation exactly similar to (5), p. 190, in which the *apparent* resistance of the arc was found by dividing the P.D.

between the carbons by the current. Thus the law of variation
of the resistance of the arc is identical, whether it is obtained
from direct measurements of the cross-sections of the vapour
film and the mist, or by dividing the observed P.D. between
the carbons by the current. It may well be, therefore, that,
instead of an arc consisting of a circuit of low resistance combined with a large back E.M.F., its *apparent* resistance—*i.e.*,
the ratio of V to A—is its *true* resistance ; or that, if there is
any back E.M.F. at all, it is very much smaller than has
hitherto been supposed.

*Both the Resistance of the Arc and the P.D. between the Carbons
     Depend, not only on the Current and the Length, but also on
     How Lately a Change has been made in Either, and on
     What that Change was.*

The whole resistance of the arc depends on the cross-sections
of the vapour film and the mist, and on the distance between
the carbons. Now, we have seen (p. 396) that when the P.D.
between the carbons is changed—increased, say—the *first*
result must be an increase of current, while the *second* is a
corresponding increase in the cross-sections of the vapour film
and the mist, causing a *diminution* of the resistance, and, consequently, of the P.D. between the carbons. *Thirdly,* if the
new current is kept constant long enough, the ends of the
carbons burn away to longer points, allowing the mist to
extend further along their sides, and so to take a smaller cross-
section, so that both the resistance and the P.D. increase
again, although they never reach such high values as they had
with the smaller current.

Fig. 130 is useful, as showing at a glance how the resistance
and the P.D. depend upon the time that has elapsed after a
change of current, when the current is kept constant after
the change. When the arc is normal in the first instance
AB, A'B' and A"B" represent the curves connecting the P.D.,
the current and the true resistance of the arc respectively
with the time. When the P.D. is increased from B to C, say,
the resistance does not alter at the first instant, but the
current rises to C', and it is then kept constant at C'. Next, the
surface of volatilisation increases in area, so that the resistance
falls to D", and the P.D. consequently falls to D. After this

the carbons begin to grow longer points, the cross-section of the mist diminishes, the resistance therefore increases to E'', and the P.D. with it to E. The arc has now become normal again, so that the curves are all now parallel straight lines, the current higher than before, and the P.D. and resistance lower.

Thus any alteration that is made and maintained in the arc sets up a series of changes in its resistance and, consequently, in the P.D. between the carbons, that cease only when the arc

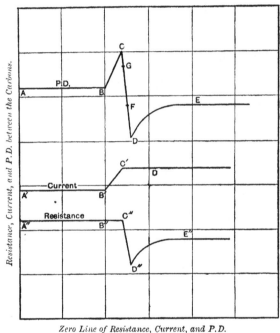

Fig. 130.—Simultaneous Time-Changes of P.D., Current and Resistance.

becomes normal again. In other words, when an arc of given length, with a given current flowing, exists between given carbons, neither the resistance nor the P.D. between the carbons has any definite value, except when the arc is and remains normal. In all other cases, each varies, within certain limits, according to the time that has elapsed since either the current or the length was altered, and according to what

change was then made. Experimental proof of the abovementioned alterations in the P.D., corresponding with current
changes, has already been given in Chapter III., Figs. 36
and 37, with the exception of the first change of P.D. in the
same direction as the change of current. This was so quick
that it was over, before it could influence the voltmeter, in
every case but when both carbons were cored. The reason
why will appear later. These changes have an important
bearing, which we will now examine, on the question of
measuring the resistance of the arc by means of a small superimposed alternating current.

This method has been employed by many experimenters, but
the results obtained have not shared the similarity of the
methods; for while Von Lang and Arons found, in 1887, that
the arc had a *positive* resistance, Messrs. Frith and Rodgers,
in 1896, found that it had a *negative* one, with solid carbons.
We shall now see the reason of this disparity, and first it may be
well to recall shortly the reasoning on which the method is based.

The equation

$$V = E + Ar$$

may be taken to represent the connection between the P.D.
between the carbons, the current and the length of the arc,
whether it has a variable E.M.F., a constant E.M.F., or none
at all. For, in the first case, E will be variable, in the second
constant and in the third zero. In any case

$$\frac{dV}{dA} = r$$

when such a small quick change is made in V *and* A *that
neither* E *nor* r *is made to vary by it.*

Instead of a single small quick change of current, the
experimenters superimposed a small alternating current on the
direct current of the arc, and measured the *average* value
of $\frac{\delta V}{\delta A}$, or its equivalent. Obviously, if the alternating current
left the resistance and any back E.M.F. that might exist in the
arc unaffected, this was a *true* measure of the resistance of the
arc. But if the alternating current changed both or either of
these, then, instead of being equal to r we should have

$$\frac{dV}{dA} = r + \frac{dE}{dA} + A\frac{dr}{dA},$$

if there is a back E.M.F., and if both it and the resistance varied with the alternating current; we should have

$$\frac{dV}{dA} = r + \frac{dE}{dA},$$

if there is a back E.M.F. and if it alone varied, and

$$\frac{dV}{dA} = r + A\frac{dr}{dA},$$

if the back E.M.F. is either non-existent or constant, and the resistance alone varied.

None of the experimenters, as far as I am aware, applied any but a few imperfect tests to see whether the alternating currents they employed affected the resistance of the arc or not, and it was, I believe, because the resistance *was* affected, in every case, that such diverse results were obtained. The low frequency of the alternations was the probable source of error, for I shall now show that, with a given root mean square value of the alternating current, the average value of $\frac{\delta V}{\delta A}$ varies, not only in magnitude, but even in sign, with its frequency.

Effect of the Frequency of the Superimposed Alternating Current on the Value and Sign of $\frac{\delta V}{\delta A}$.

In dealing with a superimposed alternating current there is, of course, no sudden increase or diminution of current such as was dealt with in Fig. 130, everything is gradual. The three changes of P.D. do not, therefore, act separately. At any moment—when the current is increasing, say—the P.D. has a tendency to rise on account of the latest increase of current, to fall on account of the diminution of resistance due to the last increase but one, and to rise on account of the re-shaping of the carbons following the last increase but two. If the frequency of the alternating current is very low indeed, so that the current changes very slowly, all three of these tendencies will be in force at each moment, and the actual change of P.D. will be the resultant of the three. If the frequency is so high that the shapes of the carbons never change at all, but so low that the area of the volatilising surface can alter, only the first two tendencies will be operative ;

while, if the frequency is so high that the area of the volatilising surface remains constant, the resistance of the arc will not alter at all, the current and P.D. will increase and diminish together and proportionately, and, unless the arc contains a variable E.M.F. which is influenced by the changes of current, $\dfrac{\delta V}{\delta A}$ will measure the true resistance of the arc.

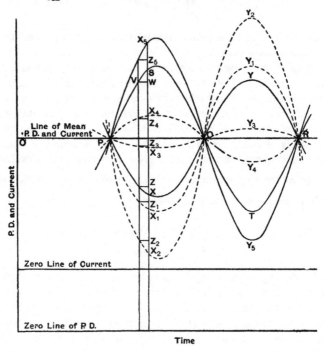

Fig. 131.—Curves showing the effect of the Frequency of an Alternating Current Superimposed on a Direct Current Arc on the Simultaneous Time-Changes of P.D. and Current.

The influence of the frequency of the alternating current on the magnitude and sign of $\dfrac{\delta V}{\delta A}$ is traced in Fig. 131, where the zero lines of P.D. and current are drawn at such a distance apart that the mean P.D. line and the mean current line are represented by the same horizontal line OPR. PR represents the time occupied by one complete alternation, whatever that time may be. If, for instance, the frequency is 50 complete alternations per second, PR represents $\frac{1}{50}$th of a second ; if the frequency is 5,000, PR represents $\frac{1}{5000}$th of a second. PSQTR represents the time change of current with any frequency. When the alternations are so slow that the arc remains normal the change of P.D., δV, for a given small change of current, SW, say, is the resultant of three such changes as BC, CD and DE in Fig. 130, and it is in the opposite direction from the change of current. The P.D. time curve is, therefore, something like PXQYR (Fig. 131), and $\frac{\delta V}{\delta A}$ is the mean of such ratios as $-\frac{ZX}{SW}$.

When the frequency is so raised that the carbons never have time to alter their shapes completely before the current changes, the third component DE (Fig. 130) is smaller than with the normal arc, so that $\dot{\delta} V$ is greater, and $\frac{\delta V}{\delta A}$ must, therefore, have a larger negative value than when the arc is normal, so that PX_1QY_1R might be the P.D. time curve in this case.

When the frequency was so high that the carbons never altered their shapes at all, but the volatilising surface underwent the maximum alteration, the third component (DE, Fig. 130) would be absent altogether, and therefore δV would undergo the greatest change it was susceptible of in the opposite direction to the change of current. PX_2QY_2R is then the P.D. time curve, $\frac{\delta V}{\delta A}$ has then its maximum negative value, and is the mean of such ratios as $-\frac{Z_2X_2}{SW}$.

With a further increase of frequency, the area of the volatilising surface would never have time to change completely, so that δV would be the resultant of two such changes as BC and CF, say (Fig. 130). $\frac{\delta V}{\delta A}$ would therefore be *smaller* than with the lower frequency last mentioned, and the P.D. time curve might again be PX_1QY_1R, or it might be PX_3QY_3R, if the frequency were high enough. When the frequency was so great that the two P.Ds. BC and CF (Fig. 130) were exactly equal, the P.D. would not alter at all when the current was changed, $\frac{\delta V}{\delta A}$ would be zero, and the straight line PQR would be the P.D.

time curve. When the frequency was further increased, the change of P.D. would be the resultant of two such changes as BC, CG only (Fig. 130); the total change of P.D. would, therefore, be in the *same* direction as the change of current, the P.D. time curve would be like PX_4QY_4R, and $\frac{\delta V}{\delta A}$ would be $+\frac{Z_4X_4}{SW}$. Finally, when the frequency was so great that the area of the volatilising surface never altered at all, the change of P.D. would be BC alone (Fig. 130), the P.D. time curve would be PX_5QY_5R, δV would be Z_5X_5, and $\frac{\delta V}{\delta A} = +\frac{Z_5X_5}{SW}$ would measure the *true* resistance of the arc, even if there is a back E.M.F. in the arc, unless that back E.M.F. varies with the current.

Thus, by applying the same alternating current, but with different frequencies, to a direct current arc, $\frac{\delta V}{\delta A}$ can be made to have any value from a fairly large negative value to the positive value which is the true resistance. It is easy to see therefore, how different experimenters might get very different values and even different signs for the resistance of the arc, when they measuerd it by means of a superimposed alternating current; and Fig. 131 shows the imperative necessity of some rigorous proof that the alternating current has not affected the resistance of the arc before any such measurements can be accepted as final. I shall presently show how such a proof can be obtained; but first it will be interesting to see how, with an arc of given length, and with a given current flowing, the value of $\frac{\delta V}{\delta A}$ is connected with the frequency of the alternating current, and what sort of frequency is required in order that the resistance of the arc shall not be affected by this current.

To find the Curve Connecting $\frac{\delta V}{\delta A}$ *with the Frequency of the Superimposed Alternating Current, and to see with what Frequency* $\frac{\delta V}{\delta A}$ *Measures the True Resistance of the Arc.*

Take an arc of 2mm, with a direct current of 10 amperes flowing.

For the arc to remain normal, when the small alternating current is superimposed on it, the frequency must be practically zero, for each alternation must take many seconds instead of only a small fraction of a second. Now, differentiating equation (3) (p. 184) we find that with the normal arc and with solid Apostle carbons

$$\frac{\delta V}{\delta A} = -\frac{11 \cdot 66 + 10 \cdot 54l}{A^2},$$

$$= -0 \cdot 33,$$

when $l = 2$ and $A = 10$.

The first point on the curve connecting $\frac{\delta V}{\delta A}$ with the frequency of the alternating current has, therefore, the co-ordinates 0 and $-0 \cdot 33$ (A, Fig. 132).

The value found for $\frac{\delta V}{\delta A}$ by Messrs. Frith and Rodgers * with the same carbons, direct current, and P.D was about $-0 \cdot 8$, more than double the normal value—which shows that the alternating current they superimposed was making the resistance of the arc vary so that the P.D. followed some such curve as PX_1QY_1R or PX_2QY_2R (Fig. 131). They also found that varying the frequency from 7 to 250 complete alternations made no difference in the value they obtained. Therefore, the curve connecting $\frac{\delta V}{\delta A}$ with the frequency must fall steeply from A, the point of no frequency, to B, the point for a frequency of 7, and must be practically horizontal from B to C. Hence, Messrs. Frith and Rodgers' observations cover the portion BC of the curve in Fig. 132.

The next point, D, is obtained from Mr. Duddell's work. In his remarkable Paper † on "Rapid Variations in the Current through the Direct-Current Arc," he said :—"I tried to record the transient rise in P.D. for the solid arc by means of an oscillograph, the sudden increase of the current being obtained by discharging a condenser through the arc. This experiment was successful, and a transient rise in P.D. was observed, *the P.D. and current increasing together, but only for about $\frac{1}{5000}th$ second*." It is clear from this that $\frac{\delta V}{\delta A}$ must at least *begin* to be positive with a frequency of 2,500 complete alternations, and

---

* *Phil. Mag.*, 1896, Vol. XLII., Plate 5.

† *Journal* of the Institution of Electrical Engineers, Vol. XXX., p. 232.

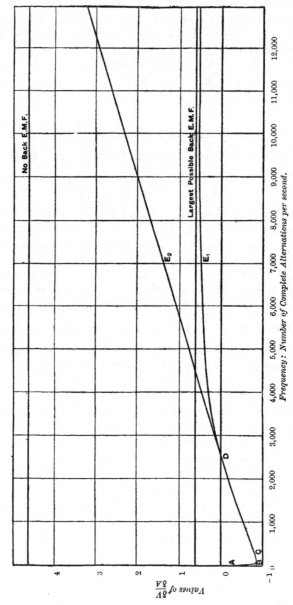

FIG. 132.—Curve connecting $\frac{\delta V}{\delta A}$ with the Frequency of the Superimposed Alternating Current.

D, where OD = 2,500, may be taken to be the point near which $\frac{\delta V}{\delta A}$ changes its sign.

To the right of D the curve must continue to rise, as indicated in Fig. 132, more and more slowly, as it approaches the horizontal line whose distance from the axis of frequency represents the value of $\frac{\delta V}{\delta A}$ that is the *true* resistance of the arc. The curve must finally become asymptotic to this line, since when once a frequency is nearly reached with which the alternating current does not practically affect the resistance of the arc, increasing the frequency will not alter the value of $\frac{\delta V}{\delta A}$.

Equation (3) (p. 184) shows that the resistance of the particular 2mm. 10 ampere normal arc under discussion cannot be greater than 4·63 ohms, nor less than 0·62 ohms ; for if there is no back E.M.F.

$$r = \frac{38 \cdot 88 + 2 \times 2 \cdot 07}{10} + \frac{11 \cdot 6\dot{6} + 10 \cdot 54 \times 2}{100} = 4 \cdot 63,$$

and if there is the largest possible back E.M.F., viz., $\left(38 \cdot 88 + \frac{11 \cdot 66}{A}\right)$ volts (for it is impossible to imagine that terms involving the length of the arc can belong to a back E.M.F.), then

$$r = \frac{2 \cdot 07 \times 2}{10} + \frac{10 \cdot 54 \times 2}{100} = 0 \cdot 62.$$

Thus the curve cannot rise higher than the horizontal line $\frac{dV}{dA}$ = 4·63, and it must rise at least as high as $\frac{dV}{dA} = 0 \cdot 62$. Consequently, as the lower curve in Fig. 132 shows, the true resistance of this particular arc could not be measured with a superimposed alternating current having a frequency of less than at least 7,000 per second, even if there were a back E.M.F. as great as 40 volts. And if, as I have suggested, the back E.M.F. is zero, or at least very much smaller than 40 volts, the frequency would have to be many times as high for $\frac{dV}{dA}$ to be on the horizontal part of the curve, *i.e.*, for the alternating current not to alter the resistance of the arc.

The Form of the P.D. Time Curve indicates whether the Resistance of the Arc is Affected by the Superimposed Alternating Current or not.

The final test as to the frequency being high enough not to affect the resistance of the arc must, of course, be the finding of two frequencies with the same root mean square value of the alternating current, but differing by many thousands of alternations per second that would give the same value of $\frac{\delta V}{\delta A}$. This would show that the horizontal part of the $\frac{\delta V}{\delta A}$ frequency curve had been found.

A very good *first* test, however, is furnished by the curve connecting the P.D. between the carbons with the time, or with the angle of the alternating wave. For this curve is unsymmetrical with respect to the corresponding current curve when the resistance *is* affected, for the following reasons :—

I have shown that the change in the area of the volatilising surface of the crater that is due to any change of current follows *after* the change of current, and requires time for its completion. If, therefore, a superimposed alternating current is affecting the resistance of a direct-current arc the P.D. required for any given current must be higher when the current is increasing than when it is diminishing. A current of 10 amperes, for instance, would require a higher P.D. when it came after 9 and before 11 amperes than when it came after 11 and before 9 amperes, because in the first case it would be flowing through an arc of which the cross-section had been made by some current *less*, and in the second by some current *greater* than 10 amperes.

I have applied this test, with very satisfactory results, to some curves made from experiments in which it is quite certain that the alternating currents must have affected the resistance of the arcs, because they had frequencies of only 47 and 115 alternations per second respectively.

The experiments formed part of a valuable series carried ou in 1896 by Messrs. Ray and Watlington, two students at the Central Technical College, in continuation of the researches of Messrs. Frith and Rodgers. The carbons were solid, and the direct-current normal arc carried a current of 10 amperes with a P.D. of 45 volts. Various P.Ds., with their corresponding currents taken from the curves, are given in Table LIV., in which the columns headed V_i are the P.Ds. with increasing currents, and those headed V_d with diminishing currents; while V_s and V_l belong to the smallest and largest currents respectively.

It may be seen at a glance that in every single instance the P.D. for the same current is higher when · the current is increasing than when it is diminishing. For instance, with the lower frequency the P.D. corresponding with a current of

Table LIV.—*Corresponding Currents and P.Ds. with Small Alternating Current Superimposed on Direct Current of 10 Amperes in the Arc.*

P.D. with Direct Current alone 45 volts.

Solid Apostle Carbons: Positive, 11mm.; negative, 9mm.

Frequency 47.					Frequency 115.				
A	V_s	V_i	V_l	V_d	A	V_s	V_i	V_l	V_d
8·425	45·6	...	...	...	8·65	46	...	...	...
9·0	...	44·8	...	44·0	9·0	...	45·6	...	44·9
9·5	...	44·2	...	43·3	9·5	...	45·2	...	44·2
10·0	...	43·7	...	42·8	10·0	...	44·6	...	43·5
10·5	...	43·1	...	42·1	10·5	...	43·9	...	42·9
11·0	...	42·4	...	41·8	11·0	...	43·2	...	42·0
11·5	...	42·0	...	41·4	11·25	...	...	42·2	...
12·0	...	41·7	...	41·2	...	...	...	...	...
12·25	...	...	41·4	...	...	...	...	...	...

11 amperes is 42·4 volts when the current is increasing, and only 41·8 volts when it is diminishing. And with the higher frequency it is 43·2 volts with increasing current and only 42·0 with a falling current. Hence, we are supplied with a very simple test as to whether the superimposed alternating current changes the resistance of the arc or not. It is only necessary to take the wave forms of P.D. and current by means of an oscillograph, and to observe whether the P.D. corre_sponding with each current is the same whether the current is increasing or diminishing. If the two P.D.'s are different the resistance is being altered, if they are alike it is not,

*How to ascertain with Certainty whether there is a Constant or
a Variable Back E.M.F. in the Arc or None, and how to
find the True Back E.M.F., if there is One.*

Returning to the equation

$$V = E + Ar ;$$

if both E and r vary $\dfrac{dV}{dA} = \dfrac{dE}{dA} + r + A\dfrac{dr}{dA} ;$

and

$$V - A\frac{dV}{dA} = E - A\frac{dE}{dA} - A^2\frac{dr}{dA}.$$

If the alternating current with which $\dfrac{dV}{dA}$ is measured is of
such high frequency that it does not alter the resistance of
the arc, and if, also, the back E.M.F. is constant, or, being
variable, the alternating current is too small to affect it, then

$$V - A\frac{dV}{dA} = E.$$

To see whether the arc has any back E.M.F. at all, therefore, it is only necessary to measure $\dfrac{dV}{dA}$ with a superimposed
alternating current of a frequency that has been found not to
affect its resistance, and to subtract $A\dfrac{dV}{dA}$ from V. If the
result is zero the arc has *no* back E.M.F. If it is not zero,
$\dfrac{dV}{dA}$ must be measured in the same way for other arcs differing
widely in current and length. If all the values of $V - A\dfrac{dV}{dA}$
thus obtained are equal or nearly so, the arc has a *constant*
back E.M.F. which is equal to this value. If $V - A\dfrac{dV}{dA}$ is *not*
the same for all the arcs, but varies according to some definite
law, then there is a *variable* back E.M.F. in the arc which
may or may not be affected by the alternating current used to
measure $\dfrac{\delta V}{\delta A}$.

Suppose, for instance, that two measuements of $\dfrac{\delta V}{\delta A}$ were
made, using the same direct current and length of arc, but
different alternating currents. If one of the alternating
currents had a root mean square value equal to 1 per cent. of

the direct current and the other a value equal to 5 per cent., one would be five times as great as the other, and yet both would be small compared with the direct current. It would, of course, be possible to make the frequency of each of these currents so great that the resistances of the arcs to which they were applied were not altered by them. Yet it would not necessarily follow that when this had been done the two values of $\dfrac{\delta V}{\delta A}$ thus obtained would be equal. For the back E.M.F. might vary, not with the *frequency* of the alternating current, but with its *magnitude.* If, therefore, it were found that E was variable it would be necessary to measure $\dfrac{dV}{dA}$ with smaller and smaller alternating currents till two were found which, while differing considerably from one another, gave the same value of $\dfrac{dV}{dA}$. Only a value obtained in this way could be accepted as measuring the *true* resistance of the arc, and $V - A\dfrac{dV}{dA}$ would then be the *true* back E.M.F. of the same arc.

The Changes introduced into the Resistance of the Arc by the use of Cored Carbons.

The preceding applies to all carbon arcs. Next let us consider the explanation of the marked effects produced by introducing a core into either or both carbons. These are, first, those such as Prof. Ayrton published at Chicago in 1893, viz.:—

(1) The P.D. between the carbons is always lower, for a given current and length of arc, when either or both of the carbons are cored than when both are solid.

(2) With a constant length of arc and increasing current, the P.D., which diminishes continuously when both carbons are solid, either diminishes less than with solid carbons, when the positive is cored, or, after diminishing to a minimum remains constant over a wide range of current, or increases again.

(3) It requires a larger current, with the same length of arc, to make the arc hiss when the positive carbon is cored than when both are solid.

Secondly, there are the facts connected with the influence of cores on the small *change* of P.D. accompanying a small

change of current, to which attention was first called by Messrs.
Frith and Rodgers in 1896 (*see* pp. 76-81).* These facts, which
were physically correct, although they were wrongly interpreted
at the time, are embodied in the following wider generalisations,
which are deduced from experiments, presently to be described,
combined with theoretical considerations :—

(1) When, on a direct current arc, an alternating current is
superimposed which is such that the resistance of the arc is
altered by it, the average value of $\frac{\delta V}{\delta A}$ is always more positive†
when either carbon is cored than when both are solid, and
most positive of all when both are cored, all other things being
equal.

(2) The frequency of the alternating current that makes $\frac{\delta V}{\delta A}$
begin to have a positive value is lower when either carbon is
cored than when both are solid, and lowest when both are cored.

(3) The value of $\frac{\delta V}{\delta A}$, with a given root mean square value of
the superimposed alternating current, depends not only on the
nature of the carbons and on the frequency of that current,
but also on the magnitude of the direct current, and on the
length of the arc.

There are two ways in which the P.D. between the carbons
may be lowered by the core : (1) by an increase in the cross-
section of the vapour film, or the mist, or both ; (2) by a lowering
of their specific resistances.   To see whether I could observe any
change in the cross-sections I have traced a series of enlarged
images of the arc, with four sets of Apostle carbons, using
(1) solid-solid ; (2) solid-cored ; (3) cored-solid ; (4) cored-cored
carbons.   The positive carbon was, as usual, 11mm. and the
negative 9mm. in diameter, and the arc was 2mm. in length in
each case, while the currents were 4, 6, 8, 10, 12, 14 amperes.
The diagrams were traced, not only when the arc was normal
in each case, but also immediately after each change of current,
so that the effect on the cross-section of the arc of both an

---

* " The Resistance of the Electric Arc," *Phil. Mag.*, 1896, p. 407.

† I call $\frac{\delta V}{\delta A}$ *more positive* in one case than in the other when it has either
a larger positive, or a smaller negative value in the first case than in the
second.

instantaneous and a normal change of current might be seen. Fig. 128 shows the first set of diagrams of the normal arc ; the others are too numerous to publish, but the mean cross-sections of the purple part—the mist—in each, measured as in Fig. 128, may be found in Table LV., those marked " non-normal " belonging to the arc *immediately* after the change of current, and those marked " normal " to the arc after all the conditions had become steady again.

Table LV.—*Mean Cross-Section of Mist between Solid-Solid, Solid-Cored, Cored-Solid and Cored-Cored Apostle Carbons, 11mm. and 9mm.*

Length of Arc, 2mm.

Current in Amperes.	Normal.				Non-Normal.			
	S.S.	S.C.	C.S.	C.C.	S.S.	S.C.	C.S.	C.C.
4	4·8	6·95	4·0	3·3	...	...	...	...
6	9·8	8·3	6·05	5·6	9·5	8·4	6·25	3·5
8	16·2	14·2	11·0	8·9	17·6	11·1	12·0	5·8
10	23·4	20·75	13·55	11·9	21·5	19·0	18·7	11·1
12	34·9	27·6	17·7	16·55	34·1	26·9	20·1	16·7
14	41·2	35·0	24·5	20·0	...	39·4	...	18·7

With a single exception, every number in each set is smaller than the corresponding number in the preceding column. Hence, with *all* these arcs the mean cross-section of the mist, for a given current, was largest when both carbons were solid, smallest when both were cored, and was more diminished by coring the positive than by coring the negative. Fig. 133, which connects the mean cross-section of the mist with the current for each pair of carbons, besides showing well this marked difference in the influence of the cores, makes it apparent that the difference increases, in every case, with the current.

We cannot measure the cross-section of the vapour film directly, but, for a constant length of arc it must be roughly proportional to the cross-section of the mist where it touches the crater. These cross-sections, which are given in Table LVI., do not, naturally, vary nearly so regularly as the mean cross-sections, but still we can judge pretty well what are the effects of the various cores. Coring the positive carbon, for instance,

E E 2

distinctly diminishes the cross-section of the vapour film; for all but one number in column (3) is less than the corresponding one in column (1), and every number in column (7) is less than the corresponding one in column (5). Coring the negative carbon, on the other hand, only seems to diminish the cross-section which the vapour film assumes immediately after a

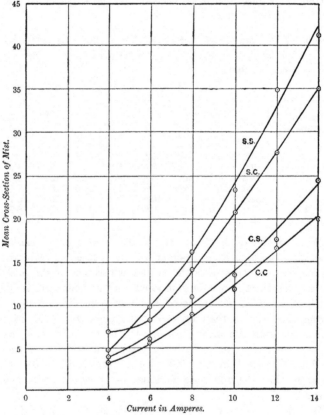

Fig. 133.—Curves connecting the Mean Cross-Section of the Arc Mist with the Current for Solid-Solid, Solid-Cored, Cored-Solid, and Cored-Cored Carbons, 11mm. and 9mm. in diameter.

Length of Arc, 2mm.

change of current, for while, in the *non-normal* section, each number in (6) is less than in (5), and in (8) two are less than in (7), one is equal, and only one is greater, in the *normal* section the numbers in (2) are sometimes less and sometimes greater than in (1), and those in (4) are nearly all greater than those in (3).

Thus, taking Tables LV. and LVI. together, we find that a core in the positive carbon keeps both the mist and the vapour film from being as large as they would be with a solid positive, both immediately after a change of current and when the arc is normal again. Coring the negative, on the other hand, while it has the same effect on the cross-section of the *mist* as coring

Table LVI.—*Cross-Section of Mist where it touches Crater, Solid-Solid, Solid-Cored, Cored-Solid, Cored-Cored Apostle Carbons, 11mm. and 9mm.*

Length of Arc, 2mm.

Current in Amperes.	Normal.				Non-Normal.			
	S.S.	S.C.	C.S.	C.C.	S.S.	S.C.	C.S.	C.C.
	(1)	(2)	(3)	(4)	(5)	(6)	(7)	(8)
4	2·9	7·8	3·2	2·9	...	...	...	...
6	6·8	9·0	5·3	5·8	10·9	6·25	6·25	3·6
8	16·0	13·0	10·9	14·4	16·0	9·0	10·9	6·25
10	23·0	26·0	12·25	21·2	19·4	16·8	15·2	15·2
12	32·5	25·0	17·6	22·1	31·4	23·0	17·6	21·2
14	39·1	36·0	23·0	24·0	...	33·6	...	19·4

the positive, only diminishes the cross-section of the vapour film immediately after a change of current. If, therefore, coring either carbon produced nothing but an alteration in the cross-section of the arc, the resistance of the arc, and, consequently, the P.D. between the carbons would be *increased* by the coring. It follows, therefore, that the diminution of the P.D. between the carbons actually observed with cored carbons must be caused by a lowering of the specific resistance of the vapour film or the mist, or both ; and this lowering must be so great that it must more than compensate for the diminution in their cross-sections.

It is easy to see how the vapour and mist from a core in the positive carbon must alter the specific resistance of the arc, but, since the negative carbon does not volatilise, there seems to be no reason why coring *it* should have the same effect.

The core, however, consists of a mixture of carbon and metallic salts; and metallic salts have a lower temperature of volatilisation than carbon, so that these salts may easily be volatilised by the *mist* touching them, and, mingling with it, lower its specific resistance.

Now take the fact that with a constant length of arc and increasing current the P.D. always diminishes less if the positive carbon is cored than if it is solid, and that the reduction of diminution is sometimes so great that the P.D. remains constant for a large increase of current, and sometimes even increases somewhat, instead of steadily diminishing, as it does when both carbons are solid.

In Chapter IV. (p. 133) this point was dealt with on the assumption that " with a given negative carbon the P.D. required to send a given current through a fixed length of arc depends principally, if not entirely, on the nature of the surface of the crater, being greater or less, according as the carbon of which that surface is composed is harder or softer." This can now be modified into the following, which is no longer a pure assumption, but must rank as a proposition :— " With a given negative carbon, current, and length of arc, the P.D. between the carbons depends principally, if not entirely, on the nature of the carbon that forms the surface of volatilisation, being higher or lower, according as that carbon is more or less free from metallic salts."

The explanation given in Chapter IV. can now be extended in the following manner :—

Every increase of current entails an enlargement of the cross-section of the arc, and a consequent tendency of the P.D. to diminish. While the current is so small that the volatilising surface does not completely cover the core, the increase of cross-section is unaccompanied by any change in the specific resistance of the arc. When the current is so large, however, that the solid carbon round the core begins to volatilise, since the resulting vapour and mist have higher specific resistances than those of the core, each increase of current is accompanied by two tendencies in the P.D.—the one to *fall*, on account of the larger cross-section, the other to *rise*, because of the higher specific resistance of the vapour and mist. The curve connecting the P.D. with the current must, therefore, be compounded of two. One, such as ABC (Fig. 134), which would connect the P.D. with the current if the positive carbon were composed entirely of

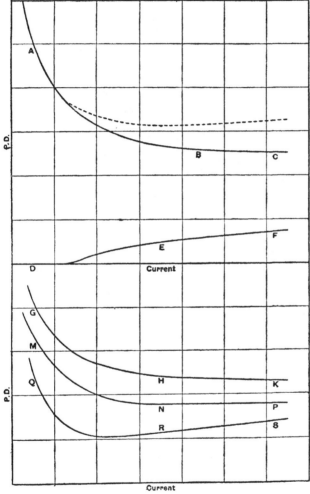

FIG. 134.—Curves exemplifying the Changes in the Curve connecting P.D. with Current caused by a Core in the Positive Carbon.

core, and the other, DEF, connecting the rise of P.D., due to the increase of specific resistance, with the current. The

curve connecting the true P.D. with the current is found by
adding each ordinate of DEF to the corresponding ordinate of
ABC as indicated in the dotted line. Whether this resulting
curve has the form GHK, or MNP, or QRS (Fig. 134), depends
evidently upon the relation between the increase of the cross-
section and the rise of specific resistance, i.e., on the relative structures and cross-sections of the core and the outer
carbon.

The fact, already obtained from Table LV., that, for the
same current and length of arc, the vapour film, and consequently the crater, is smaller with a cored than with a solid
positive carbon explains why the arc can carry such a much
larger current without hissing when the positive carbon is
cored. For, since hissing is the result of that direct contact
of the crater with the air that follows when it grows too large
to cover the end only of the positive carbon, and since this
must happen with a smaller current the larger the crater is
with a given current, it must happen with a smaller current
when the positive carbon is solid than when it is cored.

The Influence of the Core on the Value of $\frac{\delta V}{\delta A}$.

We have next to consider the influence of the cores on the
value of $\frac{\delta V}{\delta A}$, when such an alternating current is superimposed
on a direct-current normal arc that the resistance of the arc is
affected by the superposition. Here we have to deal, not with
the whole P.D. between the carbons, but with the *change* in
that P.D. that accompanies a given *change* of current, and I
shall show that the effect of the core on this change is always
to *add* a positive increment to $\frac{\delta V}{\delta A}$, the amount of which depends
on the current, the length of the arc, and the frequency of the
alternating current.

The influence of the core on the value of $\frac{\delta V}{\delta A}$ is two-fold; it
alters the amount by which the cross-sections of the vapour
film and the mist change, with a given change of current; and
it makes their specific resistances vary with the current. We
will take each separately—the change of cross-section first.
I shall call the part of $\frac{\delta V}{\delta A}$ that depends on the change of cross-

section $\frac{\delta V_c}{\delta A}$, and the part that depends on the variation in the specific resistance $\frac{\delta V_s}{\delta A}$, so that

$$\frac{\delta V}{\delta A} = \frac{\delta V_c}{\delta A} + \frac{\delta V_s}{\delta A}.$$

How the Change in the Cross-Sections of the Mist and the Vapour Film due to a Change of Current is affected by Coring either or both Carbons.

I have already pointed out (p. 402) that if, when the current is increased, the ratios of the new cross-sections of the mist and the vapour film to the old are greater than the ratio of the new current to the old, then the resistance of the arc must have been diminished more than the current was increased, and $\frac{\delta V}{\delta A}$ must be negative (provided always that the specific resistance of the arc has not been altered). Similarly, when the ratios of the cross sections are smaller than that of the current $\frac{\delta V}{\delta A}$ must be positive.

In order to see the effect of the cores on these ratios in the experiments of which the results are given in Tables LV. and LVI., Tables LVII. and LVIII. have been drawn up, in which the cross-section ratios are found by dividing the cross-section for each current by the cross-section for the next smaller current ; and the current ratios by dividing each current by the next smaller current. For the non-normal ratios the larger cross-sections were taken from the *non-normal* sets in Tables II. and III. and the smaller from the *normal,* because it is the effect of the core when the alternating current is superimposed on a *normal* arc that we are considering, and because also the non-normal numbers in these tables were found by suddenly *increasing* the current when the arc was normal. For the normal ratios both numbers were taken from the normal sets in Tables LV. and LVI. For instance, the normal cross-section for a current of 8 amperes with the + solid − cored carbons in Table LV. is 14·2, and the non-normal cross-section for 10 amperes is 19·0, while the normal cross-section for the same current is 20·75. Thus, when the current is increased from 8 to 10 amperes, the current ratio is $\frac{10}{8} = 1\cdot25$

the non-normal cross-section ratio with these carbons is $\frac{19}{14\cdot2} = 1\cdot34$, and the normal is $\frac{20\cdot75}{14\cdot2} = 1\cdot46$. In this case, therefore, $\frac{\delta V}{\delta A}$ would be negative, as far as the change in the cross-section of the mist was concerned, both when the change was

Table LVII.—*Ratio of Mean Cross-Section of Mist to Cross-Section with Next Smaller Current taken from Table LV.*

Length of Arc, 2mm.

Change of Current.	Current Ratios.	Ratios of Mean Cross Sections.							
		Normal.				Non-Normal.			
		S.S.	S.C.	C.S.	C.C.	S.S.	S.C.	C.S.	C.C.
(1)	(2)	(3)	(4)	(5)	(6)	(7)	(8)	(9)	(10)
4 to 6	1·5	2·04	1·2	1·51	1·7	1·98	1·21	1·56	1·06
6 to 8	1·33	1·65	1·71	1·82	1·6	1·8	1·33	1·98	1·04
8 to 10	1·25	1·44	1·46	1·23	1·33	1·33	1·34	1·70	1·25
10 to 12	1·20	1·49	1·33	1·31	1·39	1·46	1·30	1·48	1·40
12 to 14	1·17	1·18	1·25	1·38	1·21	...	1·43	...	1·13

Table LVIII.—*Ratios of Cross-Sections of Mist where it touches the Crater, taken from Table LV.*

Length of Arc, 2mm.

Change of Current.	Current Ratios.	Ratios of Cross-Sections at Crater.							
		Normal.				Non-Normal.			
		S.S.	S.C.	C.S.	C.C.	S.S.	S.C.	C.S.	C.C.
(1)	(2)	(3)	(4)	(5)	(6)	(7)	(8)	(9)	(10)
4 to 6	1·5	2·34	1·07	1·66	2·0	3·76	0·8	1·95	1·24
6 to 8	1·33	2·35	1·44	2·05	2·5	2·35	1·00	2·06	1·08
8 to 10	1·44	2·00	1·12	1·47	1·21	1·00	1·40	1·05
10 to 12	1·20	1·41	0·96	1·44	1·04	1·37	0·88	1·44	1·00
12 to 14	1·17	1·20	1·44	1·31	1·09	...	1·34	...	0·88

non-normal and when it was normal, for both 1·34 and 1·46 are greater than 1·25, the current ratio. The non-normal ratios show the effect of the core on the change in the resistance of the arc, and therefore on $\frac{\delta V}{\delta A}$, when the frequency of the alternations is so great that the carbons do not change their shapes ; and the normal, when it is so small that the arc remains normal throughout.

Of course, to imitate the effect of an alternating current completely it would be necessary to diminish the current suddenly as well as suddenly increasing it, but as this would only alter the signs of both δV and δA, without materially changing their relative values, it is not necessary for our purpose.

The most important point to observe in these tables is whether $\dfrac{\delta V_c}{\delta A}$ is negative or positive with each set of carbons. *i.e.*, whether the cross-section ratios are greater or less than the current ratios. Take, first, the non-normal ratios. When the positive carbon alone is cored $\dfrac{\delta V_c}{\delta A}$ is decidedly negative, for all the cross-section ratios in column (9) of both tables are greater than the corresponding current ratios in column (2). Moreover, with this particular length of arc and these currents the non-normal $\dfrac{\delta V_c}{\delta A}$ appears to be unaffected by coring the *positive* carbon alone, for the non-normal cross-section ratios in column (9) of each Table are in some cases greater and in others less than those in column (7). When the *negative* carbon alone is cored, the non-normal value of $\dfrac{\delta V_c}{\delta A}$ appears to be negative, but approaching the zero point; for in Table LVII. one cross-section ratio in column (8) is less than the corresponding current ratio, one is equal and three are greater, while in Table LVIII. three are less and two are greater. When *both* carbons are cored, the non-normal value of $\dfrac{\delta V_c}{\delta A}$ is positive; for three out of the five of the numbers in column 10 of Table LVI., and the whole of those in the same column of Table LVII., are less than the corresponding numbers in column (2).

Turning next to the normal ratios, we find that when the positive carbon alone is cored $\dfrac{\delta V_c}{\delta A}$ has still much the same negative value as when both carbons are solid, since the numbers in column (5) differ very little on the whole from those in column (3). When, on the other hand, it is the negative carbon alone that is cored there is a change, for instead of being a little below zero $\frac{\delta V_c}{\delta A}$ is decidedly negative, since in Table LVII. all but one of the numbers in column (4), and in Table LVIII. all but two are greater than the corresponding numbers in column (2). When both carbons are cored there is an even greater difference between the normal and non-normal values of $\frac{\delta V_c}{\delta A}$. For, in Table LVII. all the numbers in column (6), and in Table LVIII. all but two are greater than the corresponding current ratios, showing that $\frac{\delta V_c}{\delta A}$ is *negative* for normal changes of current, though it is positive for non-normal changes, with these carbons, currents, and lengths of arc. Thus, coring the negative carbon retards the change of cross-section that follows a change of current, for while this change follows *immediately* after the change of current when the negative carbon is solid, when it is cored an appreciable time elapses before it takes place.

To sum up the changes in the value of $\frac{\delta V_c}{\delta A}$ produced by coring one or both of the carbons, we find that while coring the positive carbon alone makes very little difference in either the normal or the non-normal change of cross-section that accompanies a given change of current, coring the negative carbon *diminishes* the change of cross-section both for normal and non-normal changes of current, but more for the second than for the first, and more when both carbons are cored than when the negative alone is cored. Thus coring the negative carbon both *diminishes* and *retards* the change in the cross-sections of the arc that accompany a change of current, since the change of cross-section is less immediately after the current is changed than it is later, after the arc has become normal for the new current. This retardation of the change of cross-section is quite sufficient to account for the fact already mentioned on p. 406, viz., that if I quickly altered the resistance in the circuit outside the arc, when both carbons were *cored*, I could sometimes see the first quick swing of the voltmeter needle in the same direction as that of the ammeter, but never when both were *solid*. For, as the resistance did not alter *directly* after the current with the cored carbons, the new current would be flowing through the old resistance for an appreciable time after the change, and so the accompanying change of P.D. in the same direction as the change of current would be able to influence the voltmeter needle.

The Change in the Specific Resistance of the Arc produced by a Change of Current when Either or Both Carbons are Cored.

We have next to consider $\dfrac{\delta V_s}{\delta A}$, the part of $\dfrac{\delta V}{\delta A}$ that depends on the changes in the specific resistances of the mist and vapour that occur with each change of current when either or both carbons are cored.

Coring the negative carbon must have a very different effect from coring the positive, for whereas, in the first case the whole of the vapour and almost the whole of the mist issues from the *uncored* carbon, the core only contributing a little metallic vapour to the mist in contact with it, in the second the whole comes from the *cored* carbon. Thus, while with the cored *negative* the vapour is always *solid-carbon* vapour, and the mist is practically *solid-carbon* mist, even with the smallest currents, with the cored *positive* the vapour and mist are both *core* vapour and mist alone until the current is large enough for the volatilising surface to cover the whole core, and only begins to have an admixture of solid-carbon vapour and mist when the current is larger than this. When, therefore, the *negative* carbon alone is cored, the specific resistance of the vapour is constant, and that of the mist increases with each small increase of current, but more and more slowly, with the *same* addition of current, the larger the original current before the addition is made. The curve connecting $\dfrac{\delta V_s}{\delta A}$ with the normal current in this case must, therefore, be of the form ABC (Fig. 135), for the specific resistance must change most when the current is just large enough for the mist to cover the whole core, and the amount by which it changes must steadily diminish as the direct current increases after that, till it becomes practically zero with very large currents, so that the curve becomes asymptotic to the axis of current.

When the *positive* carbon alone is cored, the curve is quite different. If the arc always remained perfectly central, it would be of the form DEFG (Fig. 135). The specific resistances of the vapour and mist would remain constant till the volatilising surface was large enough to cover the core, so that, until then, $\dfrac{\delta V_s}{\delta A}$ would be zero and DE would be the first part of the curve. The first increment of current that was added after this would increase the specific resistances more than any subsequent increment, because this would be the point at which the specific resistances of the existing vapour and mist and of those added would be most different. Therefore, the curve would rise suddenly at E. After this, each addition to the normal current would make the change of specific resistance due to the added small non-normal increment of current smaller and smaller, so that the curve would fall towards the axis of current as shown in FG. Finally, there

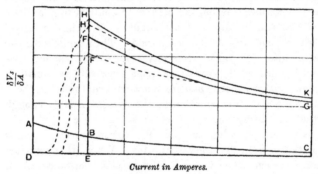

Current in Amperes.

FIG. 135.—Curves connecting $\dfrac{\delta V_s}{\delta A}$ with Current for Constant Length of Arc.

would already be so much solid-carbon vapour and mist in the arc that the addition of a little more would make practically no change, so that this curve also is asymptotic to the axis of current. The fact that the arc is never really quite central, and that the volatilising surface must therefore cover a little solid carbon long before it is larger than the core, must introduce some modifications into the first part of the curve, shortening DE and making EF rise less abruptly, something like DF'G; but these modifications are unimportant.

When both carbons are cored the curve must be like DEHK, or rather DH'K, because the effect of the metallic vapour from the negative core will be added to that of the positive core, and the change of specific resistance, when solid-carbon mist begins to be added will, therefore, be greater.

How the whole Value of $\frac{\delta V}{\delta A}$ is affected by Coring either or both Carbons.

By combining the two changes in the resistance of the arc introduced by the core—viz., that due to the difference of the changes in the cross-sections of the arc and that produced by the alterations in its specific resistance—we can see how the complete value of $\frac{\delta V}{\delta A}$ is affected by the core.

From what has been said on p. 427 it is clear that, if the cross-section ratios in Tables LVII. and LVIII. can be considered typical, $\frac{\delta V_e}{\delta A}$ never has a greater negative value when the positive carbon alone is cored than when both are solid ; never a greater negative value when the negative alone than when the positive alone is cored, and never a greater negative value when both are cored than when the negative alone is cored. But $\frac{\delta V}{\delta A}$ is zero when both carbons are solid, is greatest when both are cored, and has always some positive value, however small, when either carbon alone is cored. Consequently, when the superimposed alternating current alters the resistance of the arc, if all other things are equal, $\frac{\delta V}{\delta A}$ is more positive when either carbon is cored than when both are solid, and most positive when both are cored.

The general effect on $\frac{\delta V}{\delta A}$ of coring either or both carbons is given in the preceding paragraph, but with a given root mean square value of the alternating current $\frac{\delta V}{\delta A}$ depends, not only on the nature of the carbons, but also on the magnitude of the direct current, the length of the arc, and the frequency of the alternating current. To complete our knowledge of the influence of cores on the value of $\frac{\delta V}{\delta A}$, therefore, we must examine the effect they produce on the curves connecting each of these variables with $\frac{\delta V}{\delta A}$ when the others are constant. Take, first, the curves connecting $\frac{\delta V}{\delta A}$ with the magnitude of the direct current.

The Effect produced by Coring either or both Carbons on the Curve connecting the Non-Normal Value of $\dfrac{\delta V}{\delta A}$ with A, when the Length of the Arc is Constant.

In Tables LVII. and LVIII. the cross-section ratios for solid carbons differ less, on the whole, from the corresponding current ratios the larger the current on which the increase of 2 amperes

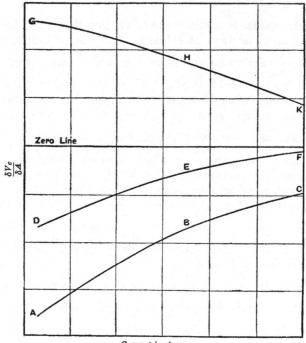

Current in Amperes.

FIG. 136.—Curves connecting $\dfrac{\delta V_c}{\delta A}$ with the Current for a Constant Length of Arc.

ABC—Solid-Solid or Cored-Solid Carbons ; DEF—Solid-Cored ; GHK—Cored-Cored.

has been superimposed. This shows that with solid carbons, when the length of the arc is constant, $\dfrac{\delta V}{\delta A}$ diminishes as the current increases. Consequently, the curve for solid carbons is of the form ABC (Fig. 136). With cored carbons the curves depend not only on $\frac{\delta V_c}{\delta A}$, which is obtained from Tables LVII.

and LVIII., but also on $\frac{\delta V_s}{\delta A}$, the curves connecting which with

A are given in Fig. 135. The curves connecting $\frac{\delta V_c}{\delta A}$ with A

cannot be obtained straight from Tables LVII. and LVIII., because the values are too irregular, but we can deduce them from what we already know. For instance, when the positive carbon alone is cored it must have the same form, ABC, as when both are solid, since the change of cross-section due to a given change of current is not materially altered by coring the positive carbon alone. Coring the negative carbon alone *diminishes* the negative value of $\frac{\delta V_c}{\delta A}$, and must diminish it

most when the current is least, for it is then that the metallic vapour from the core will be expended on the smallest quantity of hard carbon mist, and will, consequently, have most effect. Hence the curve for a cored negative and solid positive carbon must resemble DEF (Fig. 136), and the current for which $\frac{\delta V_c}{\delta A}$

becomes positive, if any, will depend upon the length of the arc and the frequency. Finally, it has been shown that with both carbons cored $\frac{\delta V_c}{\delta A}$ is even more positive than when the

negative only is cored (p. 431), so that the curve with both carbons cored must resemble GHK (Fig. 136), since the same reasoning as before shows that the cores must have least effect on both $\frac{\delta V_c}{\delta A}$ and $\frac{\delta V_s}{\delta A}$ when the current is largest.

To find the full curves connecting $\frac{\delta V}{\delta A}$ with A for each pair of

carbons, we have only to add each ordinate of each curve in Fig. 135 to the corresponding ordinate of the curve for the same carbons in Fig. 136. Curves resembling those that would be thus obtained for one length of arc and frequency of alternating current are given in Fig. 137. The exact distance of each above or below the zero line and the exact points where it cuts that line must, of course, depend upon the length of arc and frequency of alternating current for which the curves are drawn, but their relative shapes and positions must be similar to those in Fig. 137 whatever the length of the arc and the frequency.

F F

Next take the curve connecting $\dfrac{\delta V}{\delta A}$ with l, the length of the arc, when the frequency of the alternating current and the value of the direct current are both constant.

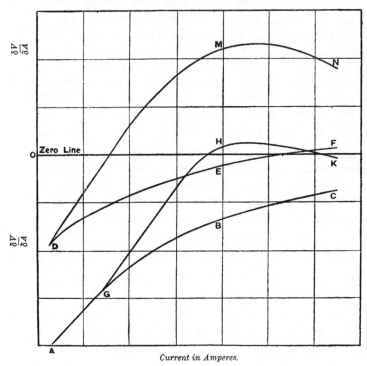

FIG. 137.—Curves connecting $\dfrac{\delta V}{\delta A}$ with A for a Constant Length of Arc when the Superimposed Alternating Current Affects the Resistance of the Arc.

ABC—Solid-Solid Carbons ; AGHK—Cored-Solid :
DEF—Solid-Cored ; DMN—Cored-Cored.

The Effect produced by Coring either or both Carbons on the Curve connecting the Non-Normal Value of $\dfrac{\delta V}{\delta A}$ with the Length of the Arc, when A is Constant.

Messrs. Frith and Rodgers found that, with a constant current of 10 amperes, curves somewhat resembling those in Fig. 139

connected $\frac{\delta V}{\delta A}$ with the length of the arc with Apostle carbons of 11mm. and 9mm. In order to see why the curves should take this peculiar form we must start, first of all, with the lowest, for which both carbons were solid.

PQ (Fig. 138) is the rise of P.D. that would accompany the increase of current δA with an arc of l millimetres if the resistance of the arc did not alter with the current. QR is the fall of P.D. due to the enlargement of the vapour film and the mist. When the current increases from A to $A + \delta A$, therefore, the P.D. actually falls from P to R. Now, the rise PQ depends only on the amount by which the current is increased, and the resistance through which that increased current has to flow, $i.e.$, on δA, A, and l; or, since A and δA are supposed to be the same for each length of arc, PQ depends simply on l, and increases directly as l increases.

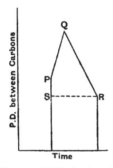

Fig. 138.—Time-Change of P.D. due to a Change of Current.

The *fall* of P.D.—QR—is more complex. It depends principally on how much of the extra carbon volatilised by the larger current remains between the carbons, and how much escapes along them. When the carbons are blunt more will remain than when they are pointed, and as the carbons get blunter, with the same current, as the arc is lengthened, so the resistance must diminish more, on account of the increase of the current, the longer the arc. But the blunting of the carbons, which is a rapid affair when the arc is short, takes place more and more slowly as it is lengthened, till at last the addition of a millimetre or so makes practically no difference. Hence, the diminution of resistance due to the addition of δA to the current increases rapidly at first, when the arc is short, and more and more slowly as the arc lengthens, till it becomes practically constant; and hence, also, QR—the fall of P.D. accompanying this diminution—increases more and

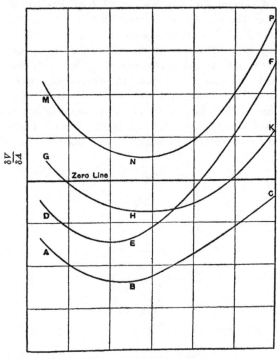

Length of Arc.

FIG. 139.—Curves connecting $\frac{\delta V}{\delta A}$ with the Length of the Arc for a Constant Current when the Superimposed Alternating Current Affects the Resistance of the Arc.

ABC—Solid-Solid Carbons ; DEF and GHK—One Carbon Solid, One Cored ; MNP—Cored-Cored.

more slowly as the arc is lengthened. Thus, while the rise of P.D.—PQ—increases at a *constant* rate as the arc is lengthened, the fall—QR—increases at a *diminishing* rate,

While the arc is so short, therefore, that QR increases more rapidly than PQ when l is increased, the whole fall of P.D. —PS—will *increase* with the length of the arc, or, since PS is $- \delta V$, and δA is the same for all the lengths of arc, $- \frac{\delta V}{\delta A}$ *increases* as the arc is lengthened. When the arc is so long that PQ increases faster than QR, $- \frac{\delta V}{\delta A}$ will *diminish* as the arc is lengthened. Between the two stages there must be a length of arc for which $- \frac{\delta V}{\delta A}$ is a maximum. The curve connecting $\frac{\delta V}{\delta A}$ with l for a constant current, with solid carbons, must therefore be of the form ABC (Fig. 139), and there seems to be no reason why, with very long arcs, $\frac{\delta V}{\delta A}$ should not actually become positive, with superimposed alternating currents of comparatively low frequency, even with solid carbons.

The curves connecting $\frac{\delta V}{\delta A}$ with l, when cored carbons are used must resemble the curve for solid carbons, ABC (Fig. 139), but must be higher up the figure (DEF, GHK) when one carbon is cored, and still higher (MNP) when both are cored. Also, since a change in the specific resistance of the arc must have more effect on the value of $\frac{\delta V}{\delta A}$, the longer the arc, the distance between the curves for cored carbons and the curve for solid carbons must increase as the arc is lengthened, as it does in Fig. 139.

The Effect produced by Coring either or both Carbons on the Curve connecting $\frac{\delta V}{\delta A}$ with the Frequency of the Alternating Current.

ABC (Fig. 140), which is copied from Fig. 132, is the curve for solid carbons. Since $\frac{\delta V}{\delta A}$ is always most positive when both carbons are cored, and more positive when one is cored than when both are solid (p.431), the curve when both carbons are cored must resemble DEF, and the curves for one carbon cored and the other solid must lie between ABC and DEF. It follows, therefore, that the frequency with which $\frac{\delta V}{\delta A}$ becomes positive, if it is not already positive, for normal changes of current (frequency (0 must be lower when one carbon is cored than when both are solid, and lowest when both are cored. Thus, with the same direct current and length of arc $\dfrac{\delta V}{\delta A}$ may be positive for all four sets of carbons, as at the points C, P, K, F, or positive for some and negative for others, as at B, N, H, E, or negative for all. Moreover, since the true resistance of the arc is greatest when both carbons are solid and least when

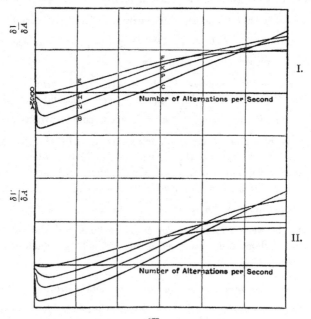

Fig. 140.—Curves connecting $\dfrac{\delta V}{\delta A}$ with the Frequency of the Superimposed Alternating Current for Solid-Solid, Solid-Cored, Cored-Solid and Cored-Cored Carbons.

both are cored, and smaller when the positive alone than when the negative alone is cored, the curve for two solid carbons must cut all the others at some fairly high frequencies, and that for two cored carbons must also cut the other two. Hence, the curves will be like I (Fig. 140) when the curve for the positive carbon alone cored is higher, with low frequencies than that for the negative alone cored, and like II (Fig. 140) when it is lower.

It is clear from these curves that, with cored carbons, the frequency with which $\frac{\delta V}{\delta A}$ first becomes positive, when its normal value is negative, may be very low indeed ; and that when its normal value is positive, as for instance, with currents over 6 amperes in such curves as those for 0mm. and 1mm. (Fig 40) $\frac{\delta V}{\delta A}$ may never become negative with any frequency. This is just what Messrs. Frith and Rodgers found by direct measurement of $\frac{\delta V}{\delta A}$; for they could get no negative value, with *any* frequency, when the P.D between the carbons was 35 volts, and the direct current was 10 amperes, with cored positive and solid negative Brush carbons ; while with some other currents and lengths of arc they found $\frac{\delta V}{\delta A}$ negative with smaller frequencies than 1·8 per second, and positive with larger, and with yet others it was negative, even with frequencies of 250 per second. Their experiments showed, then, that with a superimposed alternating current of constant root mean square value and frequency, $\frac{\delta V}{\delta A}$ was positive or negative, *with cored carbons*, according to the value of the direct current and the length of the arc. But we have seen from theoretical considerations that both the value and the sign of $\frac{\delta V}{\delta A}$ must depend on these two variables, not with cored carbons only, but with *all* carbons.

Thus, all the principal phenomena of the arc, with cored and with solid carbons alike, can, with one exception, to be presently alluded to, be explained as natural results of the variations in the specific resistances of the material in the gap that *must* exist, together with the observed variations in its cross-sections. It is quite probable, therefore, that there is neither a large back E.M.F. in the arc nor a " negative resistance," but that its resistance follows the ordinary ohmic laws, obscured only by the varying resistivities of its different parts, consequent on their varying temperatures and on the resultant differences in their physical conditions. There is only one phenomenon that these variations do not explain, viz., the fall of potential between the arc and the *negative* carbon, which has been shown (p. 225) to vary between about 8·3 and 11 volts with currents over 4 amperes. This may possibly be a true back E.M.F., which, although large compared with that of an ordinary cell is very small compared with that which has been supposed to exist in the arc.

SUMMARY.

I. All the material in the gap between the carbons of an arc cannot remain carbon vapour; the inner part must cool into carbon mist at a short distance from the crater, which is the seat of volatilisation, and the outer part must burn and form a sheath of flame in contact with the air. These different materials are indicated in images of the arc.

II. The flame has a very high specific resistance; the film of true vapour in contact with the crater a lower one; the carbon mist lowest of all. Hence the current flows principally through the vapour and mist, but meets with much greater resistance in the thin vapour film than in the thicker mass of mist.

III. The heat of the crater is principally—perhaps entirely—due to the passage of the current through the high resistance vapour film in contact with it.

IV. The end of the positive carbon acquires its characteristic shape through a race between the volatilisation of a portion of its end surface and the burning of the remainder, and of its sides. The negative carbon is entirely shaped by burning, and by deposit from the carbon mist.

V. The area of the crater, including both the part that is actually being volatilised and that which is just below the temperature of volatilisation, varies with the current, the length of the arc, and the time that has elapsed since a change was made in either. The area of the surface of volatilisation depends on the current alone.

VI. The film of vapour in contact with the crater acts exactly like a back E.M.F.

VII. With solid carbons the cross-sections of the vapour film and the mist increase more rapidly than the current. Hence, the resistance of the arc diminishes more rapidly than the current increases, which gives the arc the appearance of having a negative resistance.

VIII. Both the resistance of the arc and the P.D. between the carbons depend, not only on the current and the length of the arc, but also on how lately a change has been made in either, and on what that change was.

IX. When a small change is made in the current, it is only when the alteration is so quick and so small, that neither the resistance nor the back E.M.F. of the arc (if any) is changed by it that $\frac{\delta V}{\delta A}$ is a true measure of the resistance of the arc.

X. When the resistance of the arc is measured by superimposing a small alternating current on the direct current, the alternating current must have a frequency of many thousands of alternations per second for the resistance of the arc not to be altered by it.

XI. When the alternating current affects the resistance of the arc, $\frac{\delta V}{\delta A}$ may have any value between a large negative limit and the positive limit that is the true resistance of the arc, the value and sign of $\frac{\delta V}{\delta A}$ depending on the frequency of the alternations.

XII. The form of the P.D. time curve indicates whether the resistance of the arc is affected by the superimposed alternating current or not.

XIII. Even when it has been ascertained that the superimposed alternating current does not affect the *resistance* of the arc, measures must be taken to show that it does not, either, affect any back E.M.F. that may exist, before the average value of $\frac{\delta V}{\delta A}$ can be accepted as measuring the true resistance of the arc.

Cored Carbons.

XIV. There are two ways in which cores in the carbons affect the arc; they alter its cross-section, and they change the specific resistances of the vapour and the mist.

XV. When the sizes of the carbons, the current, and the length of the arc are all constant, the cross-section of the arc is greatest when both carbons are solid, is smaller when the negative alone is cored, smaller still when the positive alone is cored, and smallest when both are cored.

XVI. The ordinary commercial core being composed of carbon mixed with metallic salts, the resistivities of the vapour and mist from it are lower than from solid carbon, and this causes a diminution of the resistance of the arc that more than compensates for the increase of resistance due to the diminution of its cross-section. Hence the P.D. is lowered by a core in the positive carbon.

XVII. With a given negative carbon, current, and length of arc, the P.D. between the carbons depends principally, if not entirely, on the nature of the carbon that forms the surface of volatilisation, being higher or lower, according as that carbon is more or less free from metallic salts.

XVIII. The change in the shape of the P.D. current curve that takes place when a cored positive carbon is substituted for a solid one, is entirely accounted for by the fact that, as soon as the crater more than covers the core, the volatilising surface is composed of two different substances, the proportions of which depend upon the current.

XIX. A core in either carbon influences the value of $\dfrac{\delta V}{\delta A}$ when a small alternating current is superimposed, in two ways: it alters the amount by which the cross-section of the vapour film and the mist change as the current changes, and it makes their specific resistances depend upon the current.

XX. From images of the arc and carbons taken before and after changes of current, it is deduced that coring the positive carbon alone has no effect on either the normal or the non-normal change of cross-section due to a change of current. Coring the negative alone *diminishes* both changes, and also *retards* the change, and by coring both carbons both these effects are intensified.

XXI. By combining the two changes in the resistance of the arc introduced by the core—viz., that due to the difference in the changes of the cross-sections of the arc, and that produced by the alterations in its specific resistance, we find that when a superimposed alternating current alters the resistance of the arc, $\dfrac{\delta V}{\delta A}$ is more positive when either carbon is cored than when both are solid, and most positive when both are cored.

XXII. With solid and with cored carbons alike the value of $\dfrac{\delta V}{\delta A}$ depends upon the magnitude of the direct current, the length of the arc, and the frequency of the alternations.

XXIII. When the negative carbon is solid, whether the positive is solid or cored, the value of $\dfrac{\delta V}{\delta A}$ becomes more positive as A increases; but when the negative carbon is cored, $\dfrac{\delta V}{\delta A}$ increases positively to a maximum and then diminishes again.

XXIV. With all four sets of carbons, when A is constant, $\dfrac{\delta V}{\delta A}$ diminishes positively to a minimum as the arc is lengthened, and then increases again.

XXV. The frequency with which $\dfrac{\delta V}{\delta A}$ first becomes positive is smallest when both carbons are cored, smaller when the negative alone than when the positive alone is cored, and greatest when both are solid.

XXVI. The fall of potential between the arc and the negative carbon is probably due to a true back E.M.F.

APPENDIX.

I.—THE APPARENT AREA OF A DISC.

If a disc is looked at from any point—by a single eye, we will suppose for the sake of simplicity—the rays of light which make it visible form a truncated cone, of which the disc is the base and the entrance to the pupil is a section. When the centre of the pupil is in the line which is at right angles to the disc at its centre, every point on the edge of the disc is at the same distance from the centre of the pupil, and the disc is seen as a circle. When the line joining the centres of the eye and the disc makes any other angle with the plane of the disc, however, the disc has the same effect on the eye as would be produced if it were projected on a plane passing through its centre and perpendicular to the line joining that centre to the centre of the pupil. This projection is, therefore, the apparent area of the disc when looked at from the given point. The area of this projection, and, therefore, the apparent area of the disc, must depend on (1) r, the radius of the disc, (2) l, the distance between the centres of the disc and the pupil of the eye, and (3) θ, the angle between the surface of the disc and the plane perpendicular to the line joining the centre of the eye to the centre of the disc. The apparent area of the disc in terms of these three quantities may be found from Fig. 141.

Let ABC be the disc, which may be considered to be perpendicular to the plane of the paper, and to be cut by that plane through one of its diameters, ADB. Let P, a point in the plane of the paper, be the point from which the disc is viewed, and let ePe' be the diameter of the pupil. Draw FDG perpendicular to PD. Then $AD = DB = FD = DG = r$, $PD = l$, and, since AB is the diameter of the disc that intersects the plane of the paper, and FG is the line in which the plane perpendicular to PD meets the plane of the paper, AB, FG, and PD, are all in this same plane, and GDB being the angle between the disc and the plane through the centre of the disc perpendicular to PD, must be θ. Join eA, e'B, and let eA meet FG in H, and produce e'B to meet FG in K. Then HK is the projection of AB on FG, and, since AB is a diameter of the disc, HK must also be a diameter of its projection, and, being in the same plane as PD, HK must be one axis of the ellipse, into which the disc ABC is projected, and, if Q be taken bisecting HK, Q (Fig. 142) is the centre of the ellipse.

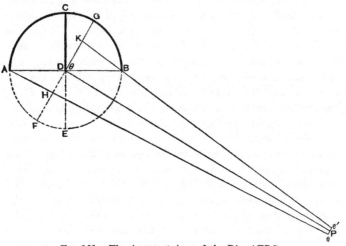

Fig. 141.—The Apparent Area of the Disc AEBC.

Next, let EDC be that diameter of the disc which is at right angles to the plane through PD—the plane of the paper. Then, since the disc is projected on to a plane which also cuts the plane of the paper at right angles through D, EDC must be common to the disc and its projection, and EDC must therefore be a chord of this projection which is bisected at right angles by PD. Let us now clear off all unnecessary lines, and deal only (Fig. 142), with PD the line joining the centre of the eye to the centre of the disc, AB the diameter of the disc that is in the same plane as PD, CDE, the chord that is common to the disc and to its projection on the plane at right angles to PD, and HQK, an axis of the projection. Thus AB, PD, and HK are all in one plane (the plane of the paper), and CDE is at right angles to that plane.

Since HK is a part only of FG, and FG = CE, which is a chord of the projection of the disc, HK must be the *minor* axis of the projection. With centre Q, and radius QH (half the minor axis), describe a circle cutting DC at R. Draw SQ perpendicular to HK, making

$$SQ : QH :: DC : DR.$$

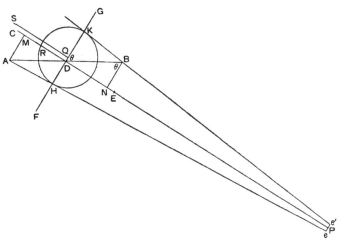

Fig. 142.—The Apparent Area of a Disc.

Then, by a well-known property of the ellipse, SQ is half the major axis of the ellipse which is the projection of the disc, and we can now find the area of this ellipse in terms of r, l and θ. For, draw AM and BN both parallel to HK. Then, by similar triangles,

$$\frac{HD}{AM} = \frac{PD}{PM} = \frac{PD}{PD + DM}.$$

But $\qquad AM = r \cos \theta, \quad DM = r \sin \theta, \quad \text{and} \quad PD = l \; ;$

$$\therefore \; HD = \frac{lr \cos \theta}{l + r \sin \theta}$$

Similarly,
$$DK = \frac{lr \cos \theta}{l - r \sin \theta};$$

$$\therefore \ HD + DK = HK = \frac{2l^2 r \cos \theta}{l^2 - r^2 \sin^2 \theta}.$$

Thus, if we call the major and minor axes of the ellipse $2a$ and $2b$, we have

$$b = \frac{l^2 r \cos \theta}{l^2 - r^2 \sin^2 \theta},$$

Also
$$a : b :: DC : DR :: r : DR,$$

and, since HRK is a semicircle,

$$DR^2 = HD. \ \ DK = \frac{l^3 r^2 \cos^2 \theta}{l^2 - r^2 \sin^2 \theta};$$

$$\therefore \ DR = \frac{lr \cos \theta}{\sqrt{l^2 - r^2 \sin^2 \theta}},$$

$$\therefore \ a : b :: r : \frac{lr \cos \theta}{\sqrt{l^2 - r^2 \sin^2 \theta}},$$

$$\therefore \ ab = \frac{r^2 l^3 \cos \theta}{(l^2 - r^2 \sin^2 \theta)^{\frac{3}{2}}},$$

and πab, which is the area of the ellipse, is

$$\frac{\pi r^2 l^3 \cos \theta}{(l^2 - r^2 \sin^2 \theta)^{\frac{3}{2}}}.$$

If l is great compared with r, this becomes

$$\pi r^2 \cos \theta.$$

Thus, when the distance of the eye from the disc is great compared with the radius of the disc, the apparent area of the disc varies as the cosine of the angle it would have to be turned through to make its plane perpendicular to the line joining the eye to its centre. This angle Mr. Trotter calls the inclination, so that we *may* say that the apparent area of the crater varies as the cosine of its inclination, since the radius of the crater is always small compared with its distance from the eye.

II.—PHOTOMETRY.

Several assumptions are made in all photometry, and it is well to have a clear understanding of what these are.

The eye is incapable of judging of quantity of light, it can only appreciate intensity—brightness—*the quantity of light emitted per unit area of the source.* Even of this, it can only judge within very narrow limits ; for if two sources have both more than a certain intensity the eye is equally dazzled by both and incapable, therefore, of distinguishing which is the brighter. For this reason it is our habit to judge of the intensity of a source of light not by regarding the source itself, but by looking at surfaces illuminated by it. When one part of a street, for instance, is lighted by an arc lamp and another by a gas jet, we realise the brightness of the pavements and walls lighted by the arc compared with the dullness of the parts illuminated by the gas-jet rather than the brilliancy of the arc itself compared with the dimness of the gas-jet.

In the scientific measurement of light we do not depart from this accustomed method of comparing the intensity of two sources. The sources are placed so that the rays they send out in given directions are perpendicular to adjacent portions of a small plane surface. If the sources are small and are sufficiently far from the surface, they may be considered to be points of light, and if the surface is also small all the rays may be considered to be perpendicular to it. When the distances between the sources and the surface are such that the two parts of the surface appear equally bright—that is, when they are *emitting or transmitting* equal quantities of light per second per unit area—we *assume* that they are *receiving* equal quantities also. When the two sources of light are of the same kind and at the same temperature the assumption is perfectly justifiable, for then the screen will absorb the same percentage of the light it receives from each source, and all other objects on which the two lights fall will do the same. When the temperatures of the two sources are widely different, however, there will be a difference in the percentage of the light of each absorbed, depending on the nature of the surface, and hence the apparent relative candle-powers of the sources will differ according to the nature of the photometer screen employed, and the relative brightness of all objects illuminated by the two sources will vary with the object. This is probably the reason for the strong objection entertained by some people to "mixed lights." The relative brightness of objects is altered by such a mixture, and this produces an unpleasant feeling of confusion.

In measuring the relative illuminating powers of different sources of light, however, we disregard this source of error, and assume that the quantity of light received by the eye from each unit area of the photometer screen is directly proportional to the quantity transmitted to that area from the source of light. The quantity thus transmitted has been found experimentally to vary inversely as the square of the distance between the source and the photometer screen. Thus, if we take some particular source of light as our standard source, and call the quantity of light received from it in some definite direction on a unit area of a screen placed at right angles to its rays at unit distance, the unit of illuminating power, we can express the illuminating power in any one direction of any source in terms of these units, and so can *measure* the illuminating power of a source in all directions. In measuring the illuminating power of a source, therefore, what we really measure is the quantity of light per unit area emitted by a special surface on which the light from the source falls, and we measure this solely by its physiological effect on the eye.

If a source emits its light equally in all directions, its illuminating power in any one direction may be called the illuminating power of the source, for it tells you how much light you can get from the source in any single direction. The total quantity of light emitted by the source can be found by multiplying its illuminating power by 4π. For if it be considered as a point of light placed at the centre of a hollow sphere of unit radius, the illuminating power is the quantity of light falling on unit surface of this sphere, and the total light emitted is the amount falling on its whole surface. Therefore

illuminating power : total light :: unit area : surface of sphere

$$:: 1 : 4\pi \; ;$$

or $4\pi \times$ illuminating power $=$ total light emitted.

In most sources of light different quantities are emitted in different directions, *i.e.*, illuminating power varies with direction, and it becomes a question, What is the illuminating power of such a source? It is evident that we cannot take the illuminating power in any one direction as *the* illuminating power of the source, for in that case we might make the illuminating power anything we liked within given limits, and there would be no possibility of comparing the relative values of two different sources. It has been suggested that the illuminating power of each source in the direction in which it is a maximum should be taken, but this would not make a fair comparison, because for one of the sources the illuminating power might have the maximum value in many more directions than in the other ; and the test would not show this, so that the light having a slightly greater maximum would be chosen as the best, whereas that having a lower maximum in many more directions would really be the more valuable. The only fair way is to compare the total quantity of light emitted by the two sources, or the mean illuminating power of each, which is, of course, the total quantity divided by 4π. This mean illuminating power is, therefore, the illuminating power that the source would have in any one direction if it emitted exactly the same quantity of light as it actually does, but equally in all directions. As a special candle is usually taken as the standard source of light, the mean illuminating power of a source is generally called its *mean spherical candle-power*, but it must not be forgotten that this is a *quantity of light*—the quantity of light that falls from the standard candle in a horizontal direction on a unit area of a surface placed at right angles to its rays at unit distance.

There are two ways of measuring the mean illuminating power of a source of light. The whole or some definite proportion of the light may be collected and then diffused equally in all directions, so that the mean illuminating power can be found by one measurement. This is the method adopted by · Profs. Ayrton and Blondel. The more usual plan is, however, to measure the illuminating power in many different directions, and from these to find the mean. The first method is of great value when employed as it was by Profs. Ayrton and Blondel to find out how the mean illuminating power of the arc was affected by changes in its length and in the current flowing ; but it would be worthless if employed to test the relative values of two totally different sources of light. For while a knowledge of the way in which the light was distributed would not be necessary for the first comparison, seeing that the *distribution* in the arc would be practically unaltered by changes in its current and length, in the second case the test would only be half made when the mean spherical candle powers of the two sources had been determined, because the relative value of the two sources might depend quite as much on the distribution of the light in each as on their mean spherical candle-powers. For this reason, in comparing two sources of light of *different* kinds, it would always be better to employ the second method.

III.—THE MEAN SPHERICAL CANDLE-POWER OF THE ARC—ROUSSEAU'S FIGURES—POLAR LIGHT CURVES.

In the ordinary vertical arc, if the carbons are well in line, if the circuit is so arranged that the current in the wires has no inductive effect on the arc, and if, also, the light is unobstructed in all directions, then the light emitted will be fairly symmetrical round the vertical axis of the two carbons. The following very pretty method of finding the mean illuminating power of such an axially symmetrical source of light, when its illuminating power in many directions in a vertical plane passing through the centre of the source has been measured, was devised by M. Rousseau.*

Archimedes proved that the area of a zone of a sphere intercepted between two parallel planes perpendicular to its axis was equal to the area of the belt of the circumscribing cylinder intercepted between the same two planes. Thus if A'E'G'C' (Fig. 143) be a sphere, and AEGC its circumscribing cylinder, the area, A'E'G'C', of the zone of the sphere intercepted between the two planes ABCD, EFGH, is equal to the belt, AEGC, of the cylinder—that is, to $2\pi r$CG, where r is the radius of the sphere.

Suppose, now, that the sphere was hollow, and that the centre of the mouth of the crater of an arc were at O, its centre. Then OA (Fig. 144) would represent the axis of the carbons. Let the lengths of the lines OB, OC, OD, &c., be proportional to the illuminating powers of the arc in directions OB, OC, OD, &c. Then OB represents the quantity of light that falls

on unit area of the sphere in the direction OB, so that, if OB
be continued to meet the sphere at E, the quantity of light
falling on a small element of the sphere at E is OB times

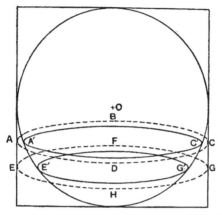

Fig. 143.—Zone of Sphere and Belt of Circumscribing Cylinder cut off by
Parallel Planes.

the area of that element. Also, since the arc emits the same
amount of light in all the directions that make an angle BOA
with the axis of the carbons, the quantity of light falling on

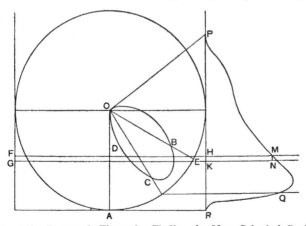

Fig. 144.—Rousseau's Figure for Finding the Mean Spherical Candle-
Power of an Axially Symmetrical Source of Light, when the Illuminating
Powers in a Number of Directions in One Plane are known.

the thin zone of the sphere intercepted between the planes FH and GK will be OB times the area of that zone, *i.e.*, OB. $2\pi r$. HK, where r is the radius of the sphere. Now make HM = OB, and complete the parallelogram HMNK. Then the total quantity of light emitted by the arc in all the directions, that make an angle BOA with the axis of the carbons is $2\pi r$ times the area HMNK, where HM and KN are infinitely near together. Now let a curve PMQR be drawn on any line PR parallel to the axis of symmetry, such that, for every point B there is a corresponding point M, where HM = OB. Then the total light emitted by the arc will be $2\pi r$ times the area of the figure PMQR, for this area includes all the elements of area similar to HMNK. To find the *average* quantity of light falling on unit area of the sphere, we must divide this quantity, $2\pi r \times$ area PMQR, by the area of the sphere, *i.e.*, by $4\pi r$. Therefore, the mean illuminating power of the arc, or its mean spherical candle-power, when the candle-power is the unit of illuminating power employed, is

$$\frac{2\pi r \times \text{area PMQR}}{4\pi r}$$

or half the area PMQR.

It is customary, when the illuminating power of the arc has been measured in many directions in one plane, and lines similar to OB, OC, &c. have been drawn representing those illuminating powers, to draw a curve through the ends of those lines, so that by means of half a dozen measurements the illuminating power in *any* direction can be found. These are the polar light curves, of which examples, drawn by Mr. Trotter, are given in Figs. 91 and 92. There is an erroneous idea—so widespread that it is essential to correct it—that the area of this curve, or the solid contents of its figure of revolution about the axis of the carbons, represents the mean spherical candle-power of the arc.

Before examining why this idea is wrong, a small practical demonstration of its impossibility may be given. Let OABC (Fig. 145) be a polar light curve, so that OA is proportional to the illuminating power of the arc in the direction OA, OB in the direction OB, &c. Now, suppose the arc to be replaced by another which has exactly half the candle-power of the first in each direction. Then it must have half the mean spherical candle-power of the first also. To obtain the new polar light curve, we must halve each radius vector of the old so that the new curve is OA′B′C′, where OA = 2OA′, OB = 2OB′, OC = 2OC′, &c. But the area of this new curve is plainly not equal to half the area of the old, as it should be if it were proportional to the mean spherical candle-power of the arc, but to a quarter of that area; and still less would the solid contents of the figure of revolution of OA′B′C′ round the axis OP be half the contents of the figure of revolution of OABC. It is quite evident, therefore, that neither the area of the polar light curve, nor the solid contents of its figure of revolution round the axis is proportional to the mean spherical candle-power of the arc; and this is the reason.

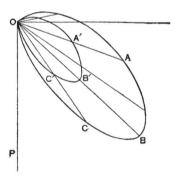

Fig. 145.—Polar Light Curves of Two Similar Sources, the one having twice the Illuminating Power of the other.

The mean illuminating power of a source of light is the total amount of light that would be received from it on the inner surface of a hollow sphere of which it occupied the centre, divided by the area of the surface of the sphere. For the numerator of this ratio—the total light that falls on the surface of the sphere—we have

$$L_1 a_1 + L_2 a_2 + \ldots = \Sigma L a,$$

where L_1, L_2, L_3, &c., are the quantities of light falling on the elements, a_1, a_2, a_3, &c., of the sphere. For the area of the sphere we have

$$a_1 + a_2 + \ldots = \Sigma a,$$

and thus for the mean illuminating power we have

$$\frac{\Sigma La}{\Sigma a}.$$

Now, each radius vector of the curve OABC gives us L for the direction in which it is drawn, but we cannot get a from this curve without drawing the Rousseau figure already described, for a is an element of the surface of a *sphere*, and the surface of the figure of revolution of OABC about OP is by no means a sphere. It is through confusing the area of this figure with the area of the sphere on which the light must fall, in order that the measurements of the light in different directions may be comparable with one another, that the error has arisen of supposing that the solid contents of this figure was proportional to the mean illuminating power of the source. *All that can be found out about the light from the polar light curve alone is the actual illuminating power of the source in each separate direction, but not its mean illuminating power.*

Another fallacy that has arisen concerning these light curves is the idea that the Rousseau figure simply gives the mean spherical candle-power plotted to rectangular co-ordinates. As there is only one mean spherical candle power for each source of light and one Rousseau's figure likewise, there can be no "plotting to rectangular co-ordinates" in the case. The area of the Rousseau figure for each arc *represents* the mean spherical candle-power of that arc.

IV.—CANDLE AND GAS SHADOW EXPERIMENTS.

When the light of an arc is sent through a candle or gas flame the lens effect is still more marked than when the crater light is simply reflected back through the arc itself, for the rim of light round the shadow is far brighter, and parts of the candle flame shadow are deeper.

Fig. 146 is taken from a photograph of the shadow of a candle flame, and shows very well the various degrees of darkness and brightness of the shadow and its rim. The darkest part of the shadow (a) corresponds with the brightest part of the flame; (b), which corresponds with the dark part of the flame round the wick, which we know to consist of unburnt gases, is bright, showing that the flame is a convex lens denser than the surrounding material. Round the whole of this bright portion, and the true shadow of the flame, is another much larger, fainter shadow of very definite form, and it is this fainter shadow that is surrounded by the rim of light previously mentioned. This outer shadow is not that of the

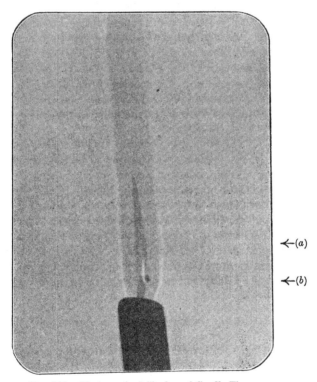

←(*a*)

←(*b*)

Fig. 146.—Photograph of Shadow of Candle Flame.

flame, as may easily be ascertained, by running the point of a thin carbon rod round the outline of the flame, and noting where this point comes in the shadow. It will be found that in all cases the point is *inside* the outer shadow, and that it just touches the inner darker one. If a small piece of cold glass is held anywhere within the region that casts the outer shadow, it will be found to be immediately covered with a thin film of moisture, while if held outside this region it remains bright. It is probable, therefore, that the outer shadow is produced by the water vapour that is evolved by the burning of a flame, and the definiteness of the shadow shows that this water vapour forms an envelope round the flame of perfectly definite shape and thickness.

It is astonishing to see to what a distance on either side and above it the flame is surrounded by a steady envelope of vapour. As Fig. 146 shows, the envelope touches the wax at the bottom, then bulges out so that it is as wide as or wider than the diameter of the candle ; and it extends upwards to some five or six times the length of the flame, before any evidence of convection currents is given. One wonders what is the path of the fresh air to the wick, as there is no indication of it in the shadow.

The limits of the vapour envelope round a gas flame are as definite on the outside as those round a candle flame, but the gas flame has no separate shadow of its own to mark the inner boundary of the envelope. Indeed, if the flame of a gas jet be gradually turned down so low that all the light-giving part is extinguished, and only a small blue bead of light is left, there is nothing in the shadow to note when the light-giving part of the flame ceases to exist ; but the small blue flame gives an astonishingly long column of water vapour, which casts a long narrow shadow, surrounded by the usual rim of light.

SUPPLEMENTARY LIST OF ORIGINAL COMMUNICATIONS CONCERNING THE ARC.*

Ann. Sci. Lomb. Veneto, 1844, Vol. XIII,, pp. 107, 169   ZANTEDESCHI.
Venezia. Atti., 1846, Vol. V., p. 519   ...   .   ZANTEDESCHI.
Annali di Fisica., 1849–50, Vol. I., pp. 57, 71, 81, 83,
    87, 141   ...   .   ZANTEDESCHI.
Wien Sitzungsberichte, 1856, Vol. XXI., p. 236   ...   ZANTEDESCHI.
Philosophical Transactions, 1881, Vol. CLXXII., p. 890   ABNEY.
Wiedemann's Annalen, 1882, Vol. XV., p. 514   ...   EDLUND.
The Electrician, 1889, Vol. XXII., pp. 534, 568, 596, 627   THOMPSON.

---

* The Italian references heading this list were kindly brought to my notice by Mr. G. Griffith, and many of the remainder by Mr. Duddell.

The Electrical Review, 1892, Vol. XXXI., p. 728 ... OLIVETTI.

Engineering, Vol. LVI., pp. 144, 223, 254 ... ... BLONDEL.

Lightning, 1894, Vol. VI., pp. 260, 298, 339 ... ... FLEMING.

Comptes Rendus, 1894, Vol. CXIX., p. 728 ... ... THOMAS.

L'Éclairage Électrique, 1894, Vol. I., p. 474 ... ... SAHULKA.

The Electrician, 1895, Vol. XXXIV., pp. 335, 364, 399, 471, 541, 610 ; 1895, Vol. XXXV., pp. 418, 635, 743 ; 1895, Vol. XXXVI., pp. 36, 225 ; 1896, Vol. XXXVI., p. 539 ... ... H. AYRTON.

Proceedings of the Royal Society, 1896, Vol. LVIII., p. 24 ... ... WILSON & GRAY.

La Société Française de Physique, 1896, July 17th, p. 243 { GUILLAUME. LE CHATELIER.

La Société Française de Physique, 1897, Feb.19th, p.12 GUILLAUME.

The Electrician, 1897, Vol. XL., p. 326 ... ...

L'Industrie Électrique, 1897, July 10th, p. 273 ... GUILLAUME.

Report of the British Association, 1897, p. 575 ... H. AYRTON.

The Electrician, Vol. XL., p. 363 ... ...

Wiedemann's Annalen, 1898, Vol. LXIV., p. 233 ... SIMON.

L'Éclairage Électrique, 1898, Vol. XV., p. 49 ... HESS.

The Physics Review, 1898, Vol. VII., p. 210 ... ... BROWN.

Proc. Amer. Acad. of Arts and Sciences, 1898, Vol. XXXIII., No. 18 ... ... CREW & BASQUIN

Report of the British Association, 1898, p. 805 ... H. AYRTON.

The Electrician, 1899, Vol. XLIV., p. 16 ... ... JARVIS SMITH.

Elektrotechnische Zeitschrift, 1899, Vol. LXVI., p. 264. WEDDING.

Journal of the Institution of Elec. Eng., 1899, Vol. XXVIII., p. 400 ... ... H. AYRTON.

Rapport du Congrès Internationale d'Electricité, 1900 p. 250 ... ... H. AYRTON.

Journal of the Institution of Elec. Eng., 1900, Vol. XXX., p. 232 ... ... DUDDELL.

L'Éclairage Électrique, 1901, Vol. XXVII., p. 379 ... CORBINO & LIGA.

Proceedings of the Royal Society, 1901, Vol. LXVIII., p. 410 ... ... H. AYRTON.

Proceedings of the Royal Society, 1901, Vol. LXVIII., p. 512 ... ... DUDDELL.

Bulletin de la Societé Internationale des Electriciens, 1901, p. 251 ... ... JANET.

Revue Generale des Sciences, 1901, pp. 612, 659 ... BLONDEL.

INDEX TO CONTENTS.

ABBREVIATIONS.

Alt. = Alternating.	Expts. = Experiments.	Pos. = Positive.
C. = Cored.	*l.* = Length of Arc.	Res. = Resistance.
Const. = Constant.	Max. = Maximum.	S. = Solid.
Cur. = Current.	Min. = Minimum.	Temp. = Temperature.
	Neg. = Negative.	

Made in the USA
Monee, IL
14 May 2021

68519499R00301